International Workshop on Discovery Physics at the LHC (Kruger2016)

Journal of Physics: Conference Series Volume 878

Kruger, South Africa
4 – 9 December 2016

ISBN: 978-1-5108-4649-4
ISSN: 1742-6588

Printed from e-media with permission by:

Curran Associates, Inc.
57 Morehouse Lane
Red Hook, NY 12571

Some format issues inherent in the e-media version may also appear in this print version.

Printed by Curran Associates, Inc. (2017)

Institute of Physics
Dirac House, Temple Back
Bristol BS1 6BE UK

Phone: 44 1 17 929 7481
Fax: 44 1 17 920 0979

techtracking@iop.org

Additional copies of this publication are available from:

Curran Associates, Inc.
57 Morehouse Lane
Red Hook, NY 12571 USA
Phone: 845-758-0400
Fax: 845-758-2633
Email: curran@proceedings.com
Web: www.proceedings.com

Table of contents

Volume 878

International Workshop on Discovery Physics at the LHC

4–9 December 2016, Kruger, South Africa

Accepted papers received: 18 July 2017
Published online: 4 August 2017

Preface

Papers

ALICE

CMS

LHCb

SHiP STAR Theory

IOP Conf. Series: Journal of Physics: Conf. Series **878** (2017) 011001 doi:10.1088/1742-6596/878/1/011001

Foreword Proceedings Kruger2016

The fourth biennial conference on "Discovery Physics at he LHC " was held December 4-9 2016 at the Kruger Gate Hotel in South Africa. In total there were about 100 participants coming from Canada, Croatia, France, Germany, Hungary, Italy, Korea, Poland, Russia, Spain, Sweden, Switzerland, UK and USA. The latest results from the Large Hadron Collider, the Relativistic Heavy Ion Collider as well the latest theoretical insights were presented. With the majestic Kruger national Park in the background this led to a very stimulating conference with many exchanges taking place. The proceedings reflect the high level of the conference.

The financial contributions from iThemba LABS, the Department of Science and Technology, the SA-CERN programme, the UCT-CERN Research Centre, the Universities of Cape Town and of the Witwatersrand are gratefully acknowledged.

Local Organizing Committee:

Zinhle Buthelezi
Jean Cleymans (chair)
Simon H. Connell
Alan S. Cornell
Thomas Dietel
Siegfried Frtsch
Nami Haasbroek
William A. Horowitz
Deepak Kar
Bruce Mellado
Sahal Yacoob

IOP Conf. Series: Journal of Physics: Conf. Series **878** (2017) 011002 doi:10.1088/1742-6596/878/1/011002

Peer review statement

All papers published in this volume of *Journal of Physics: Conference Series* have been peer reviewed through processes administered by the proceedings Editors. Reviews were conducted by expert referees to the professional and scientific standards expected of a proceedings journal published by IOP Publishing.

IOP Conf. Series: Journal of Physics: Conf. Series **878** (2017) 012001 doi:10.1088/1742-6596/878/1/012001

CERN: A global project

Rüdiger Voss
CERN, Geneva, Switzerland

Abstract. In the most important shift of paradigm of its membership rules in 60 years, CERN in 2010 introduced a policy of "Geographical Enlargement" which for the first time opened the door for membership of non-European States in the Organization. This short article reviews briefly the history of CERN's membership rules, discusses the rationale behind the new policy, its relationship with the emerging global roadmap of particle physics, and gives a short overview of the status of the enlargement process.

In 2014, CERN celebrated it 60th anniversary. On July 1, 2014, the cycle of events commemorating the birthday of this unique scientific, political and human endeavour was opened in Paris, commemorating the anniversary of the CERN Convention, which was signed at the UNESCO Headquarters by representatives of the twelve founding Members in 1953. These twelve signatures are worth commemorating indeed. For more than half a century, the Convention has stood the test of time as a masterpiece of simple and minimalistic legal language that focuses wisely on the essential cornerstones of CERNs institutional basis and governance. At the same time, it provides for the leeway that is necessary to adapt the Organization to a changing political environment, and to new scientific and technological challenges. The Convention is a testimony to the wisdom and foresight of CERN's founding fathers, on par with their vision of rebuilding peace in Europe by establishing a unique focal point that would foster scientific collaboration on an unprecedented scale, between nations that that had fought war against each other only a few years earlier. On the basis of this Convention, CERN has served as a model for other, successful European science organizations, and most recently for the SESAME synchrotron light source in the Middle East.

Some of the most intriguing aspects of the CERN Convention are in the provisions for Membership. Whereas it stipulates that "the Organization shall provide for collaboration among European States in nuclear research of a pure scientific and fundamental character ...", nowhere does it state explicitly that Membership in CERN is restricted to European States. This ambiguity is by no means fortuitous. It reflects the fact that already in the early 1950s, a possible enlargement of Membership beyond Europe was a hotly debated issue on which the provisional Council could not reach agreement. It agreed, however, on a carefully crafted compromise that left a door open to adapting the membership policy of CERN to an evolving scientific and political landscape. Indeed CERN continued to grow and counted 20 European Member States at the end of the 20th century, thanks largely due to the accession of countries from Central and Eastern Europe after the fall of the Berlin Wall.

The CERN Council debated a widening of Membership beyond Europe on several occasions in the past, but repeatedly confirmed a restrictive interpretation of the Convention, whereby Membership remained reserved for European countries. Only in 2010, it approved the most radical shift of paradigm of CERN's membership policy to date, embedded in a policy of

Figure 1. The 22 Member States of CERN. Serbia is expected to become a Member State in 2017. In addition, Cyprus, India, Pakistan, Turkey and Ukraine are Associate Member States.

"Geographical Enlargement" and opening Membership to non-European States irrespective of their geographical location [1]. At the same time, the Council introduced the new instrument of Associate Membership in order to facilitate the accession of new Members, including emerging countries outside Europe, which may not command sufficient resources to sustain full Membership in the foreseeable future. Under this new policy, Israel joined CERN as 21st, and first non-European Member State, in 2014; Romania became the 22nd member State in 2016 (Fig. 1). In parallel, a new policy of "Scientific Enlargement" allows, for the first time, for a participation in *global* accelerator projects outside Europe.

CERN's new Membership rules are not a political purpose in itself but follow a twofold science policy rationale: they underpin the unprecedented migration of the global particle physics community to Europe, reflecting the scientific attractiveness and success of the LHC; and they serve to prepare CERN for its long-term future. They mirror the globalization of particle physics, which in turn has become a prominent paradigm for the globalization of science at large.

For more than fifty years, CERN has supported free movement of scientists across countries, regions and continents with a policy of open doors, providing free access to its scientific facilities without discrimination between Member States and non-Member States. This policy had been

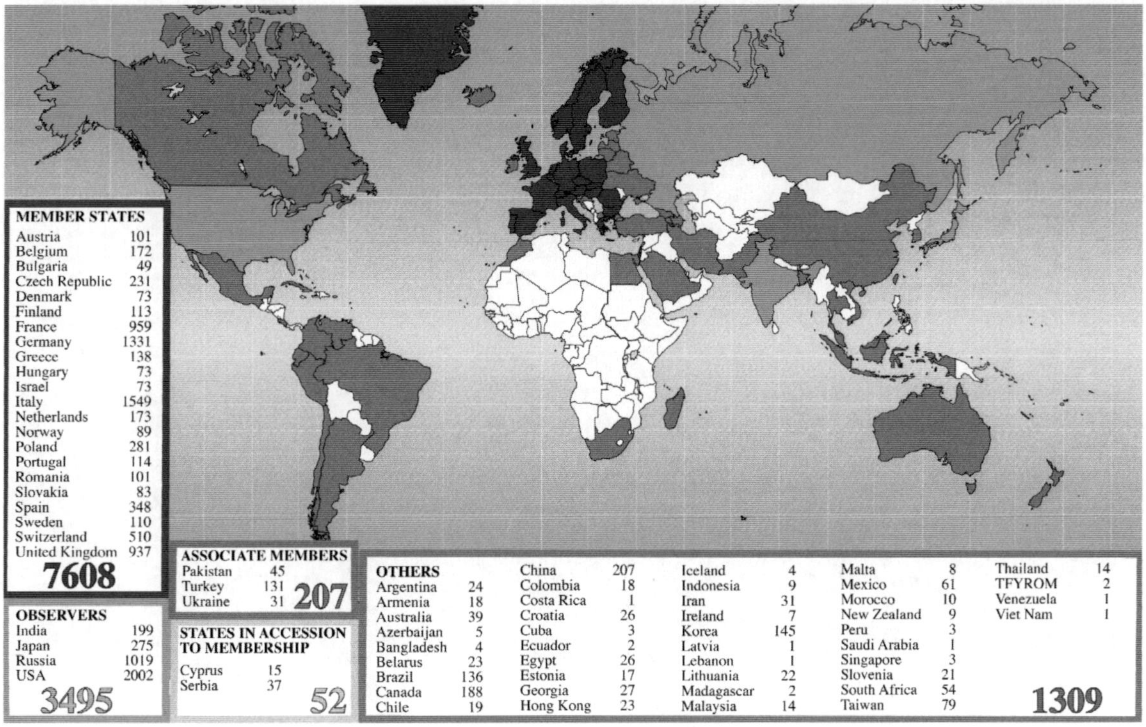

Figure 2. Countries of origin of CERN's visiting scientists (Users) by location of their home institute (September 2016). India, which is listed here as an Observer, joined as an Associate Member in January 2017.

formalized in 1980 in ICFA guidelines for free, mutual and *balanced* access of physicists from different regions to regional laboratories in other parts of the world [2]. It was shaped at a time when Europe, North America and Japan owned all major facilities for accelerator-based particle physics, when the global HEP community was strongly dominated by scientists from the same three regions, and when exchange between the same regions was healthy and balanced indeed.

Since 1990, however, the landscape has changed dramatically: starting with the demise of the SSC, the shut-down of major accelerators around the globe – or their conversion into applied science facilities – has severely broken the symmetry of scientific exchange. Since 2004, the community of CERN's visiting scientists ("Users") has grown from just above 6'000 to almost 13'000 physicists and engineers from all continents. Today, these scientists come to CERN from about 650 different institutes in more than 70 countries (Fig. 2) and represent around 100 different nationalities. This growth has been driven by non-Member States (NMS) more than by the Member States: scientists from NMS institutes now account for almost 40% of CERN's Users, and their participation is expected to increase further. Whereas the numbers are dominated by North America, in recent years important growth rates are observed for Asia and Latin America, where new players emerge on the field of international science (Fig. 3). These data convincingly illustrate particle physics' strong tradition of defying political, geographical and cultural boundaries.

From a European perspective, the unprecedented non-Member State participation in the LHC has brought about substantial scientific, technical and political benefits. It has contributed to establishing CERN as the world's leading center at the high energy frontier in the awareness

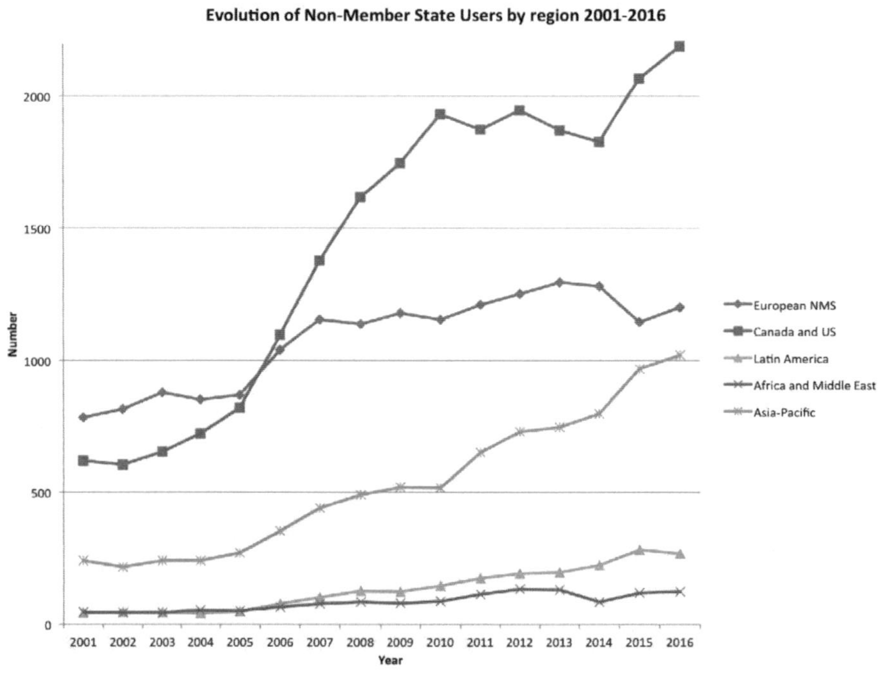

Figure 3. Evolution of the number of CERN Users from Non-Member States (NMS) grouped by region. The apparent decrease for Africa and Middle East reflects the fact that Users from Israel are not included any longer for 2014, when Israel became a Member State.

of governments and funding agencies, and of the general public. The LHC has convincingly demonstrated the potential of global partnership in basic science, and is widely perceived as a paradigm of successful world-wide cooperation on megascience projects. To take this cooperation to the next higher level, and to fully exploit its potential to the benefit of all stakeholders, CERN's enlargement policy offers a new institutional framework for a sustained global partnership in particle physics research.

More important, the Geographical Enlargement policy is a first step in preparing CERN's membership and governance for the post-LHC future. Whereas the LHC experiments today are truly global operations, the LHC accelerator was built as a predominantly European project, with a contribution from outside Europe that was technically and politically significant, but accounted for only 10% of the construction cost, and was mostly provided in-kind. This model cannot be expected to work for a large, next-generation facility in Europe. With the CLIC and FCC studies, CERN is exploring two different, challenging avenues to prepare its future, and the future of our field, after the LHC. No cost estimate exists yet for either option, but it seems inconceivable that any of them can be approved and build within the same membership, governance and funding structures which worked more than twenty years ago – successfully, but under great labour pains – for the LHC.

In one of the most significant developments in recent years, the three most important *regional* particle physics roadmaps – *i.e.* the proposal of the Japanese community to host the ILC; the 2013 Update of the European Strategy for Particle Physics [3]; and the US P5 report of 2014 [4] – have been developed in a complementary approach and with a strong focus on inter-regional collaboration, such that together they form an important milestone on the way to a *global* roadmap. CERN's enlargement policy is designed to fit seamlessly into this emerging global strategy for accelerator-based particle physics. It opens Membership in the Organization to all States irrespective of their geographical location, and introduces the new scheme of Associate Membership with reduced obligations and benefits. The important instrument of International Cooperation Agreements is maintained for partners who are not willing or able to shoulder the obligations of CERN membership in the near future. About 50 such Agreements are in place today and provide the formal framework for most of the cooperation with non-European countries; in this respect the most significant, recent development is a new Cooperation Agreement with the US signed in May 2015.

In a reciprocal move, for the first time CERN is formally allowed to participate in *global* HEP projects outside Europe; this policy, which was confirmed by the 2013 European Strategy Update, will initially enable CERN to serve as a gateway for European participation in the LBNF facility at Fermilab, and later in the ILC.

With more than ten applications for Membership or Associate Membership received in recent years from countries of different size, and from in- and outside Europe (Brazil, Croatia, Cyprus, India, Lithuania, Israel, Pakistan, Romania, Russia, Serbia, Slovenia, Turkey, Ukraine), the enlargement process has made a promising start; some of the accession procedures have been completed (India, Israel, Pakistan, Romania, Turkey, Ukraine) or are expected to conclude in the near future. Other countries which are perceived as natural candidates, while they acknowledge the promise and potential of a continued scientific and technological partnership, have so far remained absent, or are hesitant on political or financial grounds. More work, stamina, and patience will be needed to enlarge the Membership of CERN to a size that is commensurate with its future ambitions in quantity and quality. Not all States that are obvious candidates for a closer scientific and technical partnership may share today the values of a governance that is excellence-driven and consensus-oriented, and which has prevailed throughout CERN's 60-years history. In the long term, broadening the institutional base without sacrificing the traditional values of European cooperation, which have been a key ingredient in CERN's past successes, is likely to emerge as the true challenge of the enlargement process.

The partnership between CERN, its Member States and non-Member States in building the LHC and harvesting its scientific potential has developed into a solid backbone of successful cooperation of unprecedented, global dimensions. CERN wants this partnership to continue, to grow and to flourish on the basis of an enhanced institutional participation of both European and non-European countries. The policy of Geographical Enhancement is well matched to the unfolding, global roadmap of particle physics.

References

[1] CERN Council document CERN/2918/Rev., available at http://cds.cern.ch/record/1289091?ln=en
[2] ICFA Guidelines for the Interregional Utilization of Major Regional Experimental Facilities for High-Energy Particle Physics Research, http://www.fnal.gov/directorate/icfa/icfa_guidelines.html
[3] CERN-Council-S/106, available at http://cern.ch/council/en/EuropeanStrategy/esc-e-106.pdf
[4] Particle Physics Project Prioritization Panel (P5): Building for Discovery: Strategic Plan for U.S. Particle Physics in the Global Context, http://science.energy.gov/ /media/hep/hepap/pdf/May%202014/FINAL_P5_Report_053014.pdf

IOP Conf. Series: Journal of Physics: Conf. Series **878** (2017) 012002 doi:10.1088/1742-6596/878/1/012002

Relativistic Heavy Ion Physics - "Discoveries" and Future Prospects

John W. Harris

Wright Laboratory, Physics Department, Yale University New Haven, Connecticut, USA 06250

E-mail: john.harris@yale.edu

Abstract. Collisions of ultra-relativistic heavy nuclei at the Large Hadron Collider (LHC) at CERN and the Relativistic Heavy Ion Collider (RHIC) in the U.S. create energy densities where nuclear matter melts into a plasma of quarks and gluons. I will summarize the accomplishments of the ultra-relativistic heavy-ion program and present the "big questions" remaining to be answered. In the process I will elucidate how the answers are being pursued and our current understanding of the high density QCD matter created in these collisions. I will offer a brief perspective on the planned detector upgrades, the presently envisioned future experiments and collider facilities, and how they intend to address the remaining questions of the field.

1. Introduction

Since this conference covers a range of physics from nuclear and particle physics to astrophysics, it is prudent to distinguish the experimental approach taken in relativistic heavy-ion physics from the more commonly recognized approach of high-energy physics. This was encapsulated succinctly by T.D. Lee (Nobel Laureate) at the dawn of the field in his infamous quote [1]. "In high-energy physics we have concentrated on experiments in which we distribute a higher and higher amount of energy into a region with smaller and smaller dimensions. But, in order to study the question of 'vacuum', we must turn to a different direction; we should investigate bulk phenomena by *distributing high energy over a relatively large volume*." This is a concise description of the approach of relativistic heavy-ion physics to investigate Quantum Chromodynamics (QCD) at high energy density. At sufficiently high energy density (temperature), the vacuum melts to form a Quark-Gluon Plasma (QGP).

This can be seen in any number of more recent lattice QCD calculations [2] that address the behavior of QCD at high temperature. Those calculations conclude that the QCD vacuum melts to form a QGP at temperatures $T_c \geq 160$ - 190 MeV, corresponding to energy densities $\epsilon \sim 0.3$ - 1.0 GeV/fm^3. This would also then be the temperature of the phase transition from quarks to hadrons in the expansion and cooling of the early Universe. Thus, the Standard Model predicts a QCD deconfinement phase transition at T = 160 – 190 MeV, the same quark-hadron phase transition that must have occurred in the early Universe, and could occur in the cores of dense stars or in other aspects of stellar evolution. The goal of relativistic heavy-ion physics is to create the QGP in the laboratory and to study its properties.

2. An Ambitious Mission and Big Questions

Relativistic heavy-ion physics seeks to determine and understand the properties and states of matter that exist at high temperature and density. More fundamentally, it strives to explore the phase structure of a fundamental gauge theory QCD. Might our understanding of other gauge theories (like gravity!) benefit from this? There are many questions to be addressed. Is the phase diagram of QCD featureless above T_c? What are its constituents (are there quasi-particles, exotic states, others)? Can we determine characteristics of the QGP such as its transport properties, sound attenuation length, coupling strength $\alpha_s(T)$, sheer viscosity/entropy density ratio (η/s), formation time (t_f), potential excited modes, and its equation of state? Finally, are there other new phenomena or new states of matter at high energy densities [3]? Answering these questions (and more) encompasses an extremely ambitious mission for the field of relativistic heavy-ion physics!

3. Remarkable "Discoveries"

After a start in this exploration at various fixed-target facilities, the acceleration of heavy ions at the CERN Super Proton Synchrotron (SPS) facility fostered sufficient energy to adequately lead exploration in this field. The advent and use of colliders for heavy-ion physics, namely RHIC and LHC, at significantly higher energies has led to a profound increase in our understanding of the behavior of QCD at high energy density. Results from RHIC and LHC have been summarized in detail in Ref. [4] and [5], respectively. I will focus in this presentation on what in my view are the most significant results or debatably the "discoveries" of the field.[1]

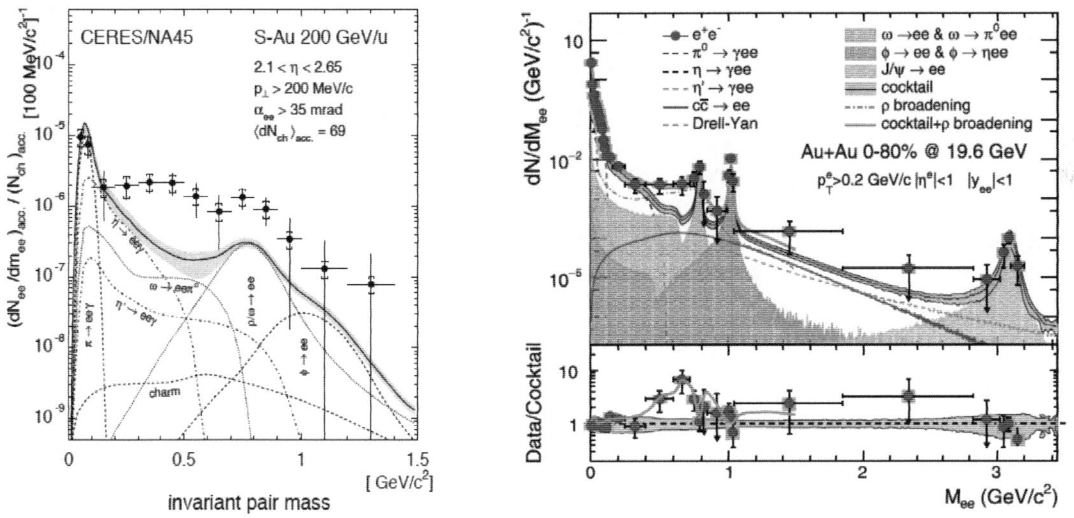

Figure 1. Dielectron invariant mass spectra. Left: E_{lab}/n = 200 GeV/n ($\sqrt{s_{NN}}$= 19.6 GeV) S-Au CERES data (circles), and contributions from hadron decays (as labeled) and their systematic error (shaded region).[6] Right: STAR data after efficiency correction, compared to decays of light hadrons and correlated decays of charm in Au-Au collisions at $\sqrt{s_{NN}}$= 19.6 GeV. [11] The data to hadronic cocktail ratio is shown in the bottom panel. Theoretical calculations incorporating a broadened ρ are also shown. Systematic uncertainties for data (green boxes) and the hadronic cocktail (gray band) are presented. See [11] for details and references therein for details on the theoretical calculations.

[1] I utilize this terminology with trepidation, and not scientifically justified in the literal sense, but I simply refer to its relation and relevance to the Conference title and beg your scientific indulgence!

4. Indications of Medium Modification and Possible Chiral Symmetry Restoration in Low-mass e^+e^- Measurements

Investigation of low-mass electron pairs in the S-Au system at $E_{lab}/n = 200$ GeV ($\sqrt{s_{NN}}=$ 19.6 GeV) by the CERES Collaboration at the SPS [6] revealed an enhancement of pairs in the mass region (200 – 700 MeV) just below the ρ mass, compared to pairs produced from hadronic decays. This is seen in Fig. 1 (left). The enhancement has been attributed in many theoretical publications to broadening of the ρ in the nuclear medium [7] and possible chiral symmetry restoration [8]. Subsequently, CERES has measured an enhancement in low-mass di-electrons in Pb-Au collisions at $E_{lab}/n = 40$ GeV [9] and NA60 an enhancement in low-mass muon pairs in 158 GeV ($\sqrt{s_{NN}}=$ 17.3 GeV) Indium-Indium collisions [10]. STAR [11] has confirmed the enhancement in Au-Au collisions at RHIC at the SPS c.m. energy, as seen in Fig. 1 (right). STAR [11] and PHENIX [12, 13] have observed enhancements of di-electrons in the same mass region in Au-Au collisions at $\sqrt{s_{NN}}=$ 200 GeV.

5. Thermal Radiation from the QGP

A thermal component of direct photons has been observed in central collisions (0-20%) $\sqrt{s_{NN}}=$ 200 MeV Au-Au by PHENIX [13, 14] at RHIC and $\sqrt{s_{NN}}=$ 2.76 TeV Pb-Pb by ALICE [15] at the LHC. After careful scaling by the number of binary collisions in next-to-leading order perturbative QCD calculations and subtraction of the predicted direct photons (or in the PHENIX case photons measured in pp interactions), an additional component of photons is observed at momenta below $\sim$ 2.5 GeV/c. These photons represent a thermal component and can be fit with an exponential to yield an inverse slope of $T = 297 \pm 12(stat) \pm 41(syst)$ MeV at the LHC and T = 221 $\pm$ 19 (stat) $\pm$ 19 (stat) MeV at RHIC. This is a strong indication of direct thermal radiation. However, it represents an integral of the radiation over the entire evolution of the system, and thus the values of T may not be strictly due to thermal radiation. A detailed comparison of the thermal photon emission data at RHIC energy [13] with a model assuming formation of a hot system with initial temperature $T_{initial}$ ranging from 300 - 600 MeV at times 0.6 – 0.15 fm/c is in qualitative agreement with the data.

6. Particles Are Formed at the Universal Hadronization Temperature

The yields of produced hadrons have been measured extensively at mid-rapidity at RHIC and LHC. The yields of the various types of hadrons have been compared to statistical hadronization (thermal) model [16] calculations, which are successful at reproducing the relative abundances of the observed hadrons. The model assumes rapid freeze-out, with typical implementations utilizing as parameters only temperature T, baryochemical potential μ_B and volume V. Shown in Fig. 2 (left) are results from the three RHIC experiments [17] and in Fig. 2 (right) those from ALICE [18] at LHC. Global thermal model fits to data are represented as a horizontal line for each ratio with the parameters T and μ_B presented in each figure. It is of interest to note that the temperatures are similar for the two c.m. energies and near that predicted by Lattice QCD as the deconfinement phase transition temperature and the quark-hadron phase transition in the early Universe.

7. High Transverse Momentum (p_T) Hadrons Are Suppressed

The products of the scattering of partons (fast quarks and gluons) in the colliding nuclei serve as probes with which to study properties of the hot QCD medium, since they are calculable in perturbative QCD. As partons traverse the medium, they lose energy through interactions or radiation leading to parton energy loss that alters the final distribution of high p_T particles, jets, and particles that contain heavy quarks.

Shown in Fig. 3 is the nuclear modification factor R_{AA} for central collisions of heavy ions for charged hadrons and pions at SPS, RHIC and LHC energies. R_{AA} is the ratio of the spectrum

IOP Conf. Series: Journal of Physics: Conf. Series **878** (2017) 012002 doi:10.1088/1742-6596/878/1/012002

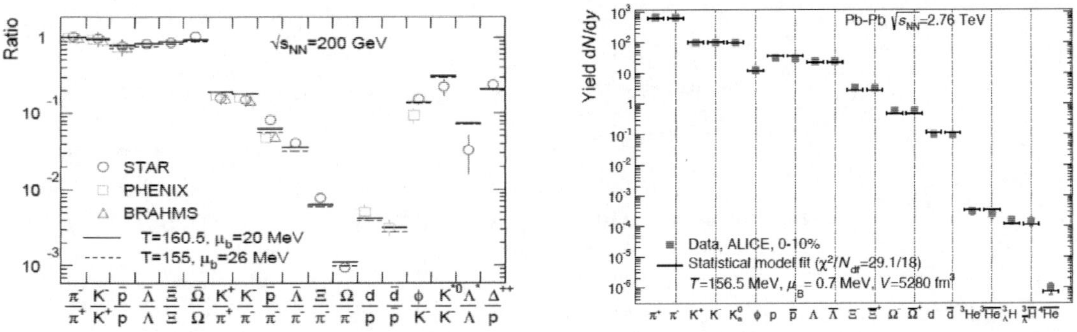

Figure 2. Left: Measured ratios (ordinate) of various hadrons (abscissa) at mid-rapidity for Au-Au central collisions by the three RHIC experiments (denoted by the different symbols in the legend).[17] Right: Measured yields per unit rapidity at mid-rapidity in central Pb-Pb collisions. A horizontal line for each ratio represents the result of a global thermal model fit to the hadron yields in each figure with parameters in each legend.[18]

of particles at high p_T in heavy-ion collisions compared to that in pp collisions, scaled by the number of binary nucleon-nucleon collisions in the nucleus-nucleus geometry and given by

$$R^i_{AA}(p_T) = \frac{d^2N^i_{AA}/dp_T d\eta}{\langle T_{AA}\rangle d^2\sigma^i_{NN}/dp_T d\eta}$$

where N^i_{AA} is the yield for particle type i in nucleus-nucleus collisions and σ^i_{NN} is the cross section for particle i in nucleon-nucleon collisions. $\langle T_{AA}\rangle$ is the ratio of the number of binary nucleon-nucleon collisions (typically calculated in the Glauber or Glauber-Gribov model) to the inelastic nucleon-nucleon cross section. R_{AA} is expected to be unity in the case where the nucleus-nucleus collisions are simply a superposition of nucleon-nucleon collisions in the absence of nuclear effects such as parton-energy loss or color screening.

A large suppression ($R_{AA} < 1$) is observed in Fig. 3 (left) for RHIC and LHC energies at intermediate p_T (between 2-20 GeV) with a gradual rise as p_T increases beyond 20 GeV.[19] Several model calculations [20, 21, 22, 23, 24] are shown, with somewhat divergent predictions. The model predictions depend on their parton energy-loss mechanisms, e.g. scattering and/or radiation and their assumed parton density. The observed suppression is in all cases a result of parton-energy loss in the dense medium.[2]

Information on parton-energy loss mechansisms may be derived from differences in the energy loss of propagating quarks and gluons by comparing the suppression of light and heavy quarks. Shown in Fig. 3 (right) is R_{AA} for D mesons (average of D^0, D^+, $D^\pm$), π^- from ALICE [26], and non-prompt J/ψ mesons from CMS [27] as a function of centrality in Pb-Pb collisions. The suppression of D mesons is consistent with that of charged particles, indicating that the energy loss of the heavier charm quark is not very different (if at all) from that of light quarks. The B-decays are less suppressed, which may indicate the emergence of the dead-cone effect [28]. QCD predicts a dead-cone effect in which the radiation pattern of gluons from partons is dependent upon the mass of the parton emitter. For quarks with mass m and energy E, radiation is suppressed for emission angles $\theta <$ m/E, resulting in heavier quarks being less suppressed than light quarks as observed for the B-mesons.

[2] Note that an enhancement is observed at SPS energies, which is commonly attributed to the Cronin effect.[25]

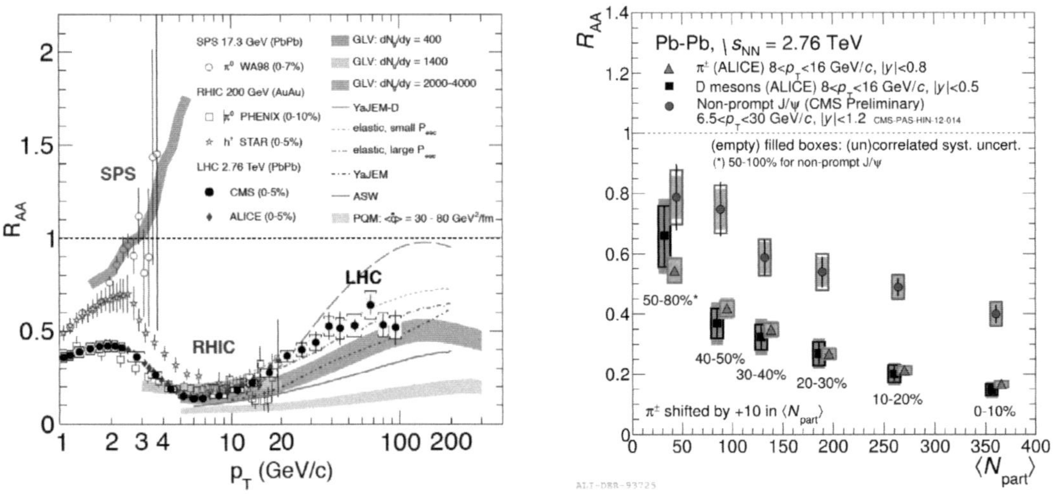

Figure 3. Left: R_{AA} as a function of p_T for central collisions of heavy ions for charged hadrons and pions at SPS, RHIC and LHC energies (see legend). Various model predictions are also shown (see text for details).[19] Right: R_{AA} for D-mesons (average D^0, D^+, $D^{\pm}$) and π^- for $8 < p_T < 16$ GeV/c from ALICE [26] and for non-prompt J/ψ mesons with $6.5 < p_T < 30$ GeV/c from CMS [27] as a function of centrality (the mean number of participant nucleons in the collision $\langle$ N_{part} $\rangle$).

8. Jets Are Quenched and Dijet Energies Modified

Additional insight into the kinematics of the initial binary parton-parton scattering and the parton energy loss is gained by measuring jets, dijets and γ-jet correlations in heavy-ion collisions and comparing to results in pp and p-A collisions. Shown in Fig. 4 (left) are the R_{AA} from particles and jets [29] (see [5] and references therein) measured in central Pb-Pb collisions. The electromagnetic probes (Z, W, γ) do not interact strongly and are thus unaffected by the medium ($R_{AA} = 1$). Charged particles are suppressed, although non-prompt J/ψ mesons are less suppressed at p_T< 20 GeV, as discussed above. B-jet suppression generally appears of similar magnitude to that of inclusive jets indicating a lack of flavor dependence in the quenching mechanism at large jet energies.

ATLAS [30] and CMS [31] have observed an energy imbalance of dijet pairs. The ATLAS results for the dijet energy asymmetry A_J are shown in Fig. 4 (right) for $\sqrt{s_{NN}}$= 2.76 TeV Pb-Pb central collisions compared to 7 TeV pp data and Monte Carlo simulations using Pythia dijets embedded in Hijing events, where A_J = (E_{T_1} - E_{T_2}) / (E_{T_1} + E_{T_2}). T_1 is defined as the highest transverse energy jet with transverse energy E_{T_1} > 100 GeV and T_2 the highest transverse energy jet in the opposite hemisphere with E_{T_2} > 25 GeV. The Pb-Pb events exhibit a large asymmetry A_J with the away-side jet having lost considerable energy compared to that observed in the pp data and Hijing + Pythia simulations. This is a clear indication of jet quenching in Pb-Pb collisions at the LHC. It is interesting to note (although not shown here) that the same measurements exhibit a peak at $\Delta\phi = \pi$ in the azimuthal angular correlation between the leading and the away-side jet with little difference in the $\Delta\phi$ widths between the Pb-Pb data and the pp data and Hijing + Pythia simulations. This latter result will require further investigation with higher statistics. Clearly a better understanding of the fragmentation and its modification in-medium and fragmentation of large energy heavy-flavor jets is necessary and the subject of further studies.

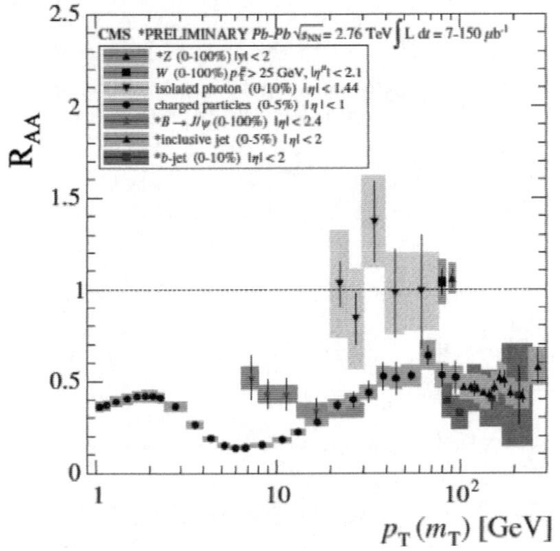

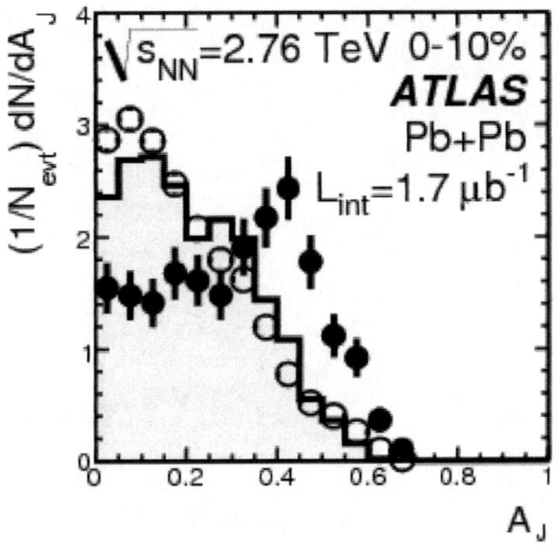

Figure 4. Left: R_{AA} for charged particles, B-meson decays to J/ψ, electromagnetic probes (Z, W, γ) and jets [29] measured in central Pb-Pb collisions (from [26] and [27]). Right: Transverse energy asymmetry of dijets (defined in text) in central Pb-Pb data (solid circles) compared to that observed in pp data (open circles) and Hijing + Pythia simulations (histogram).

9. Quarkonia - J/ψ and Υ Are Suppressed

Quarkonia, charmonium ($c\bar{c}$) and bottomonium ($b\bar{b}$), were predicted early-on [32] to be sensitive to deconfinement in the medium via the color analogue of Debye screening. The color screening of J/ψ was observed at the SPS, RHIC and LHC. Suppression of the Υ states in Pb-Pb compared to pp at LHC, seen in Fig. 5 (left), provides strong evidence of effects of the colored medium on the quarkonia [33]. The Υ(2S) and Υ(3S) states are observed to be more suppressed than the Υ ground state in the Pb-Pb data.

Theoretically, a sequential melting scheme evolved for the various quarkonium states [34], since the screening length in a deconfined medium depends on its temperature. The less tightly-bound states dissociate at lower temperatures than those that are more tightly-bound, thus producing a characteristic suppression pattern based on the binding energies of the quarkonia and the temperature of the medium. Such a pattern is observed in Fig. 5 (right), which provides a compilation of the measured suppression of the various J/ψ and Υ states as a function of their binding energies [33]. This supports the predicted color screening picture. Lattice QCD calculations are able to relate the degree of quarkonium screening to the temperature although this is complicated by the time evolution and regeneration of bound states in the medium (see [35]).

10. Collective Flow Indicative of a Strongly-Coupled Medium with Ultra-low η/s (shear viscosity/entropy density)

A clear signature of collective flow was discovered in A–A collisions as early as the 1980s at the BEVALAC [36] and later on at SPS, RHIC and LHC. The multi-particle azimuthal distributions are analyzed typically utilizing Fourier-decomposition techniques. A substantial second-order coefficient (v_2), called elliptic flow and investigated thoroughly at RHIC and LHC, is described well in a hydrodynamic description of a hot expanding fluid with a small shear viscosity to entropy density (η/s) ratio. Presented in Fig. 6 (left) and Fig. 6 (right) are results from RHIC and LHC, respectively, compared to a relativistic viscous hydrodynamic model with

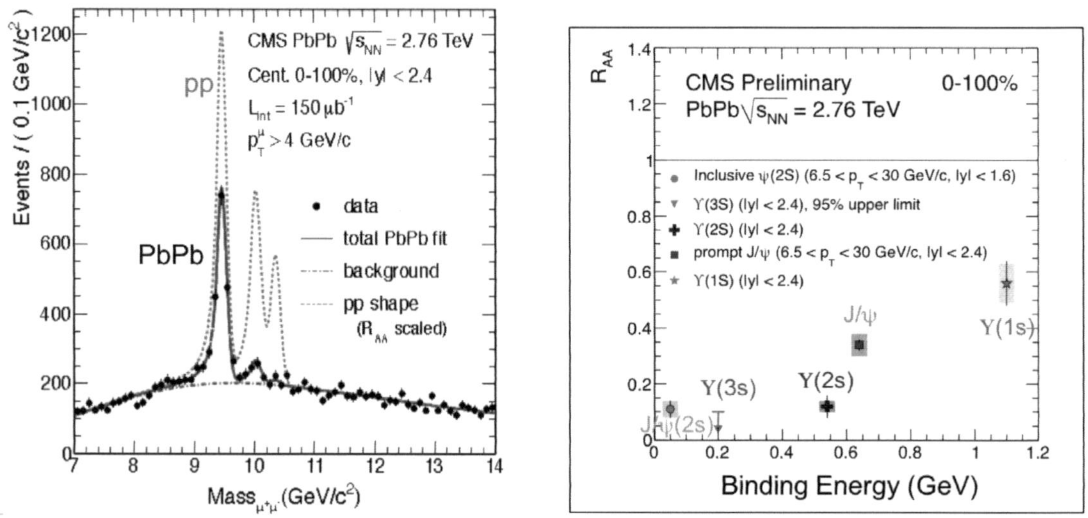

Figure 5. Left: Comparison of di-muon mass spectra measured in pp (dashed red) and Pb-Pb (solid black) collisions. Right: Suppression of Quarkonium states as a function of binding energy. (adapted from [33])

IP-GLASMA initial conditions and an evolution of matter through the quark-gluon plasma and hadron gas phases [37, 38]. Various values of η/s are shown with $\eta/s = 0.08$ matching the RHIC data from STAR and $0.08 < \eta/s < 0.16$ being closer to the LHC data from ALICE. These values are near the conjectured minimum value for an infinitely strongly-coupled theory [39, 40].

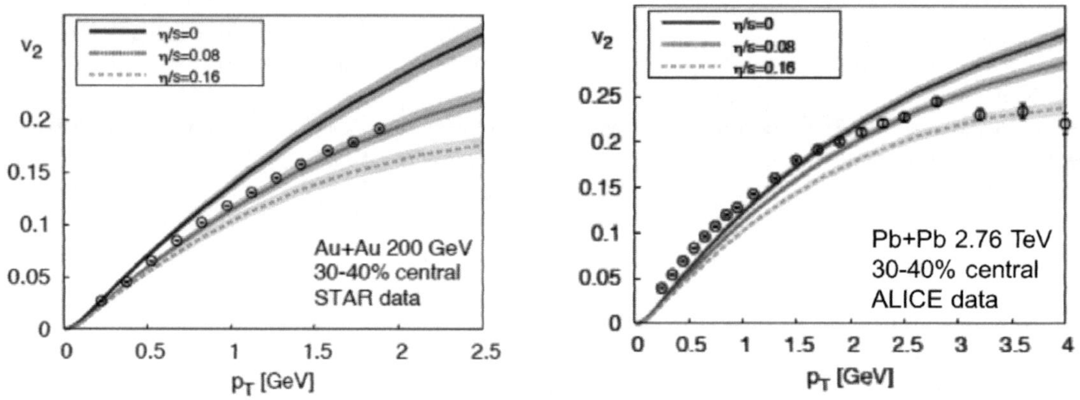

Figure 6. v_2 as a function of p_T. Left: from RHIC (STAR). Right: from LHC (ALICE). Data are compared to a relativistic viscous hydrodynamic model with various values of η/s as shown. [37, 38]

The transverse-momentum distributions of hadrons at low and high p_T are sensitive to properties of the different phases of a nuclear collision. At low p_T production of hadrons is governed by soft processes and the momentum distributions reflect properties of the medium at kinetic freeze-out. The spectra of identified hadrons (primarily pions, kaons and protons) exhibit a clear mass hierarchy where the inverse slope of the p_T distributions increases with increasing hadron mass. This is observed at RHIC and LHC energies and has been described in terms of the emission of hadrons from a thermal system that is expanding radially at a common velocity.

[41] In this scenario, heavier hadrons have larger momenta than lighter ones due to the common radial flow velocity. At p_T (> a few GeV) hadrons originate primarily from jet fragmentation in hard scattering processes that occur in the initial phase of the collision.

11. Effects in Small Systems (*p*-A, high multiplicity pp)

Proton-proton (pp) and proton-nucleus (*p*-A) collisions have been investigated in order to better understand phenomena observed in nucleus-nucleus (A-A) collisions. It is essential to identify and separate experimentally and theoretically, to the extent possible, initial-state and final-state effects. The *p*-A measurements can help to understand the initial state by establishing the extent of gluon saturation and the possible existence of a color-glass condensate at low-x in nuclei. Furthermore, the extent and impact of cold nuclear matter effects on the A-A data and their interpretation, for example in terms of color screening and jet quenching, are critical in understanding effects observed in the A-A data.

The results of measurements in small systems is the topic of another presentation[3] and thus only a summary is presented here. In measurements utilizing hard probes, no suppression of particles at high p_T nor quenching of jets is observed in *p*-A collisions. Neither are quarkonia observed to be suppressed in *p*-A, whereas cold nuclear matter effects may be expected to induce suppression. These measurements in *p*-A systems confirm that the quenching and suppression in A-A collisions are final state effects that can be interpreted from theory to result from parton energy loss and deconfinement in a high energy density environment.

One marked difference from the above conclusions that is observed in the *p*-A data is that the low p_T spectra and particle correlations in *p*-A exhibit similar effects as in A-A with regard to the spectral mass ordering of hadrons attributed to collective (radial flow) in A-A systems (see [41]). Furthermore, there are similar though weaker trends in the flow harmonic decomposition of the particle correlations in *p*-A and in high multiplicity pp data. This has generated significant interest and investigation to determine the smallest size and energy density of a droplet of QCD matter that behaves like a liquid.

12. Remaining Questions for the Field

As we investigate high density QCD phenomena in collisions of various (large and small) systems, there are still many questions to be answered! We seek to identify and separate the initial state from the final state by comparing *p*-*p*, *p*-A, A-A results to answer questions like the following. To what extent is there saturation in the initial state? Can nuclear parton distribution functions be extended from our current knowledge to describe the initial state? Does it consist of a color-glass condensate? What is the effect of cold nuclear matter on the final state observables that are being measured? How does the system evolve and thermalize from its initial state?

We seek to identify the underlying dynamical properties of the QGP that describe mechanisms of equilibration, transport and particle production (e.g. hadronization and fragmentation) and the equation of state. Can we accurately determine η/s, the coupling $\alpha_s(T)$, the formation time (t_f), sound attenuation length, the q-hat parameter (energy loss per unit length) for parton energy loss and its underlying mechanisms as a function of temperature?

Investigation of the QCD phase diagram and its understanding is fundamentally important to the field. Are there excited modes of the high energy density QCD medium, and if so what are they? What are its constituents (are there quasi-particles, exotic states, others)? Is there a critical point in the QCD phase diagram, and if so can we locate it? Is the QCD Phase Diagram featureless above T_c?

We seek to better understand quarkonium melting (suppression) at the microscopic level. What role and to what extent does quark recombination play and impact the suppression? What

[3] "Small systems in p-Pb collisions," Markert C, Proceedings of this Conference

IOP Conf. Series: Journal of Physics: Conf. Series **878** (2017) 012002 doi:10.1088/1742-6596/878/1/012002

are the effects of cold nuclear matter on the observed suppression? Is the melting/suppression sequence as a function of temperature consistent with Lattice QCD calculations as they evolve?

By identifying the dependence on the system size, multiplicity and c.m. energy of observables that can be measured in pp, *p*-A and A-A collisions, we may be able to discriminate various mechanisms and models. To make progress clearly requires close cooperation between experiment and theory. Can there be new developments in theory (lattice, hydrodynamics, parton energy loss, string theory, others) and an understanding across fields, with possibly a combination of approaches where appropriate?

13. Future

In the near-term the RHIC Beam Energy Scan 2 is scheduled for 2018-2019.[4] Its goal is to investigate the QCD phase diagram by conducting a detailed scan at low c.m. energy in a search for a critical point. It will also study fluctuation phenomena and properties of the collisions that may shed light on the location of the critical point and properties of the medium.

A new detector sPHENIX[5] is being constructed for high statistics jet and quarkonium measurements at RHIC starting in 2022. These measurements seek to understand how the properties of the QGP emerge from the underlying quark and gluon interactions, including its temperature dependence and coupling strength. Precision jet and upsilon measurements will be undertaken to probe the different length scales of the QGP to better determine and understand its properties.

In Europe NICA[6] and FAIR[7] are expected to come online in the next several years. NICA is expected to be commissioned around 2020, and will collide heavy ions at c.m. energies from 4 - 11 GeV per nucleon pair in order to investigate the QGP phase transition and critical point. FAIR is on a similar timeline with an extremely diverse experimental program. In heavy ion physics FAIR will investigate very baryon-rich matter at high densities, study in-medium modification of hadrons and search for the critical point, to name a few goals.

A detector upgrade at LHC[8] in 2018-2020 will increase capabilities allowing higher rate heavy-ion operation in Run-3 starting in 2021. Along with detector upgrades, this will allow detailed investigation of heavy flavor jets, jet substructure, high p_T particles and heavy flavors, and event-by-event correlations of hard and soft processes.

In the longer term, an Electron Ion Collider (EIC) has been designated by the U.S. Nuclear Physics Community [42] as the next construction project for nuclear physics in the U.S. for operation in the post-2025 timeframe. The EIC will allow an in-depth investigation of the gluon-dominated low-x region of QCD. Goals are to investigate the position and momentum distributions of quarks, gluons and their spins in the nucleon; to determine and investigate the existence of gluon saturation and a possible color glass condensate; and to study the response of the nuclear medium to traversing color charge.

Together these new capabilities should provide a strong and exciting physics program internationally to investigate the questions raised by the field of heavy-ion physics and further illuminate the properties of high energy density QCD.

Acknowledgments
This work was supported by the US Department of Energy under Grant DE-SC004168.

[4] "STAR Beam Energy Scan," Keane D, Proceedings of this Conference
[5] "sPHENIX: The next generation heavy ion detector at RHIC," Sarah Campbell et al.(sPHENIX Collaboration), arXiv:1611.03003.
[6] "Status of the NICA Project," Trubnikov G, Proceedings of this Conference
[7] "Physics at FAIR: Opportunities for South Africa," Stoecker H, Proceedings of this Conference
[8] "The Future Collider Program," Gray H, Proceedings of this Conference

References

[1] Lee TD 1975 *Rev. Mod. Phys.* **47** 267–75
[2] Karsch F *et al.* 2001 *Nucl. Phys. B* **605** 579–99; Bazavov A *et al.* 2009 *Phys. Rev. D* **80** 014504; Petreczky P 2012 *J. Phys. G* **39** 093002
[3] Alford M, Rajagopal K, Reddy S and Wilczek F 2001 *Phys. Rev. D* **64** 074017
[4] Arsene I *et al.* (BRAHMS Collaboration) 2005 *Nucl. Phys. A* **757** 1-27; Back B *et al.* (PHOBOS Collaboration) 2005 *Nucl. Phys. A* **757** 28-101; Adams J *et al.* (STAR Collaboration) 2005 *Nucl. Phys. A* **757** 102-83; Adcox K *et al.* (PHENIX Collaboration) 2005 *Nucl. Phys. A* **757** 184-283
[5] Averbeck R, Harris JW and Schenke B 2015 *Heavy-Ion Physics at the LHC* in *The Large Hadron Collider—Harvest of Run 1*, Ed. Schörner–Sadenius T (Switzerland: Springer) chapter 9 pp 355–420
[6] Agakichiev G *et al.* (CERES Collaboration) 1995 *Phys. Rev. Lett.* **75** 1272–5
[7] Rapp R and Wambach J 2000 *Adv. Nucl. Phys.* **25** 1–164
[8] David G, Rapp R and Xu Z 2008 *Phys. Rep.* **462** 176–217
[9] Adamová D *et al.* (CERES Collaboration) 2003 *Phys. Rev. Lett.* **91** 042301
[10] Arnaldi R *et al.* (NA60 Collaboration) 2006 *Phys. Rev. Lett.* **96** 162302
[11] Adamczyk L *et al.* (STAR Collaboration) 2015 *Phys. Lett. B* **750** 64–71
[12] Adare A *et al.* (PHENIX Collaboration) 2016 *Phys. Rev. C* **93** 014904
[13] Adare A *et al.* (PHENIX Collaboration) 2010 *Phys. Rev. C* **81** 034911
[14] Adare A *et al.* (PHENIX Collaboration) 2010 *Phys. Rev. Lett.* **104** 132301
[15] Adam J *et al.* (ALICE Collaboration) 2016 *Phys. Lett. B* **754** 235–48
[16] Becattini F and Fries R 2010 *The QCD confinement transition: hadron formation* in *Relativistic Heavy Ion Physics* (*Landolt-Börnstein—Group I Elementary Particles, Nuclei and Atoms vol 23*) Ed. R Stock (Heidelberg: Springer) chapter 4 pp 208–39
[17] Andronic A, Braun–Munzinger P and Stachel J 2006 *Nucl. Phys. A* **772** 167–99
[18] Andronic A, Braun–Munzinger P, Redlich K and Stachel J 2017 *Journal of Physics: Conf. Series* (Darmstadt) **779** 012012; Floris M 2014 *Nucl. Phys. A* **931** 103–12
[19] Chatrchyan S et al. (CMS Collaboration) 2012 *Eur. Phys. J. C* **72** 1945; *Phys. Lett. B* **715** 66–87; *Phys. Lett. B* **710** 256–77
[20] Vitev I and Gyulassy M 2002 *Phys. Rev. Lett.* **89** 252301
[21] Renk T *et al.* 2011 *Phys. Rev. C* **84** 014906
[22] Salgado CA and Wiedemann UA 2003 *Phys. Rev. D* **68** 014008
[23] Armesto N *et al.* 2005 *Phys. Rev. D* **71** 054027
[24] Dainese A, Loizides C and Paic G 2005 *Eur. Phys. J. C* **38** 461
[25] Cronin J *et al.* 1975 *Phys. Rev. D* **11** 3105
[26] Abelev B *et al.* (ALICE Collaboration) 2012 *J. High Energy Phys.* **9** 112
[27] Chatrchyan S *et al.* (CMS Collaboration) 2012 *J. High Energy Phys.* **05** 063
[28] Maltoni F, Selvaggi M and Thaler J 2016 *Phys. Rev. D* **94** 054015; Abir R, Jamil U, Mustafa MG and Srivastava DK 2012 *Phys. Lett. B* **715** 183–9
[29] Chatrchyan S *et al.* (CMS Collaboration) 2014 *Phys. Rev. Lett.* **113** 132301
[30] Aad G *et al.* (ATLAS Collaboration) 2010 *Phys. Rev. Lett.* **105** 252303
[31] Chatrchyan S *et al.* (CMS Collaboration) 2011 *Phys. Rev. C* **84** 024906
[32] Matsui T and Satz H 1986 *Phys. Lett. B* **178** 416
[33] Chatrchyan S *et al.* (CMS Collaboration) 2012 *Phys. Rev. Lett.* **109** 222301; Khachatryan V *et al.* (CMS Collaboration) 2016 (submitted to *Phys. Lett. B*) arXiv:1611.01510
[34] Digal S, Petreczky P and Satz H 2001 *Phys. Rev. D* **64** 094015; Karsch F, Kharzeev D and Satz H 2006 *Phys. Lett. B* **637** 75-80
[35] Mocsy A, Petreczky P and Strickland M 2013 *Int. J. Mod. Phys. A* **28** 1340012; Kluberg L and Satz H 2010 *Color deconfinement and charmonium production in nuclear collisions* in *SpringerMaterials—The Landolt–Börnstein Database* ed R Stock (Berlin–Heidelberg: Springer) pp 373–423
[36] Reisdorf W and Ritter H 1997 *Ann. Rev. Nucl. Part. Sci.* **47** 663–709
[37] Gale C *et al.* 2013 *Phys. Rev. Lett.* **110** 012302
[38] Gale C *et al.* 2013 *Int. J. Mod. Phys. A* **28** 1340011
[39] Policastro G *et al.* 2001 *Phys. Rev. Lett.* **87** 081601
[40] Kovtun P *et al.* 2005 *Phys. Rev. Lett.* **94** 111601
[41] Heinz U and Snellings R 2013 *Annu. Rev. Nucl. Part. Sci.* **63** 123–51
[42] Aprahamian A *et al.* (The 2015 Nuclear Science Advisory Committee) 2015 *Reaching for the horizon—the 2015 long range plan for nuclear science* INSPIRE-1398831

IOP Conf. Series: Journal of Physics: Conf. Series **878** (2017) 012003 doi:10.1088/1742-6596/878/1/012003

Hadronic Resonance production in ALICE

Christina Markert for the ALICE Collaboration
The University of Texas at Austin, Physics Department, Austin, Texas, USA
E-mail: cmarkert@physics.utexas.edu

Abstract. In heavy ion collisions a fireball of hot and dense matter is created. Short lived hadronic resonances are sensitive to the medium properties, in particular to the temperature, density and system size. Resonance yields and momentum distributions are used to gain insight into the hadronic phase, its expansion velocity and time duration. The multiplicity dependent hadronic resonance production in p-p, p-Pb and Pb-Pb collisions will be discussed within the context of the possible extended hadronic and partonic phase. The experimental results will be compared to EPOS+UrQMD model calculations to discuss the system size dependent interactions of the hadronic medium on various resonances. Small systems such as p-p and p-Pb collisions will be discussed with respect to resonance and strange particle measurements.

1. Introduction

Hadronic resonances interacting throughout the partonic and hadronic phase of heavy ion collisions will leave an imprint from the different stages of the medium on the resonance properties such as mass, width, yield and momentum distribution. Initially we have been motivated to detect invariant mass shifts and width broadenings as chiral symmetry restoration signatures from a transition between the partonic and the hadronic nuclear matter. However, further interaction within the hadronic medium until the kinetic freeze-out will dilute these early signals. This sensitivity to hadronic phase interactions are now used to determine the length and temperature dependence of the hadronic phase which changes the yields and the momentum distribution of the resonances. This can be studied in detail within the centrality dependence of the collisions where the hadronic phase is expected to live the longest in the most central collisions with the highest multiplicity of produced particles. The lifetimes of hadronic phase interactions can be extracted via microscopic model calculations EPOS+UrQMD [1] including the partonic and hadronic phase interactions. The model calculations of resonance yields and momentum distributions are compared to measurements from the ALICE experiment at the LHC which show a general agreement. The suppression of the reconstructed yield of short lived resonances have been observed at RHIC and SPS energies [2] and explained within the UrQMD model [3] by an extended hadronic lifetime. As a further consequence of the hadronic phase interactions stable particle yields may change due to annihilation processes of particles with their antiparticles. This yield changes need to be taken into account when thermal model fits are applied to calculate the chemical freeze-out temperatures. Chemical freeze-out models have successfully described the particle abundance of ground state particles with a minimum set of parameters including the temperature and the baryo-chemical potential [4, 5, 6, 7, 8, 9]. Now hadronic resonance measurements can be used to determine the existence and the lifetime of an extended hadronic phase after chemical freeze-out.

2. Resonance production in Pb-Pb and minimum bias p-p collisions

Short lived resonances are reconstructed via invariant mass through their hadronic decay particles (table 1) which are identified via their masses. This is done in the ALICE detector mainly through the velocity measurement of the Time-Of-Flight (TOF) detector or the energy loss dE/dx in the Time Projection Chamber (TPC) gas where both depend on the momentum measured in the TPC and the Inner Tracking System (ITS). Figure 1 shows the momentum distributions of the $\phi(1020)$ (left) and $K^*(892)^0$ (right) mesons after extracting the invariant mass signal and accounting for detector acceptance and efficiencies. The resonances are measured in different centralities of Pb-Pb collisions selected by the number of charged hadrons ($dN_{ch}/d\eta$) and in p-Pb and p-p collisions [11, 12, 13, 14].

Table 1. The resonances are reconstructed experimentally via their listed decay channels [17].

Resonance	decay channel	branching ratio	lifetime (fm/c)
$\rho(770)^0$	$\pi^+ + \pi^-$	1	1.335
$K^*(892)^0$	$\pi^- + K^+$	0.67	4.16
$\phi(1020)$	$K^+ + K^-$	0.489	46.26

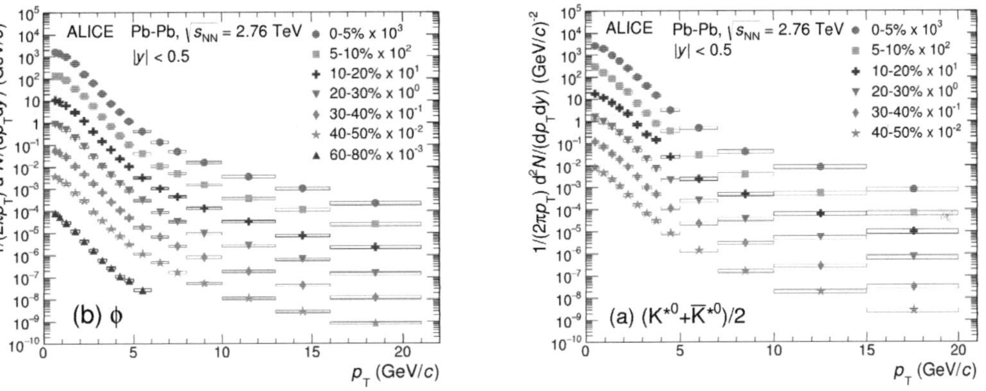

Figure 1. Production of $\phi(1020)$ (left) and $K^*(892)^0$ (right) mesons in p-p and Pb-Pb collisions at 2.76 TeV [11, 12].

The momentum integrated yield normalized to a ground state particle with same or similar quark content ($\phi(1020)$/K, $K^*(892)^0$/K and $\rho(770)^0/\pi$) versus centrality (($dN_{ch}/d\eta)^{1/3}$ used as a proxy for the system size and radius dependence [10]) are shown in figure 2. The $\phi(1020)$/K is nearly constant while the $K^*(892)^0$/K and $\rho(770)^0/\pi$ are suppresses for larger system size and for more central collisions. Within a statistical model description these particle ratios are mainly determined by the chemical freeze-out temperature T_{ch}, which is the same for different centralities as indicated in figure 3 with T_{ch}= 155 MeV. The predictions of a non-equilibrium model are shown in red triangles. The kinetic freeze-out temperature defined through the end of the elastic interactions of the system decreases with system size (figure 3 blue squares). Due to interactions in the hadronic phase and the smaller cross sections for resonance regeneration the reconstructed resonance yields for shorter lived resonances are expected to be suppressed. The observation of resonance suppression is mainly explained by the lifetime where the $K^*(892)^0$ and $\rho(770)^0$ have the shortest lifetimes (4 and 1.3 fm/c) compared to the $\phi(1020)$ (44 fm/c) (table 1). Since the $\rho(770)^0$ and the $K^*(892)^0$ show the same suppression one would expect that the $\rho(770)^0$ has a larger regeneration cross section from two pions to offset the larger re-scattering due to its three times shorter lifetime compared to the $K^*(892)^0$. EPOS+UrQMD

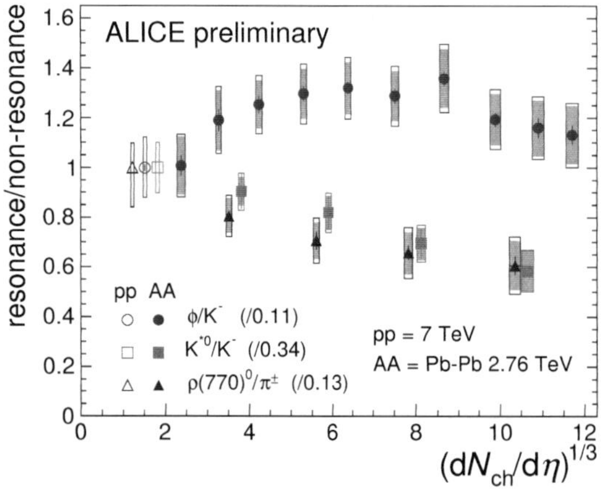

Figure 2. Resonance/non-resonance ratio as a function of charged hadrons for p-p and Pb-Pb collisions

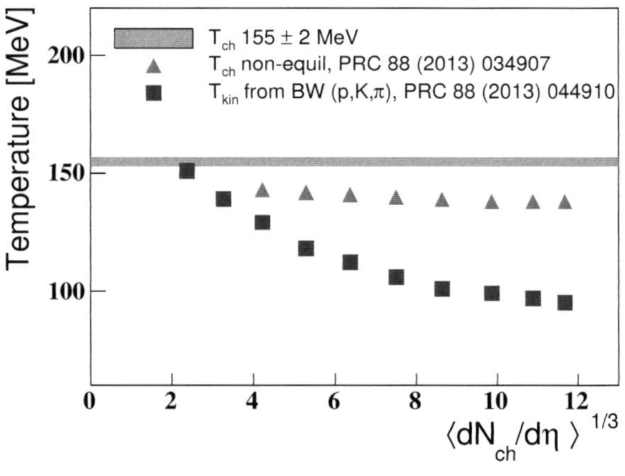

Figure 3. Chemical and kinetic freeze-out temperature as a function of charged hadrons for Pb-Pb collisions.

model predictions are able to describe the $K^*(892)^0$/K and $\rho(770)^0/\pi$ ratio versus $(\mathrm{dN}_{ch}/\mathrm{d}\eta)^{1/3}$ for Pb-Pb collisions at 2.76 TeV. Figure 4 shows the direct comparison of the $\phi(1020)$/K, $K^*(892)^0$/K as a function of charged hadrons [1].

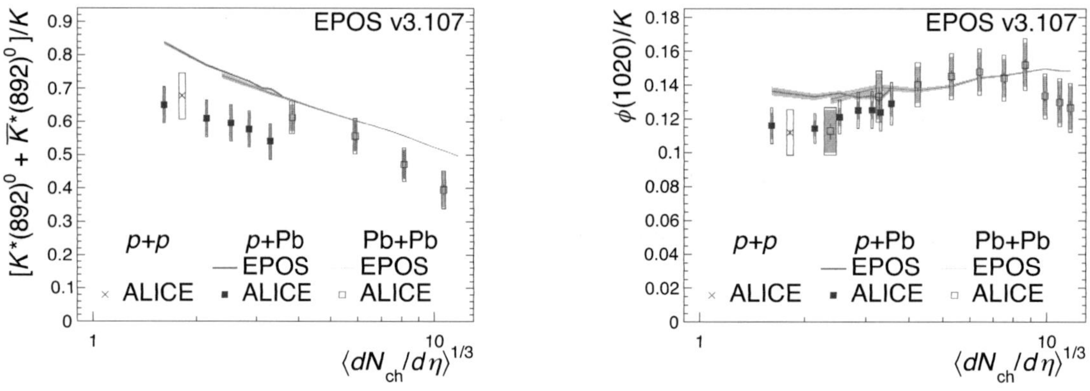

Figure 4. EPOS+UrQMD model calculations for $K^*(892)^0$/K (left) and $\phi(1020)$/K (right) in p-Pb (5.02 TeV) and Pb-Pb (2.76 TeV) collisions (red line) compared to ALICE data for p-p (7 TeV) , p-Pb (5.02 TeV) and Pb-Pb (2.76 TeV) collisions (blue points) [1].

Hadronic phase interactions predominantly occur in the low momentum region and therefore deviation from predicted momentum distributions are expected. The predicted momentum is derived from the shape of the Blast-Wave (BW) model (fit to proton, Kaon and Pion) and the resonance yield from $K^*(892)^0$/K and $\phi(1020)$/K in p-p collisions which are equal to the thermal model predictions for Pb-Pb collisions. The suppression of the $K^*(892)^0$ in the low momentum region for the most central collision 0-20% is visible in Figure 5. While the $\phi(1020)$ does not show any deviation from the predications due to its long lifetime and the reduced interactions of the decay particles in the hadronic medium. Since the BW does not describe the higher momentum $\mathrm{p}_T >$ 2 GeV/c data very well one would rather focus on the difference of the ratios

$\frac{K^*(892)^0_{data}}{K^*(892)^0_{exp}}$ and $\frac{\phi(1020)_{data}}{\phi(1020)_{exp}}$ (Figure 5 lower panel). EPOS+UrQMD model calculations are able to reproduce the suppression in the low momentum region [1].

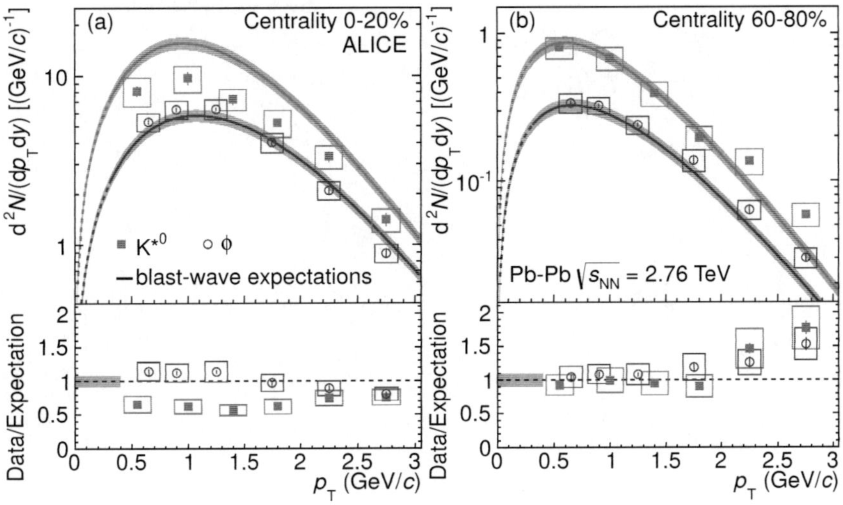

Figure 5. Transverse-momentum distributions of $K^*(892)^0$ and $\phi(1020)$ resonances in Pb-Pb collisions at 2.76 TeV along with expected distributions for central (a) and peripheral (b) collisions. The shapes of the expected distributions are given by Boltzmann-Gibbs blast-wave functions using parameters obtained from fits to $\pi^\pm$, $K^\pm$, and (anti)proton p_T distributions. The expected distributions are normalized so that their integrals from scaled p-p collisions are equal to the thermal model predictions for Pb-Pb collisions [11]. The lower panel shows the ratio of data divided by the expected value.

3. Influence of hadronic phase on stable particle yields

The statistical model successfully describes the particle yields (short resonances excluded) with a few parameters including the chemical freeze-out temperature (T_{ch}) and the baryochemical potential. The final state particles are a combination of directly produced and feed-down from weak and strong decays. Therefore the yield and the decay properties of the higher mass resonances are important. For example only about 35% of the measured Lambdas are directly produced at T_{ch} = 160 MeV [15]. The particle decay list (decay feed-down list), which are used for the statistical models include some feed-down assumptions from high mass resonances into lower mass resonances such as $\rho(770)^0$, $K^*(892)^0$, $\phi(1020)$, $\Delta(1232)^{++}$, $\Sigma(1385)^\pm$, $\Lambda(1520)$ and $\Xi(1530)^0$. These resonances have been measured in p+p and peripheral heavy ion collisions and can be directly compared to the statistical model predictions where the lifetime of the hadronic phase is at its minimum (in more peripheral collisions).

In more central collisions the extended hadronic phase changes the reconstructable resonances due to further interactions of their decay particles within the hadronic medium. This will not change the particles yields of the ground state particles but rather their momentum distribution. However, the annihilation process in the hadronic phase of particle - antiparticles in their ground states will change their yields (e.g. protons, Ξ and Ω). To determine the initial chemical freeze-out temperature ($T_{\rm ch}$) one needs to account for this effect. Therefore a microscopic model UrQMD was used to study the influence of the hadronic medium on the particle yields after chemical freeze-out [18, 19]. Figure 6 shows the ratio of the final state particle yield normalized to the yield at hadronisation for different densities. The protons and anti-protons show the largest

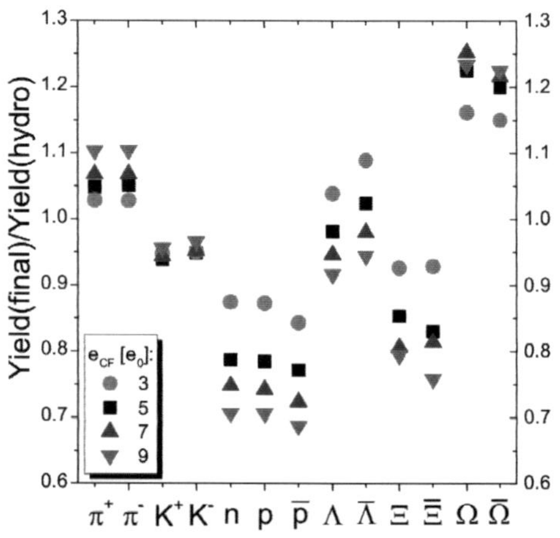

Figure 6. Ratios of particle multiplicities in the final state scaled by the particle multiplicity directly after the hydrodynamic stage for central Pb+Pb collisions at $\sqrt{s_{NN}}$=2.76 TeV. The different symbols denote different transition energy densities [16]

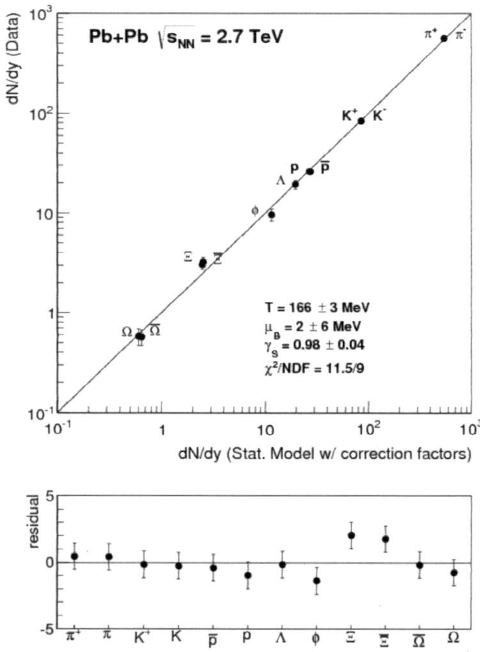

Figure 7. Statistical model fits to preliminary ALICE data for 20% central Pb-Pb collisions at $\sqrt{s_{NN}}$=7 TeV and to the same data but with modification factors from UrQMD applied in the statistical model fits [18].

suppression (absorption), the Ξ's show a smaller suppression and the Ω's show a slight increase. After correcting the hadronic yields for these effects, the thermal model fits are performed. The new temperature increases by 10 MeV and the χ^2/NDF decreases from 26 to 11 as shown in figure 7. This suggests that a systematic error of ΔT=10 MeV is present if hadronic phase interactions are not taken into account. If we understand the interactions of the hadronic phase after hadronisation and chemical freeze-out one can study the impact on correlation signatures. This was done by Steinheimer et. al. [20] who shows that about 50% of the initial fluctuations do not survive the hadronic phase in central heavy ion reactions. It seems to become more difficult to extract partonic phase (Quark Gluon Plasma (QGP)) signals or signals of the phase transition between partonic and hadronic matter when the extended hadronic phase is present. One might look into the more peripheral collisions where the hadronic phase is shorter and the kinetic freeze-out occurs at a higher temperature, or one may find the right momentum range where hadronic phase interactions are at their minimum. On the positive note one can now use resonance measurements to determine the length and expansion conditions of the hadronic phase and estimate its influence on certain observables. In small systems resonances are sensitive to the existence of existence of a possible hadronic phase. This might call into question whether an extended interacting partonic phase such as the QGP is a necessary precursor of the extended hadronic phase with high particle multiplicity.

4. Medium in small systems (p-Pb and p-p collisions)

Recent event multiplicity dependent $K^*(892)^0$ and $\phi(1020)$ measurements in small systems such as p-p and p-Pb are shown in Figure 8. The suppression of $K^*(892)^0$ follows the same trend with increasing multiplicity. The $\phi(1020)$ p-Pb is in agreement with the Pb-Pb results with no suppression due to its large lifetime. It seems that the suppression of the $K^*(892)^0$ is larger in small systems than in Pb-Pb collisions with same track multiplicity. The comparison to EPOS+UrQMD calculations in Figure 4 shows a good description of the trend with a smooth transition from small to large systems. EPOS includes in its description a QGP for the core and and single pQCD processes in the corona area. With this description a QGP is implemented by design. In this case the p-Pb collisions show the same dependence on the particle multiplicity as in Pb-Pb collisions. However, the data show a slightly larger suppression for the $K^*(892)^0$ in p-Pb than predicted by the model. And the p-p collisions show an even larger suppression (Figure 8). One can ask if the number of produced charged particles is the correct scaling variable or if the density which is larger in the small volumes of the collision systems plays a role in the scaling as well. If the suppression of the $K^*(892)^0$ resonance is caused by an extended hadronic phase then the EPOS+UrQMD model calculation estimates a lifetime of about 2 fm/c for the 0-5% highest multiplicity for the p-Pb collisions.

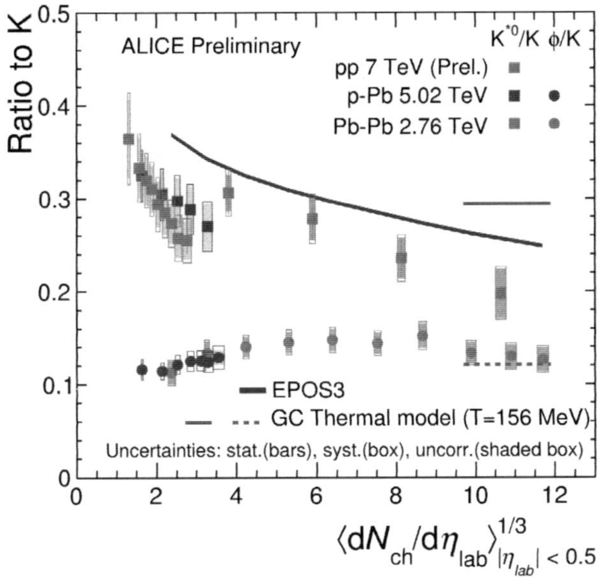

Figure 8. Resonance to stable particle ratio in p-p collisions at 7 TeV, p-Pb collisions at 5.02 TeV and Pb-Pb at 2.76 TeV for different multiplicity event classes comparison with other systems. The statistical uncertainties are shown as bars, systematic errors in boxes, and uncorrelated systematic errors in shaded boxes.

Differences in the dynamical evolution of the small systems is described by the transverse expansion velocity β_T, which drives the mean transverse momentum $\langle p_T \rangle$ and the kinetic freeze-out temperature, which is defined by the end of the elastic interactions. Figure 9 shows a steeper increase of the $\langle p_T \rangle$ for the smaller systems with increasing number of charged particles. Both resonances with similar masses as the proton, $\phi(1020)$ (80 MeV/c^2 larger) and $K^*(892)^0$ (40 MeV/c^2 smaller), show a larger $\langle p_T \rangle$ than the protons in p-p and p-Pb collisions. In Pb-Pb collisions the $K^*(892)^0$ follows the trend of the protons while the $\phi(1020)$ shows a lager $\langle p_T \rangle$ in peripheral collisions with a small increase with increasing centrality until it is in agreement with the values for protons in mid-central collisions. The number of charged hadrons in 0-5% p-Pb is the same as in 60-80% Pb-Pb collisions. However, the momentum spectra show a different expansion dynamics in this collision system. How much of the $\langle p_T \rangle$ change might be driven by a possible QGP phase in 0-5% p-Pb collisions is not clear at this moment.

Another observable in small systems is the strangeness production which shows increasing

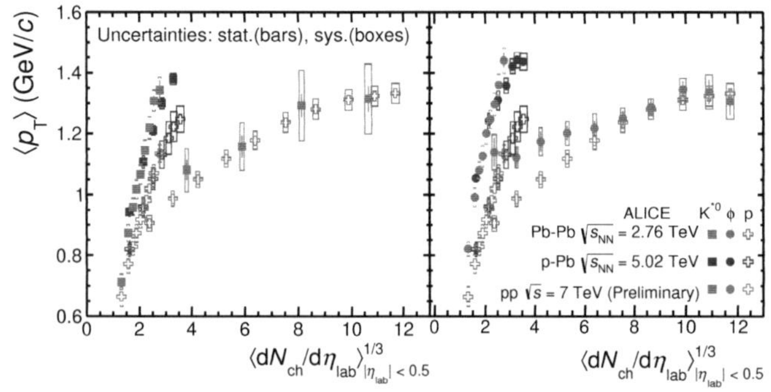

Figure 9. Mean transverse momentum $\langle p_T \rangle$ of protons for p-p, p-Pb and Pb-Pb collisions compared to $K^*(892)^0$ (left) and $\phi(1020)$ (right).

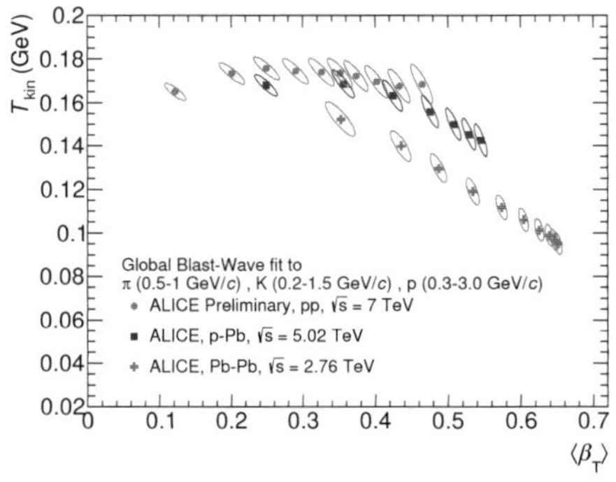

Figure 10. Expansion parameter transverse velocity β_T and kinetic freeze-out temperature T_{kin} from a Blast-Wave fit to π, K and p spectra measured in Pb-Pb collisions at 5.02 TeV.

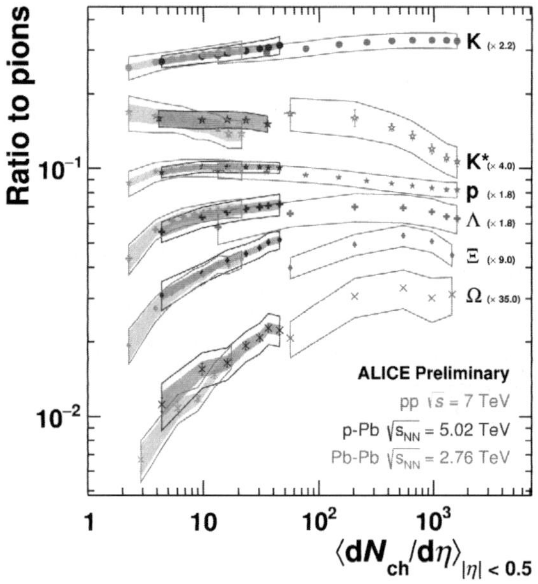

Figure 11. Particle yield ratio to pions of various identified particle species as a function of $\langle dN_{ch}/d\eta \rangle$ (measured at $|\eta| < 0.5$) in pp at 7 TeV, compared to pPb at 5.02 TeV and PbPb at 2.76 TeV. The hollow error band indicates the total systematic uncertainties, the shaded ones are the systematic uncertainties that are independent of multiplicity (latter only shown for pp and pPb).

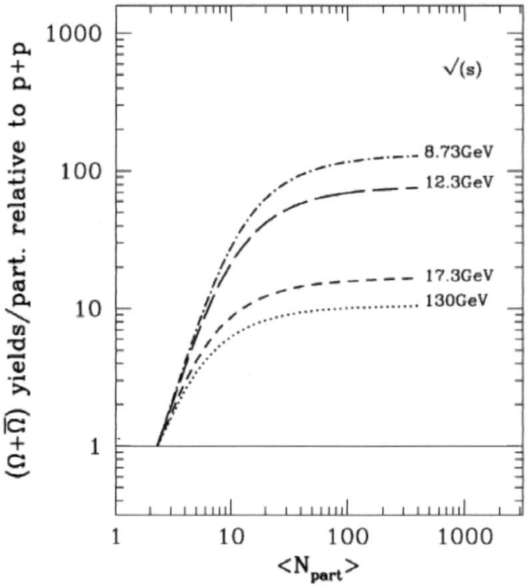

Figure 12. Centrality dependence of relative enhancement $(\Omega + \bar{\Omega})$ Yields/participant in central Pb-Pb to p-p reactions at different collisions energies.

strangeness/non-strangeness particle ratio. The strangeness production was suggested as one of the initial signatures for the QGP [21]. The increase of strangeness production as a function of centrality (charged particle multiplicity) with respect to the production in p-p collisions is shown in figure 11. This trend can be exlpained mainly driven by the strangeness suppression in p-p collisions due to a canonical description (small system size). The increase in heavy ion reactions is rather a vanishing of the strangeness suppression in small systems. This volume effect was calculated by K. Redlich for RHIC energies [22] (figure 12). Since the increase is rather a volume effect, the "real" strangeness enhancement is an enhancement in addition to the volume effect. The "real" strangeness enhancement has never been confirmed in heavy ion reactions within its statistical and systematical uncertainties. It might be also due to the difficulty of defining the turning point (volume and particle multiplicity) where the canonical system fully transfers into a grand canonical system.

The new strangeness production results in small systems from ALICE in p-p and p-Pb collisions with increasing number of produced hadrons show the same trend of the volume dependence. However, the volume effect of a canonical system also needs to be taken into account here. For sure the increase of strangeness production is an indication of the interactions of the extended hadronic phase. Whether this is driven by a extended partonic QGP phase cannot be answered at this point. Further investigation of the volume effect should help us to have a new look at this data to extract any possible signature of a QGP.

Acknowledgments

I would like to thank the organizer for inviting me to this very nice place, which stimulated very fruitful discussions in our field. The friendliness and open atmosphere allowed a deeper and critical discussion. I really enjoyed it very much and I was happy to be surrounded by this energetic and friendly group of scientists. I also would like to thank the ALICE collaboration for supporting my presentation.

References

[1] A.G. Knospe, C. Markert, K. Werner, J. Steinheimer, M. Bleicher, Phys.Rev. **C93** (2016) no.1, 014911.
[2] B.I. Abelev et al., (STAR Collaboration), Phys. Rev. Lett. **97**, 132301 (2006)
[3] M. Bleicher *et al.*, Phys. Lett. **B530**, 81 (2002);
M. Bleicher, Nucl. Phys. **A715**, 85 (2003).
[4] J. Cleymans, K. Redlich, Phys. Rev. Lett. **81** (1998) 52845286.
[5] J. Cleymans, H. Oeschler, K. Redlich, S. Wheaton, Phys.Rev. **C73** (2006) 034905.
[6] F. Becattini et al, JPG 025002 (2011)
[7] R. Rafelski, Letessier, Phys. Rev. **C83**, 054909 (2011)
[8] M. Petran, J. Letessier, V. Petracek, J. Rafelski, Phys.Rev. **C88** (3) (2013) 034907.
[9] P. Braun-Munzinger, K. Redlich, J. Stachel arXiv:nucl-th/0304013.
[10] K. Aamodt et al. Phys.Lett. **B696** (2011) 328-337
[11] B. Bezverkhny Abelev *et al.* (ALICE collaboration), Phys. Rev. **C91** 2, (2015) 024609.
[12] J. Adam et al., arXiv:1702.00555 [nucl-ex]
[13] J. Adam et al., Eur. Phys. J. **C76** (2016) no.5, 245.
[14] B. Abelev et al., Eur. Phys. J. **C72** (2012) 2183.
[15] J. Letessier, J. Rafelski, Cambridge Monographs on Particle Physics and Cosmology page 31 (2002).
[16] J. Steinheimer, J. Aichelin, M. Bleicher, Phys. Rev. Lett. **110** (2013) no.4, 042501.
[17] K. A. Olive *et al.* (Particle Data Group), Chin. Phys. **C38**, 090001 (2014)
[18] F. Becattini, M. Bleicher, T. Kollegger, T. Schuster, J. Steinheimer, R. Stock, Phys. Rev. Lett. **111** (2013) 082302.
[19] F. Becattini, E. Grossi, M Bleicher, J. Steinheimer, R. Stock, Phys. Rev. C90 (2014) no.5, 054907.
[20] J. Steinheimer, V. Vovchenko, J. Aichelin, M. Bleicher, H. Stöcker, Aug 12, 2016, arXiv:1608.03737.
[21] J. Rafelski, B. Muller, Phys. Rev. Lett. 48 (1982) 1066.
[22] A. Tounsi, K. Redlich, J.Phys. G28 (2002) 2095-2102.

Production of Nuclei, Hyper-Nuclei and Exotics in the ALICE Detector

Benjamin Dönigus (for the ALICE Collaboration)

Institut für Kernphysik, Goethe-Universität Frankfurt, Max-von-Laue-Str. 1, 60438 Frankfurt, Germany

E-mail: benjamin.doenigus@cern.ch

Abstract. Measurements of light (anti-)nuclei and (anti-)hypernuclei have been performed in different collision systems, namely pp, p–Pb and Pb–Pb, covering a large range of physics topics at the LHC using the ALICE detector setup. The results of the production of in Pb–Pb collisions is following the expectation of thermal models, whereas the production of nuclei in p–Pb collisions is tending to closer follow the expectations from coalescence models. Both models can give predictions for the production of hypothetical states such as bound states of two Λ hyperons or of a Λ and a neutron, which are expected to decay weakly. The states are searched for and no signal is found, thus upper limits are set which are significantly below the model expectations. Further, the measurement of the lifetime of the hypertriton determined in Pb–Pb collisions at the LHC gives a value significantly below the expectations of the most recent theoretical calculation. Instead this measurement is in agreement with other recent measurements in heavy-ion collisions. The combination of all measurements gives a tension of more than 2σ between the experimental lifetime and the most recent model calculation.

1. Introduction

The ALICE Collaboration has, among others, a program to investigate many different kinds of (multi-)baryon states predicted and expected by QCD and QCD inspired models [1]. One way of approaching this research topic is to start by measuring rarely observed anti- and hypermatter, which is produced copiously in Pb–Pb collisions at the LHC. With these data the model predictions, e.g. the thermal (statistical) model and coalescence approaches are tested. Ultimately, the aim is to understand the production mechanisms acting in these collisions and resulting in large production yields for (anti-)nuclei, (anti-)hypernuclei and exotica still to be discovered or excluded by setting strong upper limits.

The thermal model at the LHC has basically only one key parameter: the chemical freeze-out temperature T_{ch}. This statement is valid because the baryo-chemical potential μ_B is close to zero, and therefore often treated as being zero in thermal model fits. The nuclei themselves are subject to a strong sensitivity on T_{ch}, because of their large mass and the exponential behaviour of the expected particle yields $\mathrm{d}N/\mathrm{d}y$ on the particle mass m ($\mathrm{d}N/\mathrm{d}y \propto \exp(-m/T_{ch})$). The interesting fact to mention here, is that the binding energies of these objects are very small compared to the chemical freeze-out temperature T_{ch} [2].

Another model approach expects the constituents of a (hyper-)nucleus produced from the fireball to come close in phase space and form the (hyper-)nucleus through coalescence. In this model the produced nuclei could break apart and be created again by final-state coalescence.

IOP Conf. Series: Journal of Physics: Conf. Series **878** (2017) 012004 doi:10.1088/1742-6596/878/1/012004

2. (Anti-)nuclei

The measurement of (anti-)nuclei is performed using the time projection chamber (TPC) and the time-of-flight detector (TOF). The TOF is used in addition where the TPC is no longer sufficient to separate the different nuclei from other hadrons using the specific energy-loss $\mathrm{d}E/\mathrm{d}x$. One way to combine the information of both detectors is the measurement of the mass of the investigated particle. This is using the $\beta = \mathrm{v}/c$ from the TOF and the measured momentum, strongly constrained after a pre-selection based on the TPC measurement. A result for the measured $(m/z)^2$ for deuteron, ^{3}He and the corresponding anti-particles in Pb–Pb collisions is shown in Fig. 1. From this measurement, one can deduce a mass difference and the binding energy difference between particle and anti-particle. This test of the CPT invariance in the nuclei sector shows no hint of any violation [3].

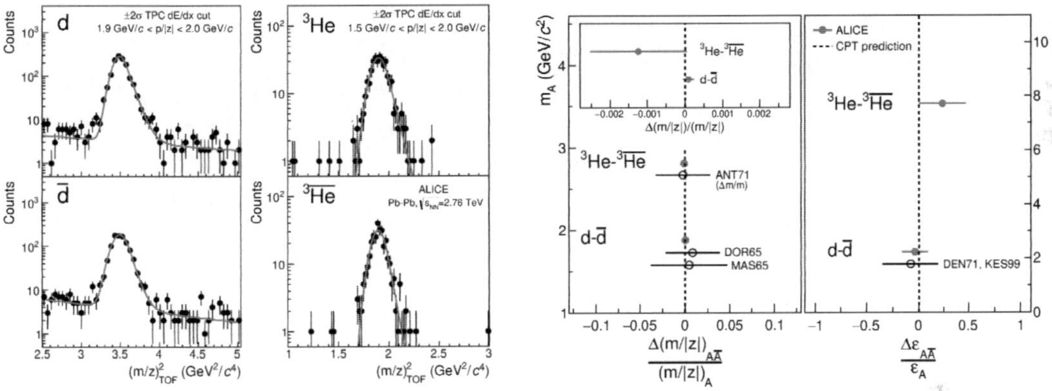

Figure 1. The left panel shows the squared ratio of mass and charge number $(m/z)^2$, distributions for d, $\bar{\mathrm{d}}$, ^{3}He and $^3\overline{\mathrm{He}}$. The right panel depicts the normalised mass difference $\frac{\Delta(m/|z|)_{A\bar{A}}}{(m/|z|)_A}$ and the nomalised binding energy difference $\frac{\Delta\epsilon_{A\bar{A}}}{\epsilon_A}$. Both figures from [3].

Furthermore, spectra are extracted dividing the signal in p_{T} slices and correcting for acceptance and efficiency. The result for deuterons in pp collisions and for five different centralities in Pb–Pb is displayed in Fig. 2, together with the ratios of the anti-nuclei to nuclei in the same centrality classes. In addition, the ratio is also shown for ^{3}He. All ratios are compatible with unity as a function of transverse momentum which is expected by thermal and coalescence models.

The spectra in Pb–Pb show a clear hardening with increasing centrality and are fitted with blast-wave distributions [4], mainly to extrapolate to the unmeasured region of the spectra and thus to extract the integrated production yield. A combined blast-wave fit (resulting in one common set of T_{kin}, $\langle\beta\rangle$ and n) to π, K, p, d and ^{3}He, shown in Fig. 3, gives a good fit for all particles (the deviation of pions at low p_{T} is due to feed-down from resonances mainly contributing in this region). This shows that nuclei are following the same radial expansion as the lighter hadrons in central Pb–Pb collisions.

The ratio of the production yields of deuterons to protons in pp, p–Pb and Pb–Pb collisions as a function of the mean multiplicity $\langle \mathrm{d}N_{\mathrm{ch}}/\mathrm{d}\eta\rangle$ is depicted in Fig. 4. It shows an increase going from pp multiplicities over p-Pb multiplicities until a saturation seems to be reached in peripheral Pb–Pb collisions. The increase with multiplicity is expected from coalescence models whereas the basically constant behaviour at higher multiplicities is expected by the thermal model.

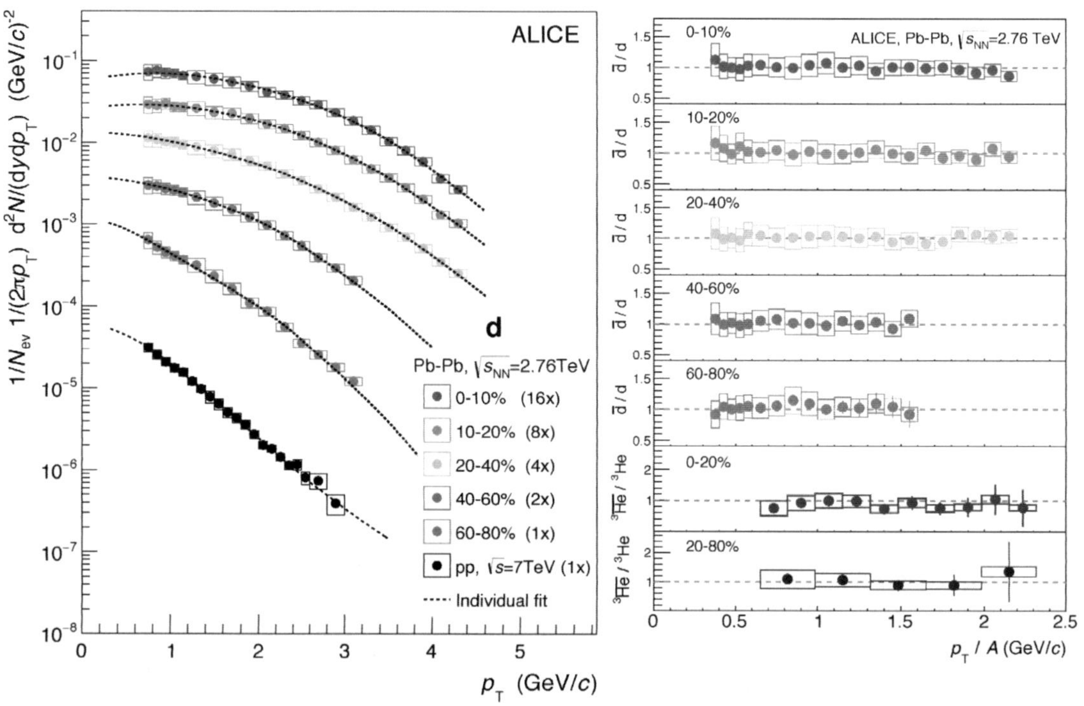

Figure 2. Invariant transverse momentum spectra for deuterons in pp and five centrality classes in Pb–Pb (left panel) and ratio between anti-nuclei and nuclei ($\bar{\mathrm{d}}$/d and $^3\overline{\mathrm{He}}$) for the different centrality classes in the lower panel. Figures taken from [5].

Figure 5 is showing the production yield of nuclei in Pb–Pb and p–Pb collisions as a function of their mass together with exponential fits. As mentioned above an exponential decrease with mass is expected from the thermal model. The penalty factor for adding a baryon to the system is about 300 in Pb–Pb and around 600 in p-Pb, which means the suppression when going from d to ^{3}He is 300 in Pb–Pb and 600 in p–Pb.

A non-central collision is creating an almond shape fireball which is basically a spatial anisotropy. This spacial anisotropy (eccentricity) will be transformed into a momentum anisotropy from the pressure gradients inside the fireball. Therefore the spectra are usually expanded into a Fourier series in azimuth and the main contribution is given by the so-called elliptic term v_2, the Fourier coefficient of the second order. The measured v_2 is displayed in Fig. 6 for π, K, p and d, which follows the mass ordering expected from hydrodynamic models. The full lines represent a blast-wave fit to π, K and p, and a prediction using this fit for the deuteron. This is describing the v_2 of deuterons rather well, whereas a simple coalescence model (dashed magenta region), scaling the v_2 of protons accordingly, fails in the description.

3. (Anti-)hypernuclei

The hypertriton is the lightest hypernucleus, consisting of a proton, a neutron and a Λ hyperon. It decays weakly, for instance into a ^{3}He and a π^-, with a lifetime close to the one of the Λ hyperon. The separation energy of the Λ is only 130 keV, which gives from simple quantum mechanical calculations a size of about 10 fm of the hypertriton. This already indicates that the

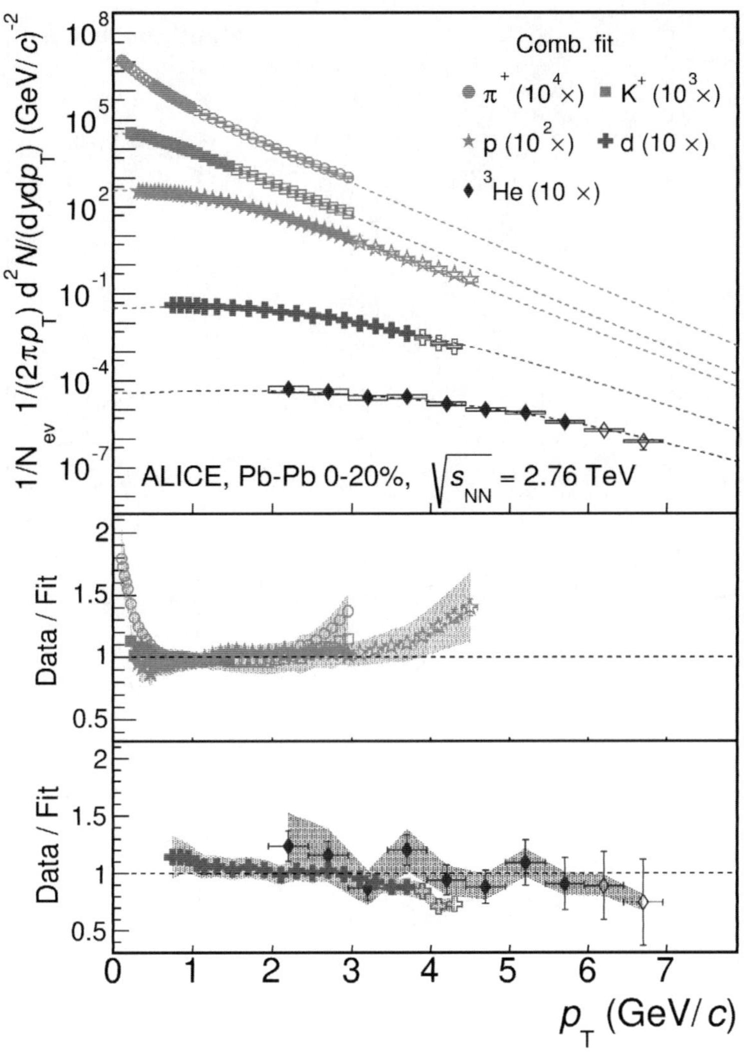

Figure 3. Invariant transverse momentum spectra for π^+, K^+, p, d, ^{3}He in 0-20% fitted with one global blast-wave function. The lower panels show the comparison of the model with the data. From [5].

lifetime should be very close to the one of the free Λ. A more sophisticated model involving also three-body forces also expects it to be close to it [6]. In addition, this model gives a branching ratio of the hypertriton into the ^{3}He+π^- of about 25%. The branching ratios (B.R.) of the hypertriton decay modes are not well known because many channels involve neutral particles such as neutrons and π^0 which are not accessible by the production experiments.

The measured $\mathrm{d}N/\mathrm{d}y\times$B.R. of the hypertriton in 0-10% central Pb–Pb collisions as a function of the branching ratio is depicted in Fig. 7, together with models describing its production. The thermal model describing the production of light hadrons and nuclei well, is in agreement with the measurement around the 25% branching ratio expected from the previously mentioned

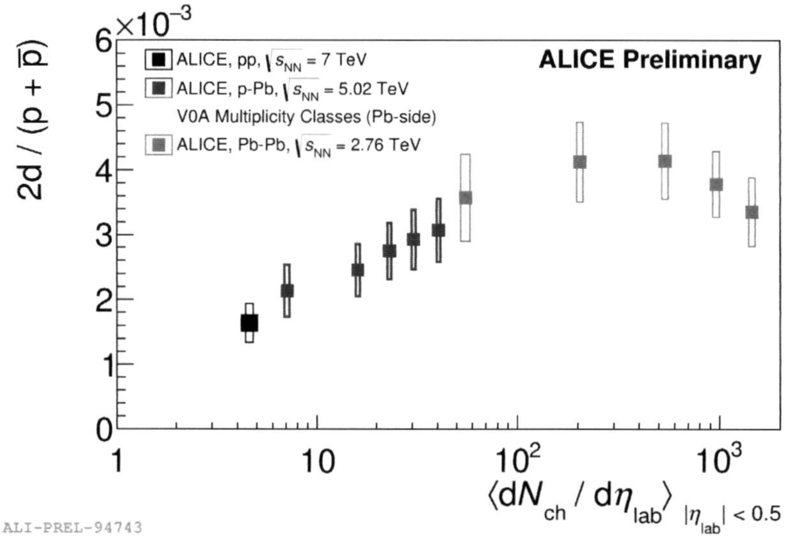

Figure 4. Ratio of the production yields of deuterons to protons ($2d/(p+\bar{p})$ as a function of the mean multiplicity $\langle dN_{ch}/d\eta \rangle$. The black point is extracted from the minimum bias pp measurement, the p-Pb results are preliminary results from a multiplicity dependent study and the Pb-Pb corresponds to the five centrality classes discussed before.

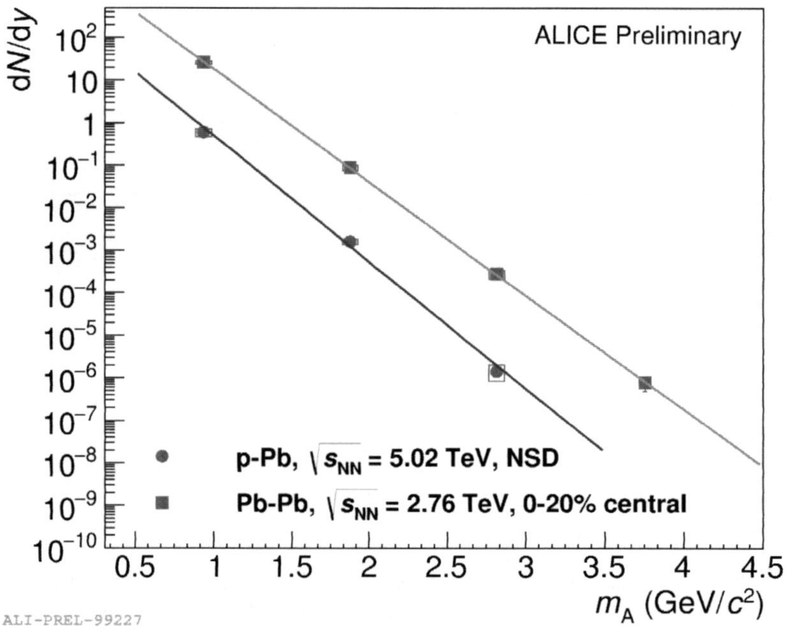

Figure 5. Production yield of different nuclei as a function of their mass. The lines indicate exponential fits through the data points at the two energies.

model.

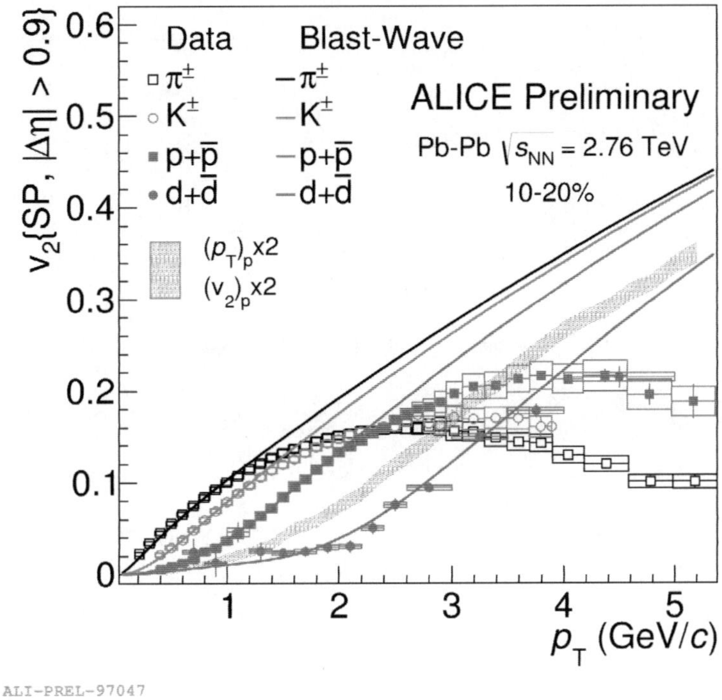

Figure 6. Elliptic flow coefficient v_2 for pions, kaons, protons and deuterons. The measured data points are compared with a blast-wave fit of pions, kaons and protons which is used as a prediction for deuterons, full lines. The dashed region indicates the prediction of a naïve coalescence model assuming a simple scaling of the v_2 of protons by a factor of 2 to get the v_2 of deuterons.

The extracted lifetime is displayed in Fig. 8 and compared with previous measurements. An average calculated following the PDG description is also shown which deviates from the free Λ hyperon lifetime by more than 2σ. The measurements before 2010 are done with emulsions whereas the latest ones are all from heavy-ion collisions. This lead to the discussion in the hyper-nuclei community that the measurements in heavy-ions might be biased because of the more rough collision environment there, which could influence the observed lifetime of the hypertriton. Therefore measurements of the lifetime in elementary collisions are desirable.

4. Exotica searches

Since the production yields of composite objects (such as light (anti-)(hyper-)nuclei) are rather well described by both production models they can be used to predict the production of exotic objects such as bound states among hyperons or hyperons and nucleons. Of particular interest is the so-called H-dibaryon, a hexaquark state composed of *uuddss* ($\Lambda\Lambda$) predicted by R. Jaffe in a bag model calculation in 1977 [8]. It is investigated in the weak decay mode $\Lambda + \mathrm{p} + \pi^-$. Another investigated bound state is a possible bound state of Λn which would decay into d $+\pi^-$. Since no signal was observed, upper limits have been extracted in a wide phase space of lifetime and branching ratio.

The upper limits of the investigated bound states are depicted in Fig. 9 as function of the

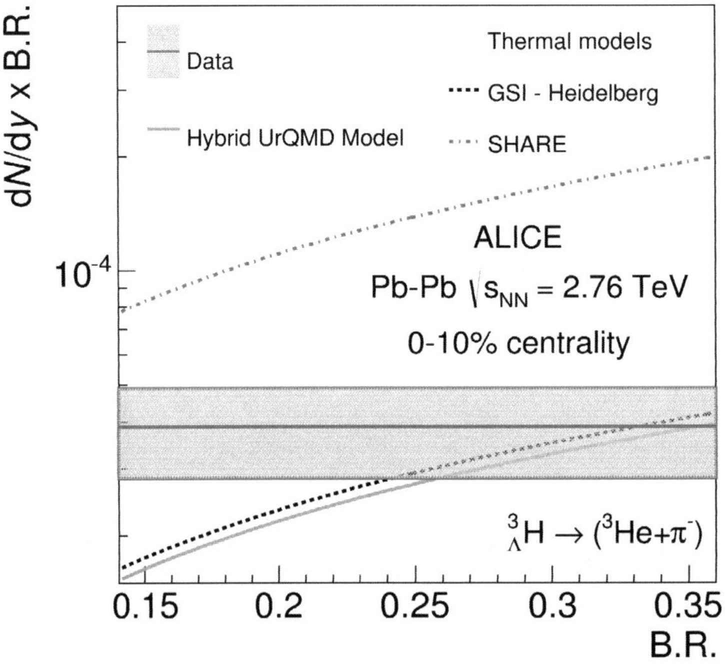

Figure 7. p_T-integrated rapidity density times branching ratio as a function of branching ratio (dN/dy×B.R. vs B.R.). The horizontal line is the measured dN/dy×B.R. and the band around it represents the quadratic sum of statistical and systematic uncertainties. The dashed lines indicate different theoretical expectations. Taken from [7].

branching ratio assuming a lifetime of the free Λ hyperon. All production models are far away from the extracted upper limits in reasonable regions of the branching ratio. Only if the branching ratios are very small, the model predictions reach down to the upper limit value.

The dependency of the upper limit on the lifetime is displayed in Fig. 10 and shows that only at very long lifetimes (about 3 m) the upper limit of the H-dibaryon comes close to the thermal model expectation. For the Λn bound state the upper limits are for all configurations under study below the model expectation.

5. Summary and outlook

The results from the production of (anti-)nuclei (in pp, p–Pb and Pb–Pb collisions) and (anti-)hypernuclei (in Pb–Pb collisions) are described rather well by the thermal and the coalescence models, but a simple coalescence model fails in describing the v_2 of the deuteron. Both models can make predictions for the production of exotic objects, such as the H-dibaryon and a Λn bound state. These exotica were searched for and no signals were observed, thus upper limits were set which are significantly below the expectations of both models. A comparison of the thermal model involving all discussed states is displayed in Fig. 11. It shows that the thermal model describes the yields of the known states very well, even the hypertriton having a Λ separation energy of only 130 keV is in good agreement. Instead, the upper limits of the exotic states are more than a factor 25 below the expectation. This makes it rather hard to accept

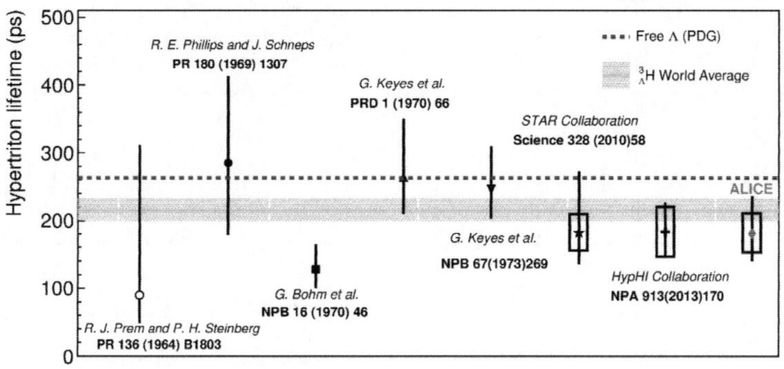

Figure 8. Measured hypertriton lifetime (red symbol) compared with previously published results. The band represents the world average of the hypertriton lifetime measurements. The dashed line indicates the lifetime of the Λ hyperon as reported by the Particle Data Group. Taken from [7].

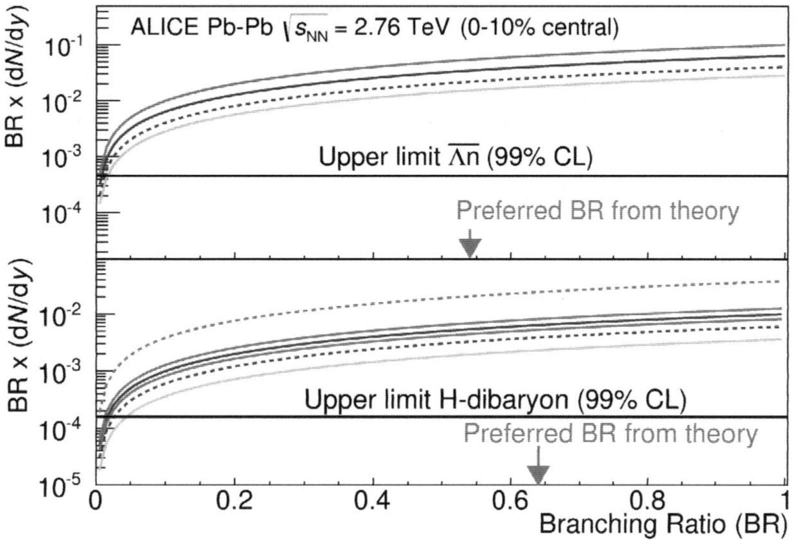

Figure 9. Extracted upper limit dN/dy as function of the branching ratio compared with different models. For details see [9].

that these particles exist in the investigated phase space.

Further, the measurement of the lifetime of the hypertriton determined in Pb–Pb collisions at the LHC gives a value significantly below the expectations of the most recent calculation. This measurement is in agreement with other recent measurements in heavy-ion collisions and in emulsions. The combination of all measurements gives a tension of more than 2σ between the experimental lifetime and the most recent model calculation. The repetition of this measurement in more elementary collisions is one aim of run 2, which has started in 2015 and will deliver significantly larger data sets in higher collision energies as before ($\sqrt{s} = 13$ TeV in pp, $\sqrt{s_{NN}} = 8$ TeV in p–Pb and $\sqrt{s_{NN}} = 5$ TeV in Pb–Pb). This data will allow also to complement the

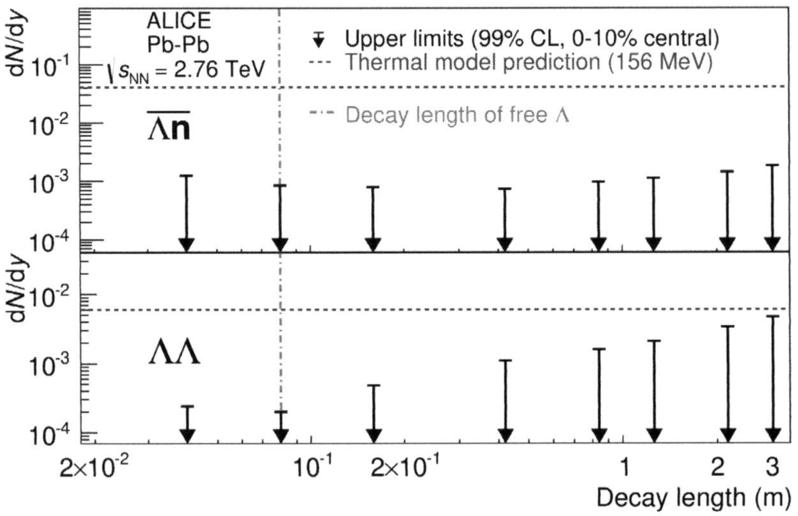

Figure 10. Upper limit dN/dy as function of the decay length compared with different models. For details see [9].

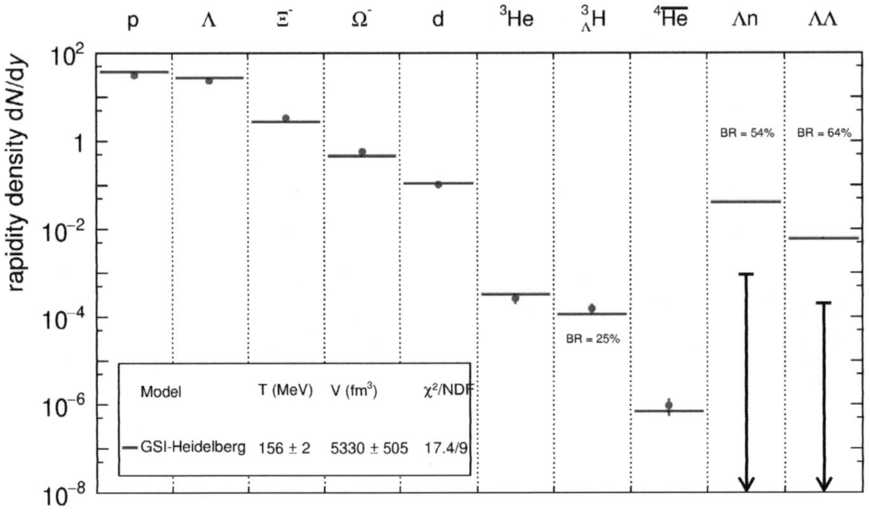

Figure 11. Comparison of production yields dN/dy of baryons, light (hyper-)nuclei and exotica with thermal model predictions. Hypertriton was corrected by 25% for the branching ratio, Λn by 54% and ΛΛ by 64%. As shown in [10].

measurements reported here and perform more systematic studies, for instance to measure the production of deuterons in pp collisions as a function of multiplicity and in particular at high multiplicities.

After 2018 in the so-called long shutdown 2 (LS2) two main detectors will be upgraded. The current silicon detectors of different technology will be replaced by pixel layers which will increase the (secondary) vertex resolution significantly [11]. The multi-wire proportional chambers of the TPC will be replaced by four layers of GEM detectors which will allow for continuous readout

IOP Conf. Series: Journal of Physics: Conf. Series **878** (2017) 012004 doi:10.1088/1742-6596/878/1/012004

at rates of 50 kHz [12]. These upgrades will allow the ALICE detector to cope with the high luminosity expected in run 3 (starting in 2021). In run 3, a data sample of 10^{10} central events is expected to be collected, which gives the possibility to study the production of the presented particles with high precision and will allow to study also hyper-nuclei with A=4 [1,11].

6. Acknowledgments
The author acknowledges the support from the German Bundesministerium fur Bildung, Wissenschaft, Forschung und Technologie (BMBF) though the project grant 05P2015 - ALICE at High Rate (BMBF-FSP202).

7. References
[1] Abelev B et al. (ALICE Collaboration), 2014 *J. Phys. G: Nucl. Part. Phys.* **41** 087001
[2] Andronic A et al., 2011*Phys. Lett. B* **697** 203
[3] Adam J et al. (ALICE Collaboration), 2015 *Nature Phys.* **11** 811
[4] Schnedermann E et al., 1993 *Phys. Rev. C* **48** 2462
[5] Adam J et al. (ALICE Collaboration), 2016 *Phys. Rev. C* **93** 02491
[6] Kamada H et al., 1998 *Phys. Rev. C* **57** 1595
[7] Adam J et al. (ALICE Collaboration), 2016 *Phys. Lett. B* **754** 360
[8] Jaffe R L, 1977 *Phys. Rev. Lett.* **38** 195 and 617
[9] Adam J et al. (ALICE Collaboration), 2016 *Phys. Lett. B* **752** 267
[10] Braun-Munzinger P, Dönigus B, Löher N, 2015 *CERN Courier* **September**
[11] Abelev B et al. (ALICE Collaboration), 2014 *J. Phys. G* **41** 087002
[12] Adam J et al. (ALICE Collaboration), 2014 *CERN-LHCC-2013-020, ALICE-TDR-016*

IOP Conf. Series: Journal of Physics: Conf. Series **878** (2017) 012005 doi:10.1088/1742-6596/878/1/012005

Identified hadron production in pp collisions measured with ALICE.

Yasser Corrales Morales for the ALICE Collaboration
University and INFN of Turin, Via Pietro Giuria 1, 10125, Turin, Italy.
E-mail: corrales@to.infn.it

Abstract. The production of identified hadrons in proton-proton collisions is frequently studied as a reference for the investigation of the strongly-interacting medium created in heavy-ion collisions. In addition, at LHC energies measurements in pp and p–Pb collisions as a function of the event multiplicity have shown some features reminiscent of those related to collective effects in Pb–Pb collisions. Thanks to its excellent PID capabilities and p_{T} coverage, the ALICE detector offers a unique opportunity for the measurement of p_{T} spectra, integrated yields ($\mathrm{d}N/\mathrm{d}y$) and mean transverse momenta ($\langle p_{\mathrm{T}} \rangle$) of identified light-flavour hadrons at mid-rapidity over a wide p_{T} range.

In this contribution, results on π, K, p, $\mathrm{K}^0_{\mathrm{S}}$, Λ, Ξ, Ω and K^{*0} as a function of multiplicity in pp collisions at $\sqrt{s}$= 7 TeV are presented. The results are compared with those measured in p–Pb and Pb–Pb collisions. A similar evolution of the spectral shape, the p_{T}-differential particle ratios and the integrated yield ratios with the charged particle multiplicity in both small and large systems is observed. The production rates of strange hadrons in pp collisions increase more than those of non-strange particles, showing an enhancement pattern with multiplicity which is remarkably similar to the one measured in p-Pb collisions.

In addition, results on the production of light flavour hadrons in pp collisions at $\sqrt{s}$= 13 TeV, the highest centre-of-mass energy reached so far in the laboratory, are also presented and the behaviour observed as a function of $\sqrt{s}$ are discussed.

1. Introduction

Heavy ion collisions, such as Pb–Pb collisions at the Large Hadron Collider (LHC) at CERN, offer a unique possibility to study in the laboratory the properties of the strongly-interacting matter under extreme conditions of pressure, energy density and temperature. In particular, the deconfined quark-gluon plasma (QGP) system which is predicted by quantum chromodynamic (QCD) calculations can be investigated. The onset of this strongly interacting phase is accompanied by some peculiar signatures in the final state. These include the collective flow of particles created in the collision and an increase in the relative production rates of strange and multi-strange hadrons with respect to minimum-bias pp and p–Pb collisions.

The measurements performed in smaller systems (e.g systems created in pp and p–Pb collisions) provide the reference data for the interpretation of the nucleus-nucleus collision results. However, the analysis of the data samples of high multiplicity pp and p–Pb collisions collected during LHC Run-1, has surprisingly produced results qualitatively resembling the collective behaviour obtained in Pb–Pb collisions at LHC and Au-Au collisions at RHIC. Observations like long-range correlation structures in pseudo-rapidity observed in two-particle angular correlations [1, 2, 3, 4] and the hardening of the transverse momentum spectra of identified particles [5] motivated several theoretical investigations [6, 7, 8, 9] as well as an effort on the experimental side in order to provide more comprehensive data from pp and p–Pb collisions. In this context, the study of identified hadrons as a function of the particle multiplicity in pp collisions allows to obtain better insight into the dynamics of small systems.

The ALICE apparatus [10] is well suited for the study of identified particles thanks to its moderate magnetic field, its high detector granularity and its excellent particle identification (PID) capabilities. In particular, PID is ensured over a wide transverse momentum range in the region at mid-rapidity via the measurement of specific energy loss in the Inner Tracking System (ITS) and in the Time Projection Chamber (TPC). Moreover, the particle velocity is measured via the Time Of Flight (TOF) detector, while the High Momentum Particle IDentification (HMPID) detector identifies particles by measuring the angle of emission of Cerenkov light emission.

2. Analysis Description

The measurements of pions, kaons and protons are performed combining the information from ITS, TPC, TOF and HMPID detectors [11]. Furthermore, a statistical PID method based on the relativistic rise of the $\mathrm{d}E/\mathrm{d}x$ in the TPC is applied to extend the measured p_T interval up to ≈ 20 GeV/c [12]. Strange (K^0_S, Λ, $\bar{\Lambda}$) and multi-strange hadron (Ξ^-, $\bar{\Xi}^+$, Ω^-, Ω^+) candidates are reconstructed via topological selection criteria and invariant mass analysis in the following channels (branching ratios): $\mathrm{K}^0_\mathrm{S} \rightarrow \pi\pi$ (69.2%), $\Lambda \rightarrow p\pi$ (63.9%), $\Xi \rightarrow \Lambda\pi$ (99.9%) and $\Omega \rightarrow \Lambda K$ (67.8%). The extraction of the raw yields were performed using the bin counting technique [13]. Finally, the production of short lived hadronic resonances (K^{*0}, $\bar{\mathrm{K}}^{*0}$) is also measured from the invariant-mass reconstruction of their decay products ($\mathrm{K}\pi$). The combinatorial background was estimated using a mixed-event technique and then subtracted from the original invariant-mass distribution. The raw yield is then extracted from a fit of the signal peak [14].

In all analyses, the transverse momentum spectra are corrected for acceptance and efficiency, which are determined through Monte Carlo simulations of the ALICE detector response. In addition, the measurement of light flavours hadrons concerns primary particles, i.e. particles produced either directly in the collision or formed by non-weak particle decays. Weak decays and contributions from particle knocked-out are removed with a data driven approach [15].

Multiplicity classes are defined on the basis of the total charge deposited in the V0A and V0C detectors located at forward ($2.8 < \eta < 5.1$) and backward ($3.7 < \eta < 1.7$) pseudo-rapidity regions, respectively. The mean pseudo-rapidity densities of primary charged particles $\langle \mathrm{d}N_\mathrm{ch}/\mathrm{d}\eta \rangle$ are measured at mid-rapidity, $|\eta| < 0.5$, for each event class [16]. In order to minimise spurious measurements originating from pile-up, events with more than one reconstructed primary vertex are rejected.

3. Results

The final transverse momentum spectra of the identified hadrons measured in pp collisions at $\sqrt{s} = 7$ TeV are reported in Fig. 1 for several event multiplicity classes, indicated by Roman numerals. The average charged particle densities in the highest (class I) and lowest (class X) multiplicity classes correspond to approximately 3.5 and 0.4 times the value in the integrated sample ($\langle \mathrm{d}N_\mathrm{ch}/\mathrm{d}\eta \rangle \approx 6.0$), respectively. For all the identified hadrons a clear hardening of the p_T spectra from low to high multiplicity events is observed. Moreover, this effect seems to be mass-dependent, being slightly more pronounced for the heavier particles. In fact, the baryon-to-meson ratio in pp collisions has been recently shown to follow a similar trend between lowest and highest multiplicity classes as the one observed in p–Pb and Pb–Pb collisions [5, 15]. In order to further explore the evolution of the different physics mechanisms related to baryon and meson production in different regions of the p_T-spectra, the ratio of proton-to-pion was studied at low-, mid- and high-p_T regions as a function of $\langle \mathrm{d}N_\mathrm{ch}/\mathrm{d}\eta \rangle$, as shown in Fig. 2. A continuity

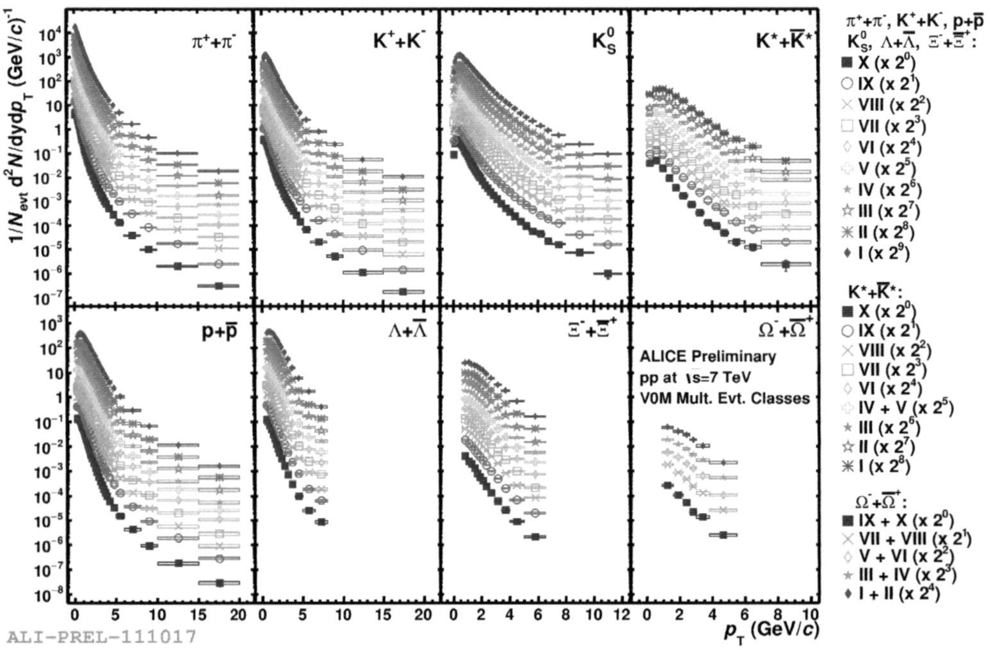

Figure 1. p_T-spectra of identified particles measured in pp collisions at $\sqrt{s}$= 7 TeV for several multiplicity classes, labelled by Roman numerals from high (class I) to low (class X) multiplicities.

across the various collision systems is observed for all the three p_T regions. The smooth evolution of the particle ratios suggests the similarity of the underlying mechanisms despite the differences in the initial colliding systems.

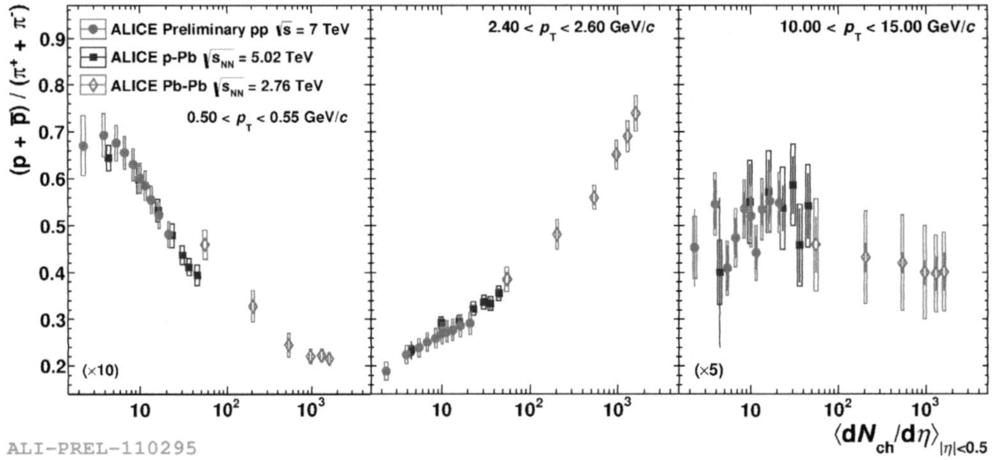

Figure 2. Proton–to–pion ratio at low (left panel), intermediate (middle panel) and high (right panel) transverse momentum as a function of the average charged particle density at mid-rapidity for pp, p–Pb and Pb–Pb collisions.

IOP Conf. Series: Journal of Physics: Conf. Series **878** (2017) 012005 doi:10.1088/1742-6596/878/1/012005

Comparisons with predictions from QCD-inspired models such as PYTHIA8 [17], DIPSY [18, 19, 20] and HERWIG7 [21, 22] have also been performed. The colour reconnection mechanism in PYTHIA8 as well as the colour ropes implementation in DIPSY seem to qualitatively reproduce the general trend observed at low and intermediate p_{T} (see Fig. 3). An alternative description is also provided by EPOS LHC [23], which implements collective radial flow using hydrodynamics. In this case, the predictions reproduce the data only qualitatively at low p_{T}.

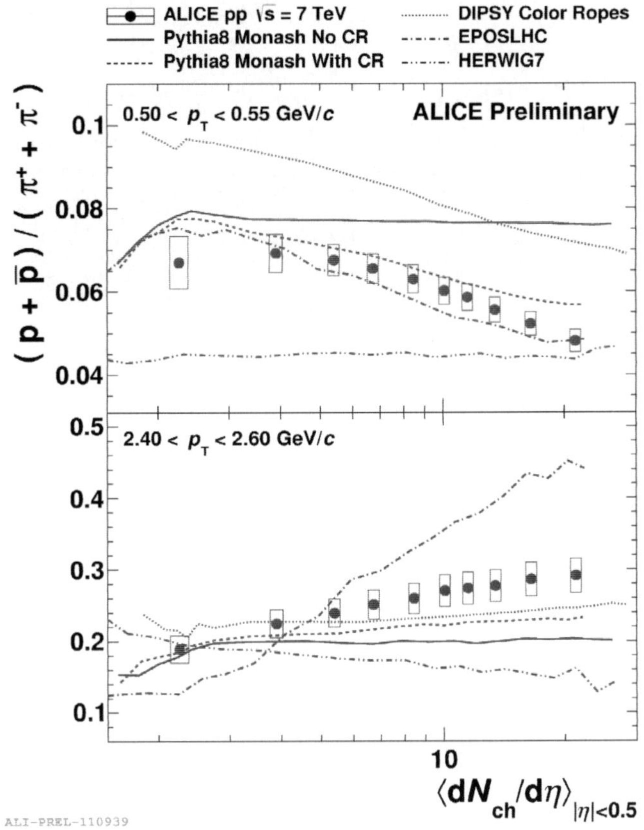

Figure 3. Proton-to-pion ratio at low (top panel) and intermediate (bottom panel) transverse momentum as a function of the average charged particle density at mid-rapidity for pp collisions at $\sqrt{s} = 7$ TeV (symbols) compared to model predictions (lines).

From the p_{T}-integrated yields of strange and multi-strange hadrons to pions, shown in Fig. 4 for pp, p–Pb and Pb–Pb collision systems, a multiplicity-dependent increase in the relative yield for particles having non-zero net strangeness content is observed. Moreover, the observed increase seems to dependent on the number of constituent strange quarks, as the increase in strangeness is larger for multi-strange hadrons. The comparison of the ratios with the predictions from Monte Carlo models shows that the increase is reproduced by the DIPSY event generator, while it is unable to reproduce the p/π ratio shown in Fig. 5. The EPOS model manages to describe only the magnitude of the effect and overestimates the rise while PYTHIA misses both trend and magnitude with increasing differences as the strangeness content increases.

The ratios between the p_{T}-integrated yields of particles with equal amount of strange valence quarks, such as $\Lambda/\mathrm{K}^0_{\mathrm{S}}$ and p/π, respectively, are reported in Fig. 5 as a function of the charged particle multiplicity for pp and p–Pb collisions. The trends for both ratios are independent of the event multiplicity, thus suggesting that the increase in the production of strange particles is related to the strangeness content rather than due to a difference in the particle masses. Comparisons to Monte Carlo models show poor agreement, especially when trying to reproduce all the observations simultaneously.

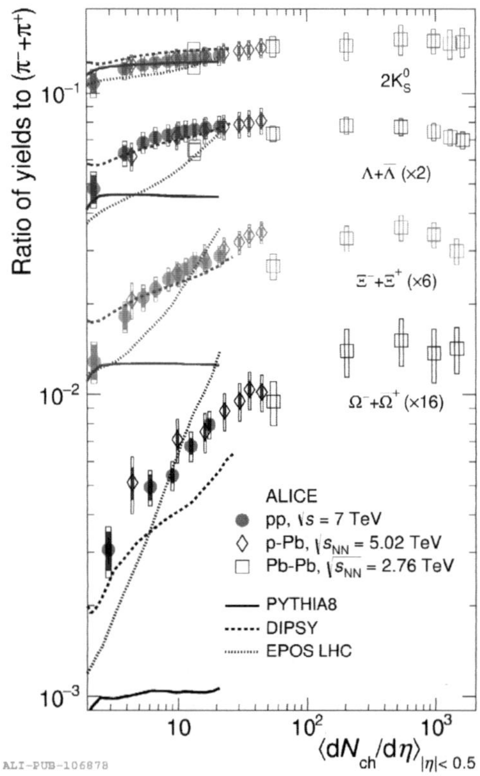

Figure 4. Ratios of K^0_S, Λ, Ξ and Ω yields to the one of π for pp, p–Pb and Pb–Pb collisions as a function of the charged multiplicity density compared to Monte Carlo models.

The evolution of particle ratios with the collision energy is another important aspect which may reveal essential details concerning the underlying particle production mechanisms. The p_T-integrated $(p+\bar{p})/(\pi^+ + \pi^-)$ and $(K^+ + K^-)/(\pi^+ + \pi^-)$ ratios measured by ALICE in pp collisions as a function of $\sqrt{s}$ are shown in Fig. 6. Results from E735 at $\sqrt{s} = 0.3$, 0.54, 1.0 and 1.8 TeV [25, 24], from PHENIX at $\sqrt{s} = 62.4$ and 200 GeV [26] and CMS at $\sqrt{s} = 0.9$, 2.76 and 7 TeV [27] are also shown and compared for both particle ratios. The results at 13 TeV confirms both the trend and the saturation behaviour observed at 900 GeV.

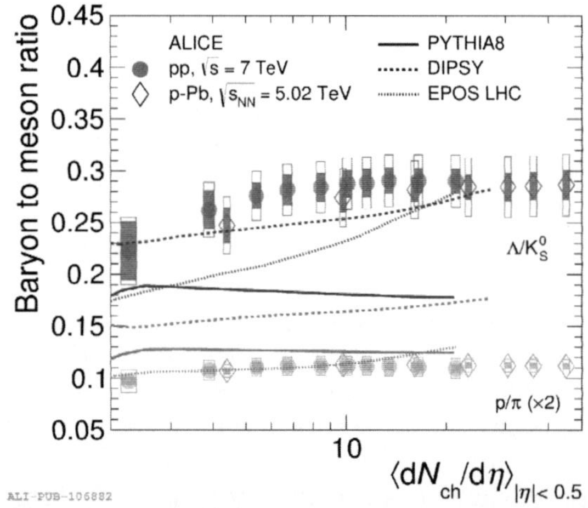

Figure 5. Λ/K^0_S and p/π ratios for pp and p–Pb as a function of multiplicity compared to Monte Carlo models.

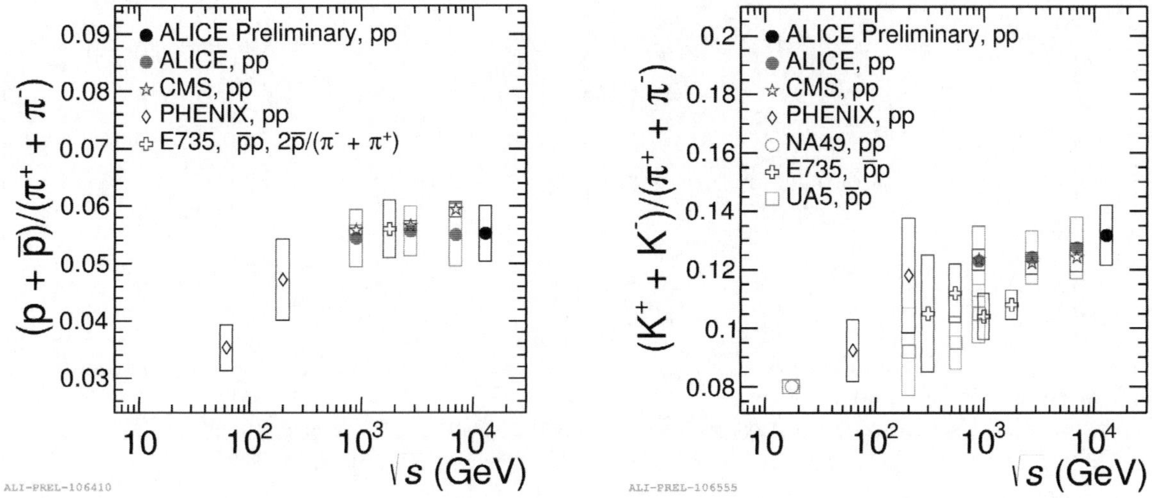

Figure 6. $(\mathrm{p}+\bar{\mathrm{p}})/(\pi^+ + \pi^-)$ and $(\mathrm{K}^+ + \mathrm{K}^-)/(\pi^+ + \pi^-)$ in pp and p$\bar{\mathrm{p}}$ collisions as a function of $\sqrt{s}$. ALICE measurements in pp collisions at $\sqrt{s}$ = 0.9, 2.76, 7 (red points) and 13 (black point) TeV are shown. Errors bars in ALICE results represent the quadratic sum of statistical and systematics uncertainties.

The hyperon-to-pion ratio as a function of $\sqrt{s}$ measured in ALICE is reported in Fig. 7. In comparison with the kaon- and proton-to-pion ratio the strange- and multi-strange-to-pion ratios show a hint of increase at 13 TeV with respect to the lower collision energies.

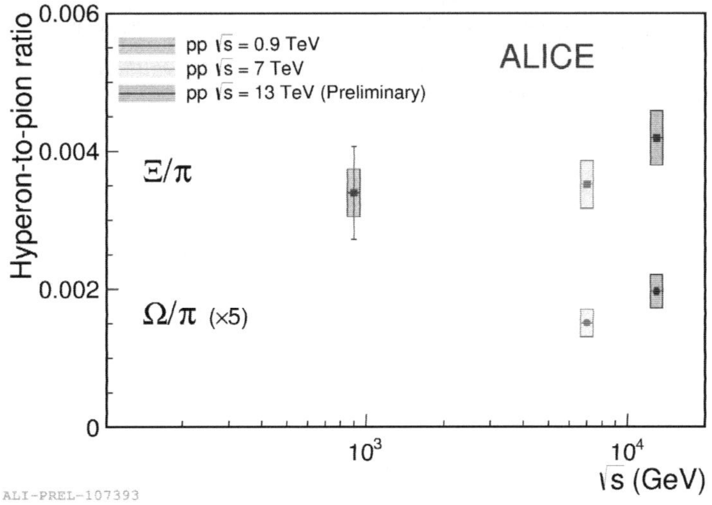

Figure 7. Hyperon to pion ratio in pp collisions as a function of the collision energy.

The K^{*0} and ϕ meson production has been also measured in pp collisions at $\sqrt{s}$ = 13 TeV. The p_{T}-integrated $\mathrm{K}^{*0}/\mathrm{K}$ and ϕ/K ratios are shown in Fig. 8. However, no energy evolution is observed in the range from RHIC to the highest LHC energy.

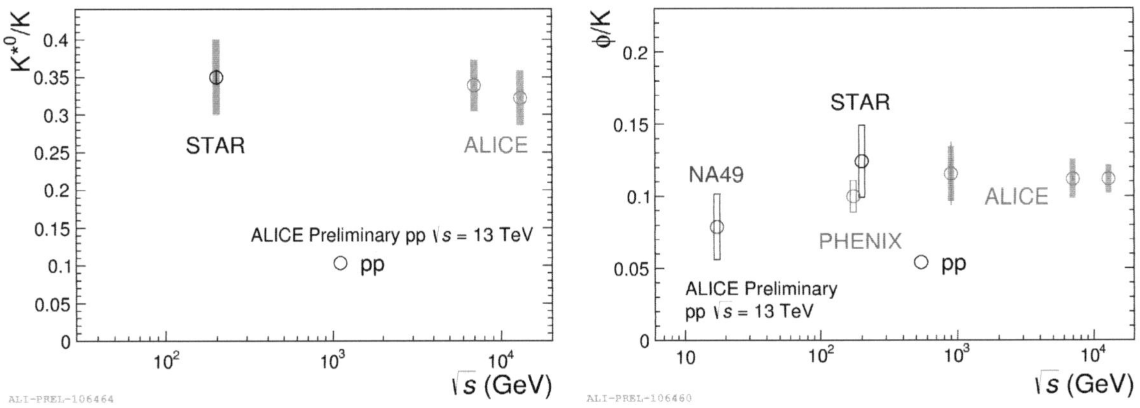

Figure 8. K^{*0}/K and ϕ/K ratio in pp collisions as a function of the collision energy. Bars represent statistical uncertainties. Boxes represent the total systematic uncertainties.

4. Summary and Outlook

The ALICE Collaboration has measured a comprehensive set of identified hadrons in pp collisions at $\sqrt{s} = 7$ TeV as a function of the charged particle multiplicity. A mass-dependent hardening of the spectral shapes was observed for high multiplicity collisions. In addition, a multiplicity-dependent increase in the production of strange hadrons relative to the production of pions was found. The effect seems to be related to the strangeness content rather than to the particle mass and it is qualitatively reproduced by Monte Carlo generators such as DIPSY, while the same model is not in quantitative agreement with the measurement for p/π.

We have also presented the measurement of the production of light flavour hadrons in pp collisions at 13 TeV, the highest centre-of-mass energy ever reached in the laboratory. Above $\sqrt{s} = 900$ GeV the $(p+\bar{p})/(\pi^+ + \pi^-)$ and $(K^+ + K^-)/(\pi^+ + \pi^-)$ ratios are energy independent within the systematic uncertainties. Furthermore, the hyperon-to-pion ratios at 13 TeV suggest an increase with respect the same results at lower energies. This effect may be related to the enhanced production of strange and multi-strange hadrons as a function of event multiplicity already observed in pp collisions at 7 TeV, as the mean charged particle multiplicity density also increases with collision energy [24]. Moreover, the relative production of strange resonances (K^{*0}/K and ϕ/K) remains constant within a wide range of centre of mass energies. Multiplicity dependent studies also in pp collisions at 13 TeV will help to disentangle the role of collision energy and event multiplicity in the particle production dynamics of pp collisions.

References

[1] CMS Collaboration, 2010 *JHEP* **09** 091 (*Preprint arXiv:1009.4122*)
[2] CMS Collaboration, 2013 *Phys. Lett. B* **718** 795
[3] ALICE Collaboration, 2013 *Phys. Lett. B* **719** 2941 (*Preprint arXiv:1212.2001*)
[4] ATLAS Collaboration, 2016 *Phys. Rev. Lett.* **116** 172301 (*Preprint arXiv:1509.04776*)
[5] ALICE Collaboration, 2014 *Phys. Lett. B* **728** 2538 (*Preprint arXiv:1307.6796*)
[6] Dusling K. and Venugopalan R., 2013 *Phys. Rev. D* **87** 094034 (*Preprint arXiv:1302.7018*)
[7] Ortiz Velasquez A. et al., 2013 *Phys. Rev. Lett.* **111** 042001 (*Preprint arXiv:1303.6326*)
[8] Bozek P. and Broniowski W., 2013 *Phys. Rev. C* **88** 014903 (*Preprint arXiv:1304.3044*)
[9] Ortiz Velasquez A. et al., 2016 *Preprint arXiv:1608.04784*
[10] ALICE Collaboration, 2008 *JINST* **3** S08002
[11] ALICE Collaboration, 2015 *Eur. Phys. J. C* **75** 226
[12] ALICE collaboration, 2011 *Eur. Phys. J. C* **71** 1594 (*Preprint arXiv:1012.3257*)
[13] ALICE Collaboration, 2016 *Preprint arXiv:1606.07424*
[14] ALICE Collaboration, 2012 *Eur. Phys. J. C* **72** 2183 (*Preprint arXiv:1208.5717*)

IOP Conf. Series: Journal of Physics: Conf. Series **878** (2017) 012005 doi:10.1088/1742-6596/878/1/012005

[15] ALICE Collaboration, 2012 *Phys. Rev. Lett.* **109** 252301
[16] ALICE Collaboration, 2013 *Phys. Rev. Lett.* **110** 032301 (*Preprint arXiv:1210.3615*)
[17] Sjostrand T., Mrenna S. and Skands P. Z. 2008, *Comput. Phys. Commun.* **178** 852867 (*Preprint arXiv:0710.3820*)
[18] Flensburg C., Gustafson G. and Lonnblad L., 2011 *JHEP* **08** 103 (*Preprint arXiv:1103.4321*)
[19] Bierlich C., Gustafson G., Lnnblad L. and Tarasov A., 2015 *JHEP* **03** 148 (*Preprint arXiv:1412.6259*)
[20] Bierlich C. and Christiansen J. R., 2015 *Phys. Rev. D* **92** 094010 (*Preprint arXiv:1507.02091*)
[21] Bahr M. et al., 2008 *Eur. Phys. J. C* **58** 639707 (*Preprint arXiv:0803.0883*)
[22] Bellm J. et al., 2016 *Eur. Phys. J. C* **76** 196 (*Preprint arXiv:1512.01178*)
[23] Pierog T. et al,. 2015 *Phys. Rev. C* **92** 034906 (*Preprint arXiv:1306.0121*)
[24] E735 Collaboration, 1992 *Phys.Rev. D* **46** 2773-2786.
[25] E735 Collaboration, 1993 *Phys.Rev. D* **48** 984-997.
[26] PHENIX Collaboration, 2011 *Phys. Rev. C* **83** 064903.
[27] CMS Collaboration, 2012 *Eur. Phys. J. C* **72** 2164.

Nuclear modification of light flavour and strangeness at LHC energies with ALICE

Ramona Lea for the ALICE Collaboration

University and INFN Trieste, Via Valerio 2, 34127 Trieste, Italy

E-mail: ramona.lea@cern.ch

Abstract. Thanks to its unique particle identification capabilities the ALICE detector is able to identify light-flavour, strange and multi-strange hadrons, including π, K, p, K^0_S, Λ, Ξ and Ω, over a wide range of transverse momentum, from pp and p-Pb interactions up to central Pb–Pb collisions. The latest results on the nuclear modification factor, R_{AA}, as a function of the Pb–Pb collision centrality, is shown for various particle specie at $\sqrt{s_{NN}} = 2.76$ TeV centre-of-mass energy. For each particle specie, the R_{AA} is compared with the nuclear modification factors R_{pA} in p-Pb collisions to asses the presence of hot nuclear matter effects affecting the high-p_T particle production in Pb–Pb collisions. The results on the R_{AA} of charged hadrons at $\sqrt{s_{NN}} = 5.02$ TeV, the highest energy ever reached in the laboratory for heavy-ion collisions, is also shown.

1. Introduction

One of the main goals of ultra relativistic heavy-ion collisions is the study of strongly interacting matter where the formation of the Quark-Gluon Plasma (QGP), a deconfined state of quarks and gluons, is expected [1]. Hard partons that propagate through this matter are predicted to loose energy via (multiple) scattering and gluon radiation. As a result, transverse momentum (p_T) spectra of final state hadrons and jets will be modified with respect to what is expected from a simple superposition of incoherent proton-proton collisions. This change, quantified by the nuclear modification factor, R_{AA}, is used to study parton energy-loss mechanisms and medium properties. Disentangling energy-loss signatures from initial state nuclear effects which may also modify p_T spectra, requires a comparison of R_{AA} to R_{pA}, the nuclear modification factor for proton-nucleus collisions. In this paper, an overview of ALICE results on the nuclear modification factor for identified hadrons at $\sqrt{s_{NN}} = 2.76$ TeV and charged particles at $\sqrt{s_{NN}} = 5.02$ TeV are presented.

A full description of the ALICE detector and its performance can be found in [2]. The ALICE sub-detectors that played an essential role for the analysis presented here are the V0 detectors [3] – covering the pseudorapidity ranges $2.8 < \eta < 5.1$ (V0-A) and $-3.7 < \eta < -1.7$ (V0-C) – the Inner Tracking System (ITS) [4] – composed of six cylindrical layers of silicon detectors with radii between 3.9 and 43 cm from the beam axis – the Time Projection Chamber (TPC) [5], which is the main tracking detector, and the Time-Of-Flight (TOF) detector [6]. At high momenta (> 1 GeV/c) identification is complemented by a small acceptance ring imaging Cherenkov detector, the HMPID [2].

The nuclear modification factor $R_{\rm AA}$ is defined as:

$$R_{AA} = \frac{1}{\langle T_{\rm AA}\rangle} \frac{{\rm d}N_{\rm AA}/{\rm d}p_{\rm T}}{{\rm d}\sigma_{pp}/{\rm d}p_{\rm T}} \tag{1}$$

where $N_{\rm AA}$ and σ_{pp} represent the yield in nucleus-nucleus collisions and the cross section in pp collisions, respectively. $\langle T_{\rm AA}\rangle = \langle {\rm N_{coll}}\rangle \;/\; \sigma_{inel}^{NN}$ is the nuclear overlap function, which corresponds to the ratio of the number of binary nucleon-nucleon collisions $\langle {\rm N_{coll}}\rangle$ estimated with a Glauber model [7] and the inelastic nucleon-nucleon cross section σ_{inel}^{NN}. If the Pb–Pb collisions are a simple superpositions of elementary nucleon-nucleon collisions, the $R_{\rm AA}$ is expected to be equal to unity, while $R_{\rm AA} < 1$ indicates a suppression of charged-particle production compared to binary-collision scaling.

2. $R_{\rm AA}$ of identified hadrons at $\sqrt{s_{\rm NN}}$ = 2.76 TeV

ALICE employs several complementaries particle identification techniques. Specifically, charged pions, kaons and protons at high transverse momentum are identified via the specific ionization dE/dx measured in the TPC. For $p_{\rm T} < 4$ GeV/c, the combined PID of the HMPID, the TOF detector and dE/dx in the ITS and TPC are used. Neutral strange and multi-strange particles ($\rm K^0_S$, Λ, $\overline{\Lambda}$) and multi-strange baryons (Ξ and Ω) are reconstructed via topological selection criteria and invariant-mass analysis of their characteristic weak decays. Specifically, tracks reconstructed by the TPC and the ITS are combined to select candidates satisfying a set of geometrical criteria. In addition, particle identification of the decay products is performed via a selection based on the specific energy loss in the TPC. Particle yields as a function of $p_{\rm T}$ are determined in various centrality intervals using an invariant mass analysis. Acceptance and efficiency corrections are calculated using Monte Carlo simulations. For more details, refer to [8–10].

2.1. Results

The transverse momentum spectra of charged pions, kaons and protons as well as $\rm K^0_S$ and hyperons in Pb–Pb collisions at $\sqrt{s_{\rm NN}}$ = 2.76 TeV have been published in [8,11,12]. At low $p_{\rm T}$ a flattening of the spectra is observed, which is more pronounced for central collisions and heavier particles. The mass dependence of this effect is characteristic of radial flow during the collective expansion of the fireball. In Figure 1 the $R_{\rm AA}$ as a function of $p_{\rm T}$ for multi-strange baryons is compared to the one for lighter hadrons for two different centrality intervals, 0-10% (left) and 60-80% (right). At high $p_{\rm T}$ the $R_{\rm AA}$ of pions, kaons, protons and Ξ baryons is significantly smaller than unity. The suppression is stronger for the 10% most central collisions (left panel) than for peripheral collisions (60-80% central, right panel). At $p_{\rm T} > 10$ GeV/c, the $R_{\rm AA}$ of π, K, and p is similar [11]. The absence of a specie dependence may indicate vacuum-like parton fragmentation for the partons that will originate high-$p_{\rm T}$ hadrons in the final state. Differences at low $p_{\rm T}$ ($p_{\rm T} \leq 6$ GeV/c) may be attributed to the mass-dependent flow effects. The nuclear modification factor for Ξ is similar to the one for p, while $R_{\rm AA}$ for Ω shows a pronounced difference with respect to Ξ and exceeds unity, which might be the result of larger strangeness enhancement compared to Ξ [8]. In peripheral collisions the difference between the $R_{\rm AA}$ of Ω and Ξ is strongly reduced, as can be observed in the right panel of Figure 1. In Figure 2 the $R_{\rm pPb}$ as a function of $p_{\rm T}$ in minimum-bias p–Pb collisions at $\sqrt{s_{\rm NN}}$ = 5.02 TeV is shown for π, K, p, Ξ and Ω. In this case, there is no suppression for π, K and p at high $p_{\rm T}$($>$ 6 GeV/c) [13], as found for unidentified particles [14]. This observation leads to the conclusion that the observed suppression in Pb–Pb collisions is a hot matter effect. Finally, in the so-called Cronin region ($2 < p_{\rm T} < 6$ GeV/c) an increase of the $R_{\rm pPb}$ with the mass and the strangeness content of the particles is visible.

IOP Conf. Series: Journal of Physics: Conf. Series **878** (2017) 012006 doi:10.1088/1742-6596/878/1/012006

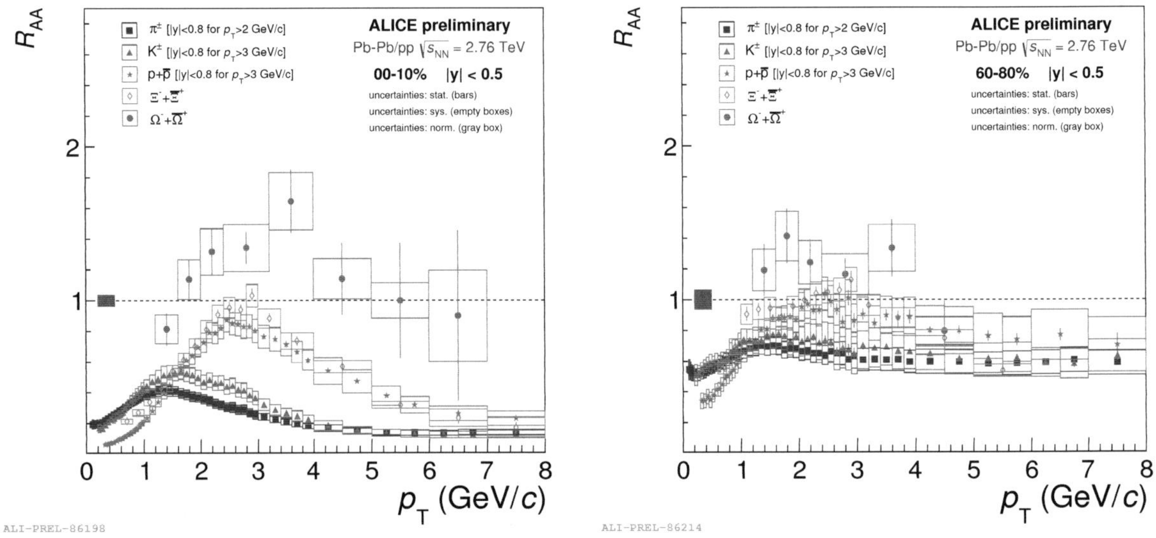

Figure 1: Nuclear modification factor R_{AA} for charged π, K, p, Ξ and Ω particles in 0-10% central (left) and 60-80% central (right) Pb–Pb collisions at $\sqrt{s_{\mathrm{NN}}}$ = 2.76 TeV. Bars represents statistical uncertainties, open boxes systematic uncertainties while gray boxes are normalization uncertainties.

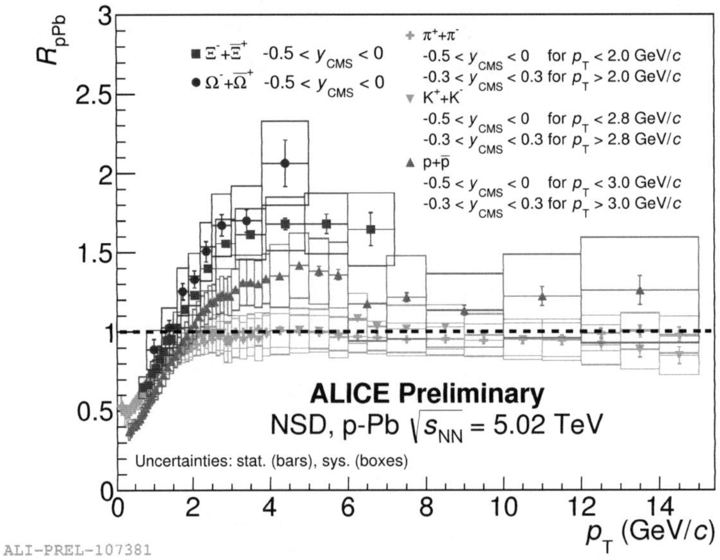

Figure 2: R_{pPb} as a function of p_{T} in minimum-bias p–Pb collisions of π, K, p, Ξ and Ω. Bars represents statistical uncertainties, while open boxes systematic uncertainties.

3. R_{AA} of charged hadrons at $\sqrt{s_{NN}} = 5.02$ TeV

The results presented in this section are based on the analysis of 25×10^6 pp collisions and 3.3×10^6 Pb–Pb collisions at $\sqrt{s_{NN}} = 5.02$ TeV collected by ALICE during 2015, which represent approximately 25% and 3% of the full collected statistics, respectively. Events are triggered using the V0A and V0C detectors, which in Pb–Pb collisions are also used to determine the centrality of the collision [15]. For the measurement of the charged particle spectra, only primary particles are selected. Primaries are defined as all the prompt charged particles produced in the collisions including all decay products, except charged particles from weak decay of light flavour hadrons.

All the corrections applied to the data are calculated using PYTHIA8 [16] Monte Carlo simulations for pp and HIJING [17] for Pb–Pb collisions where the particle transport in the ALICE detector is simulated via GEANT3 [18]. To account for the differences in particle composition between event generators and data, the charged-particle reconstruction efficiency was calculated from particle-dependent efficiencies weighted by the relative particle abundances measured in pp collisions at 7 TeV and Pb–Pb collisions at 2.76 TeV. The correction for contamination by secondary particles produced from weak decay or secondary interaction is important only at low p_T ($\sim 10\%$), while the correction for p_T resolution is important at high p_T ($\sim 2\%$).

3.1. Results

The transverse momentum distributions have been measured for pp and Pb–Pb collisions at the same energy per colliding nucleon pair $\sqrt{s_{NN}} = 5.02$ TeV. The left panel of Figure 3 shows the p_T spectrum of charged particles for pp collisions at $\sqrt{s}= 5.02$ TeV in the p_T interval $0.15 \leq p_T \leq 40$ GeV/c and $|\eta| < 0.8$. The measurement is compared to Monte Carlo calculations performed with PYTHIA8 (Monash-2013 tune) [16] and EPOS LHC [19]. Both models describe the data reasonably well at intermediate p_T, but both overestimate the production at low and high p_T by a similar amount ($\sim 20\%$). The Pb–Pb transverse momentum distributions for six centralities are shown on the right panel of Figure 3. On the same panel the pp references scaled with the nuclear overlap function $\langle T_{AA} \rangle$ are shown as dashed lines.

The nuclear modification factor can be calculated as the ratio of the two distributions. The R_{AA} for six different centralities as a function of p_T is shown in Figure 4. The solid symbols correspond to the measurement at $\sqrt{s_{NN}} = 5.02$ TeV while the open symbols represent the published data at $\sqrt{s_{NN}} = 2.76$ TeV [20]. For the six centrality intervals, the measurements for both energies show a similar behavior within systematic uncertainties. In central collisions (0-5%) a strong suppression is observed, with a minimum at $p_T \sim 7$ GeV/c and a significant rise for $p_T > 7$ GeV/c. For peripheral collisions (60-80%) R_{AA} exhibits a moderate suppression (~ 0.7).

Figure 5 shows the comparison of the measured R_{AA} at $\sqrt{s_{NN}} = 5.02$ TeV with different theoretical models. On the left panel the measurement in the 0-5% centrality interval is compared with the model proposed by Djordjevic at al. [21] (long dashed lines) and by Majumder et al. [22] (dashed lines). Djordjevic et al. model uses a dynamical energy loss formalism and takes into account interactions of the probe with the medium. Both the radiative and the collisional energy loss mechanisms are implemented in the model. Additionally, the formalism also takes into account finite magnetic mass and running coupling. Majumder et al. model is based on a higher twist (HT) radiative energy loss calculation and uses a viscous hydrodynamical evolution of the system. The model parameters are adjusted from soft hadron yield and elliptic flow. The two models succeed in describing the R_{AA} measurements at high p_T. On the right panel of Figure 5 the measurement in the 0-10% centrality interval is compared with the predictions from the model of Vitev et al. [23], which is based on the soft-collinear effective theory (SCET) [24], an effective theory of QCD for jet physics. Also in this case the model describes the data within the model uncertainties.

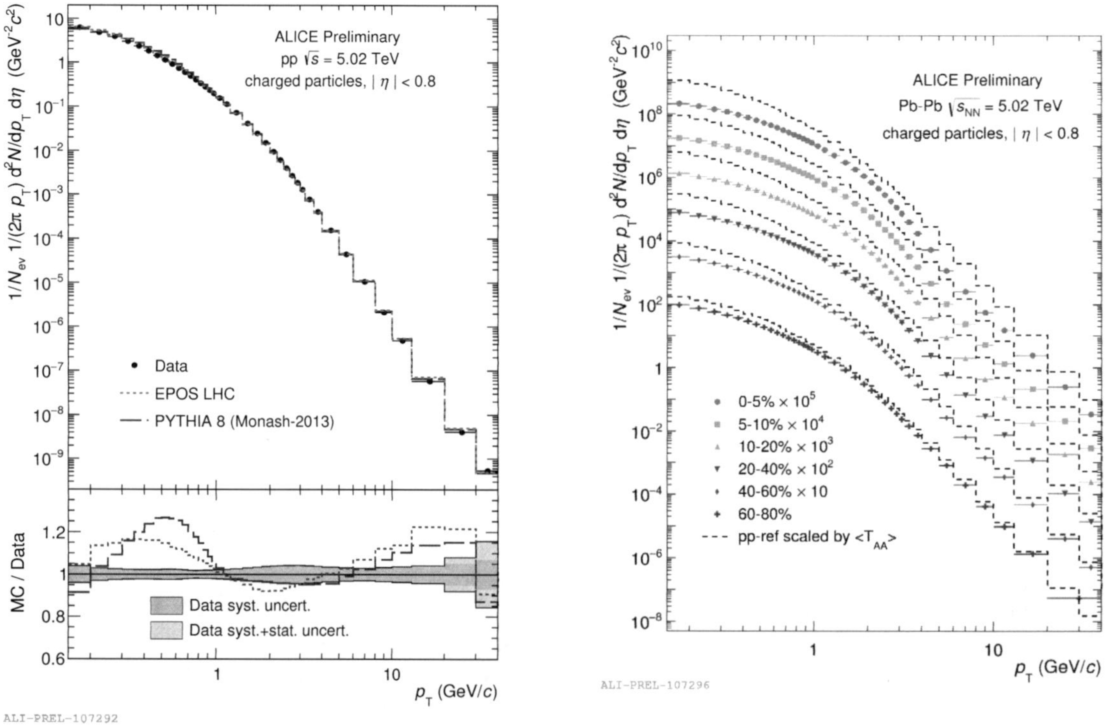

Figure 3: Left: Transverse momentum distribution of charged particles for pp collisions at $\sqrt{s}$=5.02 TeV, compared to PYTHIA8 and EPOS calculations [16, 19]. Right: Transverse momentum distribution of charged particles for Pb–Pb collisions at $\sqrt{s_{\mathrm{NN}}} = 5.02$ TeV for six different centrality classes. The dashed lines represent the measured pp collisions scaled by $\langle T_{\mathrm{AA}} \rangle$.

4. Conclusions

The production of charged pions, kaons and protons, neutral strange mesons and hyperons was measured with the ALICE detector in pp, p–Pb and Pb–Pb collisions. In central Pb–Pb collisions, R_{AA} of light hadrons at high p_{T} is smaller than unity, indicating a suppression attributed to parton energy loss. The nuclear modification factor of charged pions, kaons and protons is similar for $p_{\mathrm{T}} > 10$ GeV/c, hinting at vacuum-like fragmentation of the leading parton. The R_{AA} of Ξ is consistent with the R_{AA} of p for p_{T} larger than 6 GeV/c, whereas the R_{AA} of Ω is significantly higher. The difference between the R_{AA} of Ω and Ξ is much less pronounced in peripheral collisions. The absence of a suppression at high p_{T} in R_{pPb} indicates that the suppression in R_{AA} is due to hot nuclear matter effect.

The nuclear modification factor of unidentified charged particles in Pb–Pb collisions at $\sqrt{s_{\mathrm{NN}}} = 5.02$ TeV was measured and it was found to be similar to the measurement at 2.76 TeV. A maximal suppression was observed around 6-7 GeV/c for the most central collisions (0-5%) which becomes smaller as centrality decreases. The R_{AA} was compared with three models and they succeed in describing the general features of the measurement for a transverse momentum p_{T} larger than 5 GeV/c.

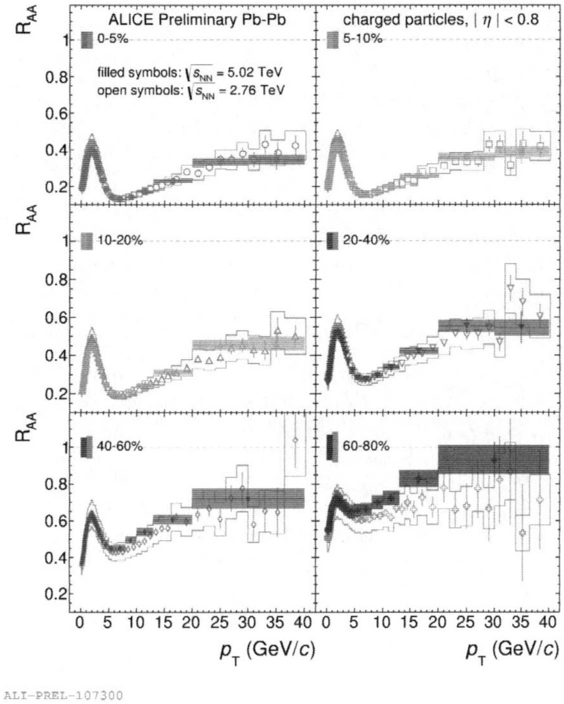

Figure 4: The nuclear modification factor (R_{AA}) as a function of p_{T} for six centrality intervals. Filled symbols represent the measurements at $\sqrt{s_{\mathrm{NN}}} = 5.02$ TeV, while empty symbols corresponds to the R_{AA} at $\sqrt{s_{\mathrm{NN}}} = 2.76$ TeV [20]. The normalization uncertainty is shown for both energies as a box around unity.

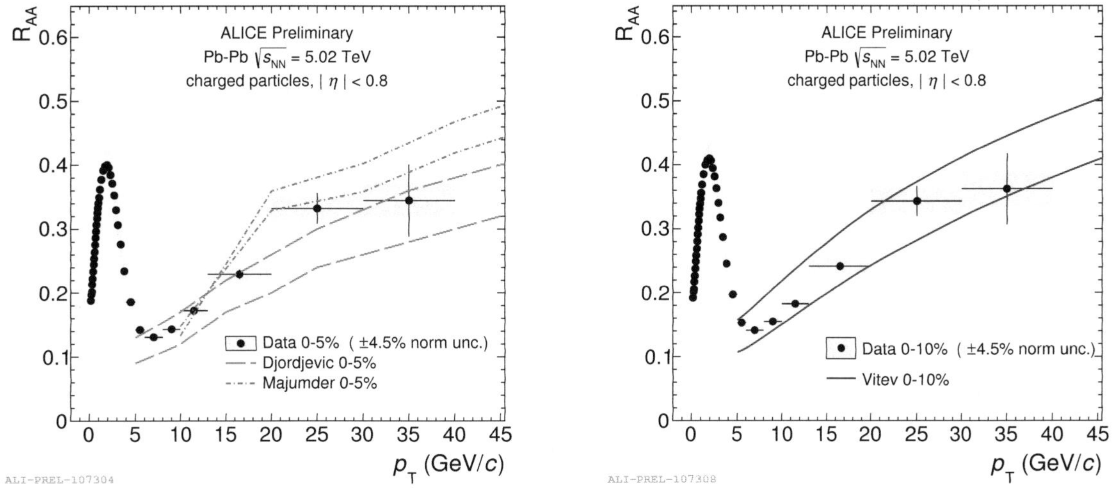

Figure 5: The nuclear modification factor measured at $\sqrt{s_{\mathrm{NN}}} = 5.02$ TeV for 0-5% (left) and 0-10% (right) centrality intervals compared to different theoretical models calculations [21–23].

References

[1] H. Stoecker and W. Greiner, Phys. Rept. **137** (1986) 277.
[2] B. B. Abelev *et al.* [ALICE Collaboration], Int. J. Mod. Phys. A **29** (2014) 1430044
[3] E. Abbas *et al.* [ALICE Collaboration], JINST **8** (2013) P10016
[4] K. Aamodt *et al.* [ALICE Collaboration], JINST **5** (2010) P03003
[5] J. Alme *et al.*, Nucl. Instrum. Meth. A **622** (2010) 316
[6] A. Akindinov *et al.*, Eur. Phys. J. Plus **128** (2013) 44.
[7] M. L. Miller, K. Reygers, S. J. Sanders and P. Steinberg, Ann. Rev. Nucl. Part. Sci. **57** (2007) 205
[8] B. B. Abelev *et al.* [ALICE Collaboration], Phys. Lett. B **728** (2014) 216 Corrigendum: [Phys. Lett. B **734** (2014) 409]
[9] B. B. Abelev *et al.* [ALICE Collaboration], Phys. Lett. B **728** (2014) 25
[10] J. Adam *et al.* [ALICE Collaboration], Phys. Lett. B **758** (2016) 389
[11] J. Adam *et al.* [ALICE Collaboration], Phys. Rev. C **93** (2016) no.3, 034913
[12] B. B. Abelev *et al.* [ALICE Collaboration], Phys. Rev. Lett. **111** (2013) 222301
[13] J. Adam *et al.* [ALICE Collaboration], Phys. Lett. B **760** (2016) 720 doi:10.1016/j.physletb.2016.07.050 [arXiv:1601.03658 [nucl-ex]].
[14] B. Abelev *et al.* [ALICE Collaboration], Phys. Rev. Lett. **110** (2013) no.8, 082302
[15] B. Abelev *et al.* [ALICE Collaboration], Phys. Rev. C **88** (2013) 044910
[16] P. Skands, S. Carrazza and J. Rojo, Eur. Phys. J. C **74** (2014) no.8, 3024
[17] X. N. Wang and M. Gyulassy, Phys. Rev. D **44** (1991) 3501.
[18] R. Brun, F. Bruyant, F. Carminati, S. Giani, M. Maire, A. McPherson, G. Patrick and L. Urban, CERN-W5013.
[19] T. Pierog, I. Karpenko, J. M. Katzy, E. Yatsenko and K. Werner, Phys. Rev. C **92** (2015) no.3, 034906
[20] B. Abelev *et al.* [ALICE Collaboration], Phys. Lett. B **720** (2013) 52
[21] M. Djordjevic, B. Blagojevic and L. Zivkovic, Phys. Rev. C **94** (2016) no.4, 044908
[22] A. Majumder and C. Shen, Phys. Rev. Lett. **109** (2012) 202301
[23] Y. T. Chien, A. Emerman, Z. B. Kang, G. Ovanesyan and I. Vitev, Phys. Rev. D **93** (2016) no.7, 074030
[24] G. Ovanesyan and I. Vitev, Phys. Lett. B **706** (2012) 371

IOP Conf. Series: Journal of Physics: Conf. Series **878** (2017) 012007 doi:10.1088/1742-6596/878/1/012007

Searches for Exotic Physics in ATLAS using Substructure Techniques

J. Katharina Behr
On behalf of the ATLAS collaboration
E-mail: katharina.behr@cern.ch

Abstract. The significant increase of the centre-of-mass energy of the Large Hadron Collider (LHC) from 8 to 13 TeV has allowed the LHC experiments to explore previously inaccessible kinematic regimes in their search for phenomena beyond the Standard Model. The sensitivity of these searches depends crucially on the efficient reconstruction and identification of hadronic decays of highly energetic (boosted) objects, the decay products of which are typically collimated into a single large jet with a characteristic substructure. In this contribution, the searches conducted by the ATLAS experiment on data recorded during 2015 and 2016 that rely on jet substructure techniques to identify signatures of interest are reviewed. A particular emphasis is placed on recent developments in the rapidly evolving field of boosted object tagging.

1. Introduction

In the first two years of the first data-taking period in proton-proton (pp) collision at the $\sqrt{s} = 13$ TeV (Run-2), the ATLAS [1] collaboration has already presented a diverse programme of searches for new phenomena beyond the Standard Model (BSM), probing the predictions of numerous models over an unprecedented kinematic regime. These searches are well-motivated by the known shortcomings of the Standard Model (SM), which explains neither why the Higgs boson is light (hierarchy problem), nor provides a candidate for dark matter. Extensions of the SM include theories with warped extra dimenions (Randall-Sundrum models), new strong interactions, vector-like quarks (VLQs), or models with a second Higgs doublet (2HDMs).

Top quarks, Higgs bosons, and W/Z bosons play a key role in many BSM searches as numerous models predict new particles that couple most strongly to the Higgs and electroweak sector and the third quark generation. At the LHC, a large fraction of these most massive SM particles are produced with large Lorentz boost. If the transverse momentum of a massive particle exceeds twice its rest mass, it is referred to as a *boosted object*. The efficient reconstruction and identification of boosted object decays is of key importance for BSM searches targeting large BSM particle masses and interactions at the highest accessible energies. Their key property is the angular collimation of their decay products, which for a two-body decay $X \to ab$ is given by:

$$\Delta R(a,b) \approx 2m^X/p_T^X \tag{1}$$

where $\Delta R(a,b)$ denotes the angular separation between the decay products.[1] As a consequence,

[1] $\Delta R = \sqrt{\Delta\eta^2 + \Delta\phi^2}$ with pseudorapidity η = -ln tan($\theta/2$). θ (ϕ) is the polar (azimuthal) angle of the ATLAS standard coordinate system.

IOP Conf. Series: Journal of Physics: Conf. Series **878** (2017) 012007 doi:10.1088/1742-6596/878/1/012007

hadronically decaying boosted objects can be reconstructed as a single jet with large radius parameter R and a characteristic substructure.

2. Jet Substructure

In ATLAS, hadronically decaying boosted objects are typically reconstructed as anti-k_t [2] $R = 1.0$ jets built from locally calibrated calorimeter clusters. The most powerful discriminating variable between jets initiated by top quarks, Higgs bosons, or W/Z bosons and background jets from light quarks or gluons is, naturally, the jet mass, which is calculated from the sum of the four-momenta of the (massless) jet constituents. A number of variables that describe the internal structure of the jet are used as additional discriminating variables. The two most commonly used jet substructure variables in early Run-2 ATLAS searches are n-subjettiness [3], which describes how compatible the jet substructure is with the n-subjets hypothesis, and variables based on energy correlation functions calculated for the jet constituents [4, 5, 6].

A key issue for analyses at high-luminosity hadron colliders is the presence of contaminations from additional pp interactions in the same and subsequent bunch-crossings, referred to as pile-up. These soft, wide-angle contaminations dilute the jet substructure and bias the jet mass, shifting it towards larger values. In order to restore the discriminating power of these variables, *jet grooming* procedures have been developed. In ATLAS, the current standard grooming algorithm is *trimming* [7], in which the constituents of the ungroomed large-R jets are re-clustered using the k_t algorithm [8, 9] with radius parameter $R = 0.2$. All k_t $R = 0.2$ subjets that carry a transverse momentum of less than 5% of that of the large-R jet are rejected and the trimmed large-R jet is built from the constituents of the remaining subjets.

3. Searches with Boosted Top Quarks

Searches for massive resonances decaying into a $t\bar{t}$ pair have been the first BSM searches to rely on jet substructure techniques, both at the Tevatron and the LHC. Today a wide and still growing range of dedicated top tagging algorithms or *top taggers* is on the market, with different taggers suited for different kinematic regimes and final states, as illustrated for example in Section 6 of Ref. [10]. For searches relying on early Run-2 data, the ATLAS collaboration has developed a simple baseline top tagger [11]. It relies on the trimmed jet mass and the n-subjettiness ratio τ_{32} [3] to identify jets from boosted hadronic top-quark decays, which are reconstructed from trimmed anti-k_t $R = 1.0$ jets. The requirements placed on the two discriminating variables depend on the jet transverse momentum and are optimised to yield the best rejection of background jets from gluons and light quarks for a given signal efficiency. Two working points with an overall signal efficiency of 50% and 80%, respectively, are provided.

This baseline top tagger is used in the search for massive resonances decaying to $t\bar{t}$ in the single-lepton-plus-jets (ℓ+jets, $\ell \in \{e, \mu\}$) final state [12]. The search relies on $\int \mathcal{L} = 3.2$ fb^{-1} of data collected in 2015 and targets the kinematic regime where the hadronically decaying top quark can be reconstructed as a single large-R jet. The presence of a single isolated electron or muon from the semileptonically decaying top quark allows efficient suppression of the background from multijet production. This enables the 80% top tagging working point to be chosen, thus allowing for a larger signal efficiency compared to the tighter 50% top tagging working point. The dominant background component in the signal region is the irreducible background from SM $t\bar{t}$ production which accounts for more than 80% of the total background contribution. No significant deviation of the data from the SM expectation is observed in the reconstructed $t\bar{t}$ invariant mass spectrum (Figure 1a). Upper limits on the cross-section times branching ratio to $t\bar{t}$ are derived for a leptophobic top-colour Z'_{TC2} boson (Figure 1b). The limits at 95% confidence level (CL) range from 300 pb to 0.05 pb for masses from 0.4 TeV to 5 TeV. A Z'_{TC2} boson of width $\Gamma/m = 1.2\%$ (3%) is excluded for masses in the range 0.7 TeV $< m_{Z'} <$ 2.0 TeV (0.7 TeV $< m_{Z'} <$ 3.2 TeV).

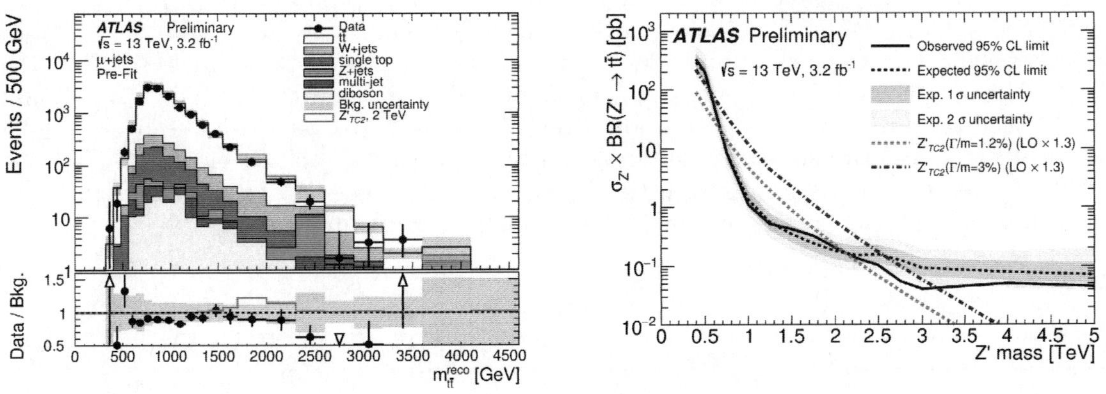

(a) (b)

Figure 1: (a) The reconstructed $t\bar{t}$ invariant mass distribution $m_{t\bar{t}}^{\mathrm{reco}}$ in the μ+jets channel. The ratio of the data to the SM expectation is given in the lower panel. The expected signal distribution for a Z'_{TC2} boson with width $\Gamma/m = 1.2\%$ is indicated by a red line. (b) Observed and expected upper limits on the cross-section times branching ratio to $t\bar{t}$ at 95% CL [12].

4. Searches with Boosted W/Z Bosons

Jet substructure techniques play a key role in all searches involving hadronic vector boson decays on $\sqrt{s} = 13$ TeV data that were published by the ATLAS collaboration to date. As in the case of boosted top quarks (Section 3), the ATLAS collaboration has developed a set of baseline W/Z taggers for searches with early $\sqrt{s} = 13$ TeV data [13]. Boosted W/Z bosons, like boosted top quarks, are reconstructed from trimmed anti-k_t $R = 1.0$ jets. The jet mass, naturally, is the most important discriminating variable for the identification of W/Z bosons. Only jets within a mass window around the W/Z boson mass are selected as W or Z boson candidates. The mass window is determined using Monte Carlo simulations and is defined as the smallest window around the peak in the trimmed jet mass distribution for W or Z boson jets, respectively, containing 68% of the jets. In addition, a jet transverse momentum dependent requirement is placed on the energy correlation ratio $D_2^{\beta=1}$ [5] to further suppress the background from multijets production. Two working points with an overall 25% and 50% signal efficiency, respectively, are provided.

4.1. Searches for diboson resonances

The baseline W/Z taggers are used in a number of resonance searches such as the search for resonances decaying to a pair of hadronically decaying vector bosons (WW, ZZ, and WZ) [14], for massive WW and WZ resonances in the $\ell\nu qq$ final state [15], for massive ZW and ZZ resonances in the $\ell\ell qq$ and $\nu\nu qq$ final states [16], and for massive resonances decaying to $Z\gamma$ [17].

Diboson searches in the purely hadronic final state especially benefit from the use of jet substructure techniques, as these not only allow for an effective suppression of the large background from multijet production but also reduce the problem of the combinatorial background that arises in resolved topologies due to incorrect assignments of the small-R jets to the vector boson candidates. Hence the search for massive diboson resonances in the fully hadronic final state focuses on the boosted regime where the events exhibit a simple dijet topology. Signal events are required to have two boson-tagged jets with $p_T > 250$ GeV ($p_T > 450$ GeV for the leading jet) and a dijet invariant mass greater than 1 TeV. To further suppress the background from multijet production, the boson candidate jets must meet the requirement $N_{\text{trk}} < 30$, where N_{trk} denotes the number of charged-particles tracks from the primary vertex that are matched to the ungroomed jet via ghost-association [18, 19, 20]. This requirement is motivated both by the larger fraction of gluon-initiated jets in background events as well as the different fragmentation scales for signal and background events that both lead to a larger track multiplicity in background jets. No significant deviation from the SM expectation is found in the dijet invariant mass distribution with $\int \mathcal{L} = 15.5\ \text{fb}^{-1}$ of data (Figure 2a) and upper limits at 95% CL are derived on the production cross section times branching ratio to WZ and WW final states of resonances arising within Heavy Vector Triplet (HVT) models (Figure 2b), and to WW and ZZ final states of a bulk Randall-Sundrum graviton.

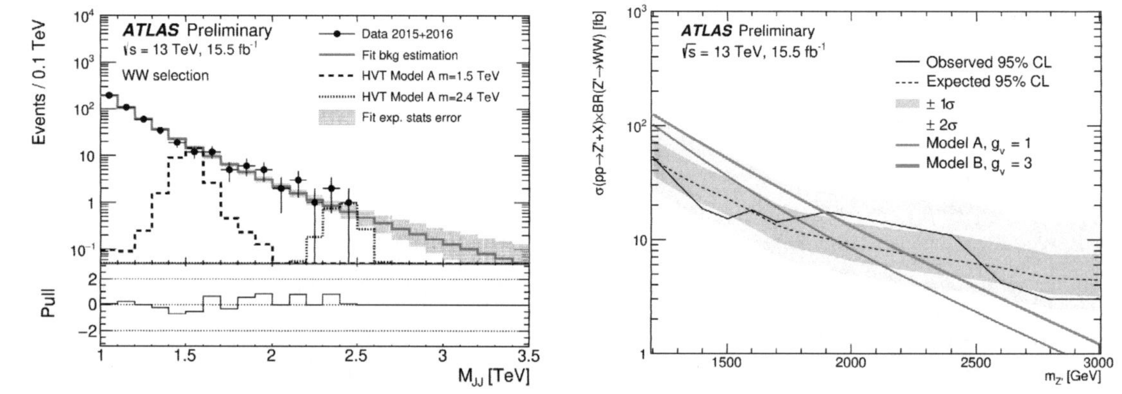

(a) (b)

Figure 2: (a) The dijet invariant mass distribution for the WW selection. The ratio of the data to the SM expectation is given in the lower panel. (b) Observed and expected upper limits at 95% CL on the cross-section times branching ratio to WW for a Z' boson in a HVT model [14].

4.2. Search for dark matter in final states with missing transverse energy and a W/Z boson

The ATLAS baseline boson tagger is also used in a search for dark matter produced in association with a hadronically decaying W/Z boson [21]. The results are interpreted in the context of an effective field theory and a simplified model with an s-channel mediator. In the latter case, the W/Z boson is radiated from one of the initial-state partons and recoils against the mediator that decays to a pair of dark datter particles, giving rise to missing transverse momentum E_T^{miss}. Final states in which the W/Z boson can be reconstructed as a single jet allow one to probe large values of E_T^{miss}. No dark matter signal is observed in $\int \mathcal{L} = 3.2\ \text{fb}^{-1}$ of data. In the case of the effective field theory, lower limits at 95% CL are derived on the mass scale M_* of the effective interaction. For the simplified model, upper limits at 95% CL on the signal strength μ are set in the two-dimensional plane of the dark-matter (m_χ) versus the mediator mass (m_{med}).

5. Searches with Boosted Higgs Bosons decaying to a $b\bar{b}$ Pair

The reconstruction and identification of boosted Higgs bosons decaying to a $b\bar{b}$ pair, described in Ref. [22], follows the same approach that is used in the case of boosted hadronically decaying W and Z bosons. It relies on trimmed anti-k_t $R = 1.0$ calorimeter jets and both the trimmed jet mass and the energy correlation variable $D_2^{\beta=1}$ have been found to be powerful discriminating variables against light-quark and gluon jets. In addition, b-tagging naturally plays a key role in the identification of $h \to b\bar{b}$ decays. In the boosted regime, it is crucial that the b-tagging

IOP Conf. Series: Journal of Physics: Conf. Series **878** (2017) 012007 doi:10.1088/1742-6596/878/1/012007

algorithms be optimised for high and stable performance at large b-hadron momenta and in dense decay topologies where the decay products of the two b-hadrons may overlap in the detector. The b-quark jet identification relies on anti-k_t $R = 0.2$ track jets that are ghost-associated to the selected large-R calorimeter jet. A tagging algorithm that is based on a Boosted Decision Tree (BDT) and takes as input various jet properties and properties computed by three other b-tagging algorithms is used to identify b-quark jets [23, 24].

5.1. Search for resonant and non-resonant Higgs pair production in the $b\bar{b}b\bar{b}$ final state

Boosted $h \to b\bar{b}$ decays play a key role in the search for resonant and non-resonant Higgs pair production in the $b\bar{b}b\bar{b}$ final state. The search relies on two separate analysis strategies, one targeted at low invariant di-Higgs masses m_{hh} where the four b-quark jets are well separated in the detector, the other targeted at the fully merged regime where the two Higgs bosons are reconstructed as a single large-R jet with $p_T > 250$ GeV ($p_T > 450$ GeV for the leading jet) and selected to have a trimmed jet mass consistent with the Higgs boson mass. The selected events are classified into three categories based on whether each of the two Higgs candidate jets has two associated b-tagged track jets (4b-tag category), one Higgs candidate has two b-tagged track jets and the other one (3b-tag category), or whether each Higgs candidate jet has exactly one associated b-tagged track jet (2b-tag category). While one would expect the 4b-tag category to be the natural choice for $hh \to b\bar{b}b\bar{b}$ final states, the introduction of the latter two categories serves to recover signal acceptance for large values of m_{hh} where the two b-quark jets from a Higgs boson decay are so highly collimated that they can no longer be resolved even with the narrow $R = 0.2$ track jets, as illustrated in Figure 3a. No requirement is placed on $D_2^{\beta=1}$ or any other substructure variable as the b-tagging requirements alone allow for a sufficient suppression of the dominant background from multijet production.

No significant deviation from the SM expectation is found in $\int \mathcal{L} = 13.3$ fb^{-1} of $\sqrt{s} = 13$ TeV data [25]. The exclusion limits are derived for both resonant and non-resonant Higgs pair production. In the former case, the results are interpreted in a bulk Randall-Sundrum (RS) model with $k/\bar{M}_{\text{Pl}} = 1.0$. The upper limits at 95% CL on the cross-section times branching ratio to $b\bar{b}b\bar{b}$ for a RS graviton range between 1000 and 2 fb for graviton masses in the range between 300 and 3000 GeV, as illustrated in Figure 3b. In the case of non-resonant Higgs pair production the upper limit at 95% CL is 330 fb, which is the most stringent limit to date.

5.2. Search for dark matter in final states with missing transverse energy and a Higgs boson

Boosted $h \to b\bar{b}$ decays also play an important role in the search for dark matter produced in association with a single Higgs boson [26]. In contrast to the mono-V signature (Section 5.1), Higgs radiation off the inital-state partons is Yukawa-suppressed. Hence the Higgs boson would be part of the interaction that produces the dark matter, thus providing a direct probe of the interaction between dark matter and the SM sector. The search results are interpreted in the context of a simplified model in which the Higgs boson is radiated off an s-channel vector mediator, and in the context of a Two-Higgs-Doublet Model with an extra vector mediator Z' (2HDM+Z') that decays into a Higgs boson and a heavy pseudoscalar A, which in turn decays into a pair of dark matter particles. Signal events are selected via both a **resolved** approach, in which the two b-jets from the Higgs boson decay are reconstructed as two separate small-R jets, and a **boosted** approach, where the Higgs boson is reconstructed as a trimmed anti-k_t $R = 1.0$ jet with two associated b-tagged track jets. No dedicated mass or substructure requirements are imposed. The resolved approach is used in the regime 150 GeV $< E_T^{\text{miss}} <$ 350 GeV, while the boosted approach is used to probe the region with $E_T^{\text{miss}} > 350$ GeV. No deviation from the SM expectation is observed in the E_T^{miss} distribution with 3.2 fb^{-1} of data. Exclusion limits are derived in the m_χ-m_{med} plane in the case of the simplified model, and in the m_A-m_{med}

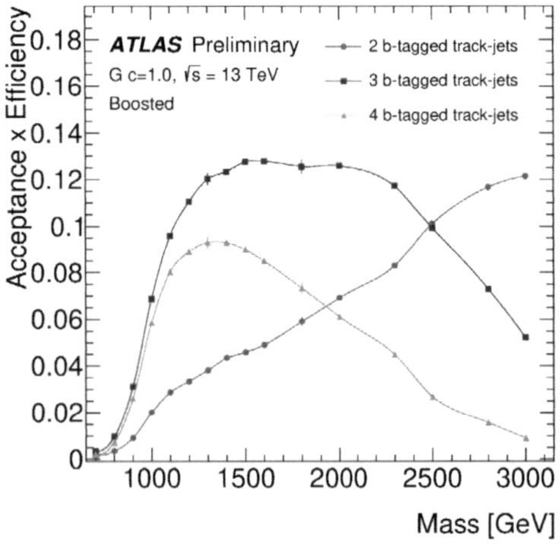

(a)

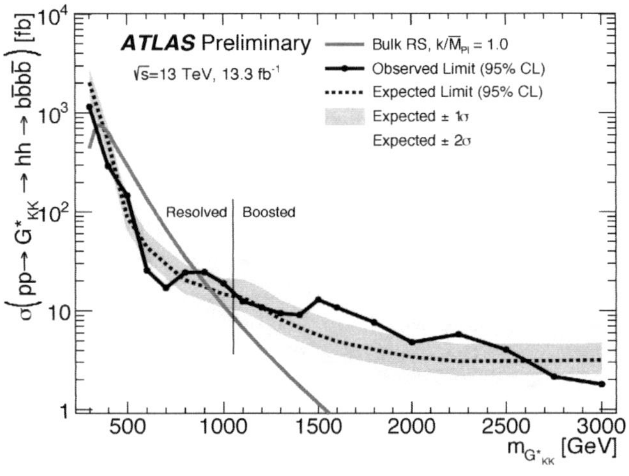

(b)

Figure 3: (a) Acceptance times efficiency for the $2b$-, $3b$-, and $4b$-tag regions in the search for Higgs pair production as a function of the mass of a hypothetical resonance. (b) Observed and expected upper limits at 95% CL on the cross-section times branching ratio for a RS graviton decaying to $b\bar{b}b\bar{b}$ via a pair of Higgs bosons [25]. The limits are taken from the analysis (resolved/boosted) that yields the tighter limits for a given signal hypothesis.

plane for $m_\chi = 100$ GeV in the case of the 2HDM+Z' benchmark.

6. Searches involving various different Boosted Objects

The ATLAS collaboration has published a variety of searches in final states that involve two or more different boosted objects. These include searches for massive resonances decaying to a W or Z and a Higgs boson in the $q\bar{q}^{(')}+b\bar{b}$ [27] and the $\nu\nu, \ell\nu, \ell\ell+b\bar{b}$ final states [28], searches for pair production of vector-like top quarks T decaying via $T\bar{T} \rightarrow WbWb$ [29], or $T\bar{T} \rightarrow Zt + X$ [30], and a search for new phenomena in $t\bar{t}$ final states with additional heavy-flavour jets [31].

For the sake of brevity, only the latter will be discussed in further detail. The search targets a variety of signal processes, including pair production of VLQs, four-top ($t\bar{t}t\bar{t}$) production in the SM and via hypothetical BSM processes, as well as associated production of heavy flavour quarks with heavy neutral (A/H) and charged ($H^{\pm}$) Higgs bosons in a 2HDM, namely $t\bar{t}A/H \rightarrow t\bar{t}t\bar{t}$, $b\bar{b}A/H \rightarrow b\bar{b}t\bar{t}$, and $tbH^{\pm} \rightarrow b\bar{b}t\bar{t}$. The search strategy is optimised for $T\bar{T}$ production where at least one of the VLQs decays to a top quark and a Higgs boson. Given that $h \rightarrow b\bar{b}$ is the dominant Higgs boson decay mode with a branching ratio of $\approx 56\%$, the signal is characterised by large jet- and b-tag multiplicities. To provide sensitivity to a large range of decay modes over a large kinematic regime, 20 orthogonal signal regions are defined, belonging either to the ℓ+jets final state, characterised by the presence of an isolated electron or muon, large E_T^{miss}, and at least six small-R jets, or the jets+E_T^{miss} final state, characterised by the presence of at least seven small-R jets and large E_T^{miss}. The regions are further distinguished based on the number of b-tags and the number of large-R jets.

Unlike the large-R jets in the searches discussed in the previous sections, the large-R jets used in this search are *re-clustered* jets [32] built from fully calibrated anti-k_t $R = 0.4$ jets with $p_T > 25$ GeV that pass a requirement on the jet vertex fraction [33] that rejects pile-up jets. These preselection requirements on the small-R input jets effectively act like trimming. The re-clustered jets are required to have $p_T > 300$ GeV and a mass[2] greater than 100 GeV in order to select jets associated with hadronically decaying Higgs bosons or top quarks.

No significant deviation from the SM expectation is found in any of the signal regions. Upper limits on the signal production cross-section are placed for various benchmark scenarios of $T\bar{T}$ production as well as on the other models discussed above. For a T-quark doublet (singlet), T-quark masses below 1160 GeV (1020 GeV) are excluded at 95% CL (Figure 4a).

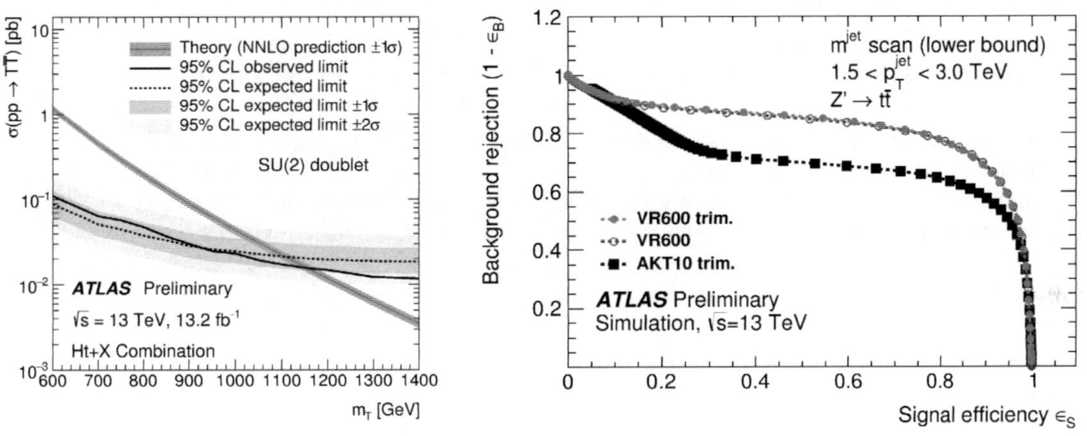

(a) (b)

Figure 4: (a) Observed and expected upper limits on the cross-section for $T\bar{T}$ production for a T-quark doublet benchmark [31]. (b) Background rejection versus signal efficiency for the variation of a lower selection requirement on the jet mass of trimmed (red) and ungroomed (blue) variable-R and trimmed anti-k_t $R = 1.0$ jets (black). Signal jets are top-quark-initated jets from simulated $Z' \to t\bar{t}$ events; background jets are taken from simulated dijet events [34].

7. Novel Substructure Techniques for Future Searches

The field of jet substructure is evolving fast with new techniques being explored to further increase the sensitivity of searches (and increasingly also measurements) in the boosted regime. For the sake of brevity, only two examples are highlighted here. The first one is related to the energy correlation variables discussed in Section 2. Recently, a new set of discriminating variables, based on generalised energy correlation functions, has been proposed [35]. It includes a class of variables optimised for good discrimination power after the removal of soft radiation and specifically designed to replace variables such as $D_2^{\beta=1}$ for groomed jets.

The second example concerns the reconstruction of the boosted object itself. All searches involving boosted objects to date rely on jets with a fixed angular size, set by the R-parameter of the jet algorithm, while in fact the collimation of the decay products, described by Equation 1, implies that the jets associated with hadronically decaying top quarks, W/Z and Higgs bosons will continue to shrink with increasing transverse momentum of the mother particle. For example, the two b-hadrons from the decay of a Higgs boson with a transverse momentum of 1 TeV are expected to be within $\Delta R < 0.25$, implying that a jet with a significantly smaller size than the traditionally used $R = 1.0$ jets would be sufficient to reconstruct the boosted Higgs boson decay. Variable-R jets [36] provide a more natural approach to the reconstruction

[2] The mass is calculated from the sum of the massive four-vectors of the small-R constituent jets.

of boosted objects as their effective size scales inversely with their transverse momentum [34]. Thus the jet shrinks around the increasingly collimated decay products, minimising the jet area and reducing the impact of contaminations from pile-up or initial-state radiation. Significant performance improvements are found with variable-R jets for a number of discriminating variables used in boosted object taggers, most notably the jet mass, as illustrated in Figure 4b.

8. Conclusion

The ATLAS collaboration has published results from a wide range of searches for BSM physics in final states with boosted objects, making use of a variety of substructure techniques to enhance the signal significance. No significant deviation from the SM expectation has been found to date and the lower exclusion limits on the masses of a number of hypothetical BSM particles have been pushed further into the TeV regime. As data-taking at $\sqrt{s} = 13$ TeV continues in 2017 and beyond at increasing luminosities, searches will be able to further explore the kinematic ranges characterised by final states with extremely boosted particles where data statistics is still poor. The challenge now lies in resolving substructure at scales below the calorimeter resolution and at unprecedented levels of pile-up. The field of jet substructure still continues to evolve at a fast rate with applications in a growing number of BSM searches and even measurements.

References

[1] ATLAS Collaboration (ATLAS) 2008 *JINST* **3** S08003
[2] Cacciari M, Salam G P and Soyez G 2008 *JHEP* **0804** 063 (*Preprint* 0802.1189)
[3] Thaler J and Van Tilburg K 2011 *JHEP* **03** 015 (*Preprint* 1011.2268)
[4] Larkoski A J, Salam G P and Thaler J 2013 *JHEP* **06** 108 (*Preprint* 1305.0007)
[5] Larkoski A J, Moult I and Neill D 2014 *JHEP* **12** 009 (*Preprint* 1409.6298)
[6] Larkoski A J, Moult I and Neill D 2016 *JHEP* **05** 117 (*Preprint* 1507.03018)
[7] Krohn D, Thaler J and Wang L T 2010 *JHEP* **1002** 084 (*Preprint* 0912.1342)
[8] Ellis S D and Soper D E 1993 *Phys.Rev.* **D48** 3160–3166 (*Preprint* hep-ph/9305266)
[9] Catani S, Dokshitzer Y L, Seymour M and Webber B 1993 *Nucl.Phys.* **B406** 187–224
[10] ATLAS Collaboration 2013 ATLAS-CONF-2013-084 https://cds.cern.ch/record/1571040
[11] ATLAS Collaboration ATL-PHYS-PUB-2015-053 http://cds.cern.ch/record/2116351
[12] ATLAS Collaboration ATLAS-CONF-2016-014 http://cds.cern.ch/record/2141001
[13] ATLAS Collaboration ATL-PHYS-PUB-2015-033 http://cds.cern.ch/record/2041461
[14] ATLAS Collaboration ATLAS-CONF-2016-055 http://cds.cern.ch/record/2206137
[15] ATLAS Collaboration ATLAS-CONF-2016-062 http://cds.cern.ch/record/2206199
[16] ATLAS Collaboration ATLAS-CONF-2016-082 http://cds.cern.ch/record/2206275
[17] ATLAS Collaboration 2017 *Phys. Lett.* **B764** 11–30 (*Preprint* 1607.06363)
[18] Cacciari M and Salam G P 2008 *Phys. Lett.* **B659** 119–126 (*Preprint* 0707.1378)
[19] Cacciari M, Salam G P and Soyez G 2008 *JHEP* **04** 005 (*Preprint* 0802.1188)
[20] ATLAS Collaboration 2013 *JHEP* **09** 076 (*Preprint* 1306.4945)
[21] ATLAS Collaboration 2016 *Phys. Lett.* **B763** 251–268 (*Preprint* 1608.02372)
[22] ATLAS Collaboration ATLAS-CONF-2016-039 http://cds.cern.ch/record/2206038
[23] ATLAS Collaboration 2016 *JINST* **11** P04008 (*Preprint* 1512.01094)
[24] ATLAS Collaboration ATL-PHYS-PUB-2016-012 https://cds.cern.ch/record/2160731
[25] ATLAS Collaboration ATLAS-CONF-2016-049 http://cds.cern.ch/record/2206131
[26] ATLAS Collaboration 2017 *Phys. Lett.* **B765** 11–31 (*Preprint* 1609.04572)
[27] ATLAS Collaboration ATLAS-CONF-2016-083 http://cds.cern.ch/record/2206276
[28] ATLAS Collaboration (ATLAS) 2017 *Phys. Lett.* **B765** 32–52 (*Preprint* 1607.05621)
[29] ATLAS Collaboration ATLAS-CONF-2016-102 http://cds.cern.ch/record/2219436
[30] ATLAS Collaboration ATLAS-CONF-2016-101 http://cds.cern.ch/record/2217232
[31] ATLAS Collaboration ATLAS-CONF-2016-104 http://cds.cern.ch/record/2220371
[32] Nachman B, Nef P, Schwartzman A, Swiatlowski M and Wanotayaroj C 2015 *JHEP* **02** 075
[33] ATLAS Collaboration (ATLAS) 2016 *Eur. Phys. J.* **C76** 581 (*Preprint* 1510.03823)
[34] ATLAS Collaboration ATL-PHYS-PUB-2016-013 https://cds.cern.ch/record/2199360
[35] Moult I, Necib L and Thaler J 2016 *JHEP* **12** 153 (*Preprint* 1609.07483)
[36] Krohn D, Thaler J and Wang L T 2009 *JHEP* **06** 059 (*Preprint* 0903.0392)

Neutron induced radiation damage of plastic scintillators for the upgrade of the Tile Calorimeter of the ATLAS detector.

J E Mdhluli[a,b,1], H Jivan[a], R Erasmus[a], Yu I Davydov[c], V Baranov[c], S Mthembu[a] B Mellado[a], E Sideras-Haddad[a,b], O Solovyanov[d] C Sandrock[a], G Peter[a], S Tlou[a], N Khanye[a],B Tjale[a]

[a] University of the Witwatersrand, 1 Jan Smuts Avenue, Braamfontein 2000, Johannesburg.
[b] DST-NRF Centre of Excellence in Strong Materials.
[c] Joint Institute for Nuclear Research, Dubna, Russia.
[d] Institute for High Energy Physics, Protvino 142281, Russia

E-mail: [1]joyemmie@gmail.com

Abstract. With the prediction that the plastic scintillators in the gap region of the Tile Calorimeter will sustain a significantly large amount of radiation damage during the HL-LHC run time, the current plastic scintillators will need to be replaced during the phase 2 upgrade in 2018. The scintillators in the gap region were exposed to a radiation environment of up to 10 kGy/year during the first run of data taking and with the luminosity being increased by a factor of 10, the radiation environment will be extremely harsh. We report on the radiation damage to the optical properties of plastic scintillators following irradiation using a neutron beam of the IBR-2 pulsed reactor in Joint Institute for Nuclear Research (JINR), Dubna. A comparison is drawn between polyvinyl toluene based commercial scintillators EJ200, EJ208 and EJ260 as well as polystyrene based scintillator from Kharkov. The samples were subjected to irradiation with high energy neutrons and a flux density range of 1×10^6–7.7×10^6. Light transmission, Raman spectroscopy, fluorescence spectroscopy and light yield testing was performed to characterize the damage induced in the samples. Preliminary results from the tests done indicate a minute change in the optical properties of the scintillators with further studies underway to gain a better understanding of the interaction between neutrons with plastic scintillators.

1. Introduction

The ATLAS detector is at one of the four collision points for protons on the Large Hadron Collider (LHC) ring together with CMS, ALICE and LHCb detectors. It is used to search for new particles through high energy proton-proton collisions at the LHC of CERN and also perform unprecedented precision measurements. The detector consists of six different detecting subsystems that are arranged in layers around the collision point to record the path, momentum and energy of the particles in order for the particles to be individually identified[1]. Amongst the detecting subsystem, there is the calorimeter subsystem that consists of the tile calorimeter. The tile calorimeter is a hadronic calorimeter responsible for detecting hadrons. It consists of a central barrel and 2 extended barrels with each barrel containing 64 modules that are made of a matrix of steel plates and plastic scintillator tiles sandwiched between them. The steel plates act as an absorber medium that converts the incoming hadrons into a "shower" of particles. The plastic scintillator tiles then absorb the energy of the particles and fluorescence to emit light. The light from the scintillators is passed through wavelength shifting (WLS) optical fibres and is detected by photomultiplier tubes. The signal is further processed using readout electronics in order to digitize the data for further analysis [2] [3].

In addition to the plastic scintillators found within the modules of the calorimeter, additional scintillators are found in the gap region. The gap region is the area between the central and extended barrels. During the first run of data taking, the scintillators in the gap region were exposed to a radiation environment of up to 10 kGy/year. It is predicted that during the high luminosity (HL)-LHC run time, the scintillators in the Gap region will sustain a significantly large amount of radiation damage and will require replacement. This prediction has led to the comparative study of proton induced radiation damage on plastic scintillators conducted by H Jivan [2] [4], C Pelwan [5] and S Liao [6].

To further extend the project on the comparative study, neutron irradiation and damage assessment studies on the same plastic scintillator samples is being conducted. Neutron interaction with matter makes this study interesting since unlike with proton irradiation where the interaction with the plastics was through direct ionization, with neutron irradiation the interaction will be through indirect ionization. Neutron bombardment on materials creates a collision cascade within the material that results to point defects and dislocations. The collisions cause a massive transfer of kinetic energy to the lattice atom that has been displaced from its lattice site, becoming what is known as the primary knock-on atom (PKA).The knock-on atoms then lose energy with each collision and that energy therefore ionizes the material[7]. The study follows the same experimental procedures as that of the proton induced radiation damage study with a few minor changes to better suite the study.

2. Scintillation mechanism

Four plastic scintillator grades were under study, three of which were obtained from ELJEN Technologies and one from ISMA, Kharkov.The plastic scintillator grades under study are the EJ200, EJ208, EJ260 and the UPS-932A type. The three plastic scintillators obtained from ELJEN technologies are composed of a polyvinyl toluene base and 3% added organic fluors [1], the fourth plastic scintillator grade is the UPS-932A type obtained from ISMA, Kharkov and is composed of polysterene base. Polyvinyl toluene consists of long chains of vinyl toluene molecules that encompass a benzene ring (C_6H_4) bonded to a methyl group (CH_3) and a vinyl group ($CH_2 - CH-$). Information on the exact composition of the doped fluors are not made public by the manufacturers. Table 1 below has some important properties of the Eljen samples listed.

Table 1. Properties of Eljen scintillators [8].

	EJ200	EJ208	EJ260
Light Output (% Anthracene)	64	60	60
Wavelength of maximum emission (nm)	425	435	490
Rise time (ns)	0.9	1.0	-
Decay time (ns)	2.1	3.3	9.2
Refractive index	1.58	1.58	1.58

Organic scintillators have a basic scintillation mechanism that involves Foster energy transfer and self-absorption. They consists of one or two dopants [9]. The benzene ring within the scintillators determines the scintillation mechanism. The chemical bonds found within a benzene ring are: σ-bonds that are in the plane with bond angle 120° and are from $sp3$ hybridization. The

IOP Conf. Series: Journal of Physics: Conf. Series **878** (2017) 012008 doi:10.1088/1742-6596/878/1/012008

other chemical bonds found are π-orbitals which are out of plane and overlap. The π-electrons are completely delocalized.

The molecule undergoes vibrational relaxation to the S_{10} state after the scintillator has absorbed a photon or excitation by ionization. The S_{10} excited state radiatively decays to the vibrational sub-levels of the ground state. The lifetime of the S_{10} state is in the nanoseconds time range. The short lifetime allows for the fluorescence emission spectrum to be roughly a "mirror image" of the absorption spectrum. The Stokes shift is observed when the emitted photons have less energy than the $S_{00} - S_{10}$ phase transition. There is no $S_2 - S_0$ emission, thus there is an internal non-radiatively de-excitation occurring within the scintillator taking place in the picoseconds time range. The excited triplet state cannot decay to the ground state as a result angular momentum selection rules, it therefore results in a delayed fluorescence and phosphorescence [9].

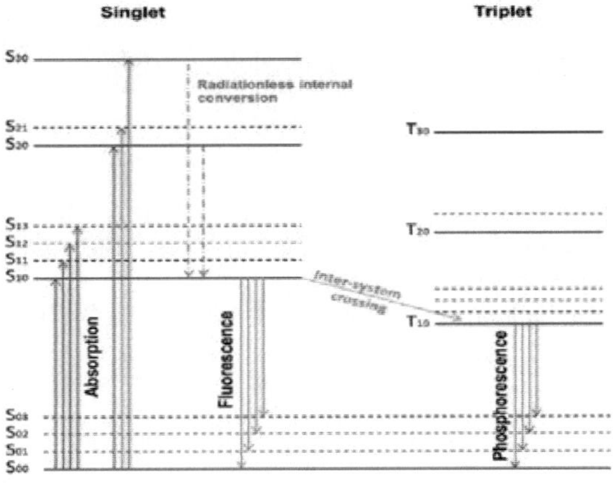

Figure 1. Energy level diagram of an organic molecule with π-electron structure, adapted from[9].The singlet states with spin 0 are labelled S_{ab} and the triplet states of spin 1 are labelled T_{ab} where a indicates the excitation level and b indicates the vibrational state of that level.

3. Experimental Procedure

Several samples of each plastic scintillator grade were cut and polished to dimensions 20 mm by 20 mm by 6 mm thickness at the Dzhelepov Laboratory of Nuclear Problems (DLNP) at JINR. Special sample holders were made to accommodate the samples for irradiation. Channel 3 of the IBR-2 pulsed reactor at the Frank Laboratory of Neutron Physics (FLNP) at JINR was used to irradiate the samples [10]. The samples were subjected to irradiation with a beam of fast neutrons with a wide energy range of up 10 MeV for 337 hours. The reactor was operating at an average power of 1875 kW. The samples were placed at three different positions from the reactor core to expose them to different neutron fluences in order to achieve various doses. The neutron flux density ranged between $1 \times 10^6 - 7.7 \times 10^6 n/cm^2/s$.

The Monte Carlo N-Particle (MCNP) 5 [11] code was used to simulate neutron transport through plastic scintillators and to determine the dose rate. The Monte Carlo method is used to simulate statistical processes theoretically, in particular complex problems that cannot be solved/modelled using computer codes that use deterministic methods. The MCNP code is used to simulate neutron, photon and electron or coupled neutron/photon/electron transport. The code is in 3-D and is capable of tracking up to 34 particles and 4 light ions. The code takes into account the absorption and the moderation of the neutrons as they travel though matter. Table 2 below shows the neutron flux density and neutron fluences at the various positions.

Table 2. Neutron flux density and neutron fluences at the various positions.

Sample position	**Flux density** $(n/cm^2/s)$	**Fluence** (n/cm^2)
1	1×10^6	1.2×10^{12}
2	3.6×10^6	3.6×10^{12}
3	7.7×10^6	9.4×10^{12}

Light spectroscopy was conducted using the Varian Carry 500 spectrophotometer to characterize the optical properties of the irradiated samples due to the damage of the neutron irradiation. The light transmission of the samples was measured relative to the transmission in air over a laser wavelength range between 200-800 nm.

The Raman spectroscopy techniques was used to investigate any structural damage undergone by the scintillators after irradiation. The Horiba Jobin-Yvon Raman spectrograph consists of an Argon laser used to provide a 515 nm excitation wavelength. Raman spectra were obtained for the un-irradiated control samples as well as on the irradiated samples to gauge the changes in the samples.

Fluorescence testing was measured with the aid of LabrRAM HR Raman spectroscopy setup. A laser of wavelength of 229 nm with a power ranging from $3 \sim 5\ mW$ was employed to provide enough energy for molecular excitation, which results in prompt light emission through luminescence. The laser is guided through a series of mirrors and optics in the setup, and is incident on the sample through the microscope aperture. As the sample fluoresces, the light emitted in the direction back up the aperture is collected by a detector and a differential wavelength spectrum is obtained.

Light yield testing was conducted at CERN. The samples were excited with a Sr-90 source that undergoes beta decay. As the sample flourescence to emit light, the light is passed through four WLS optical fibres and detected by a PMT. The signal from the PMT is further processed through electronics and is digitized in the computer. For better contact between the sample and the fibres, profiles were placed along the sides of the sample to ensure the stability of the fibres.

4. Results and Discussion

Below are the spectra of the EJ200 samples. All the other scintillator grades behaved in the same manner as the EJ200 and therefore their results will not be discussed in detail.

Figure 2 shows the light transmission spectra of the EJ200. It is observed that at a wavelength of 400 nm the absorptive edge falls away completely and that the overall transmission of the grade remains unchanged even after irradiation. The transmission loss is observed at wavelength 450 nm as this corresponds to the peak absorption wavelength of the wavelength shifting optical fibres coupled with these scintillators within the Tile Calorimeter. Even at this wavelength, there is still no change observed in the samples with increasing doses.

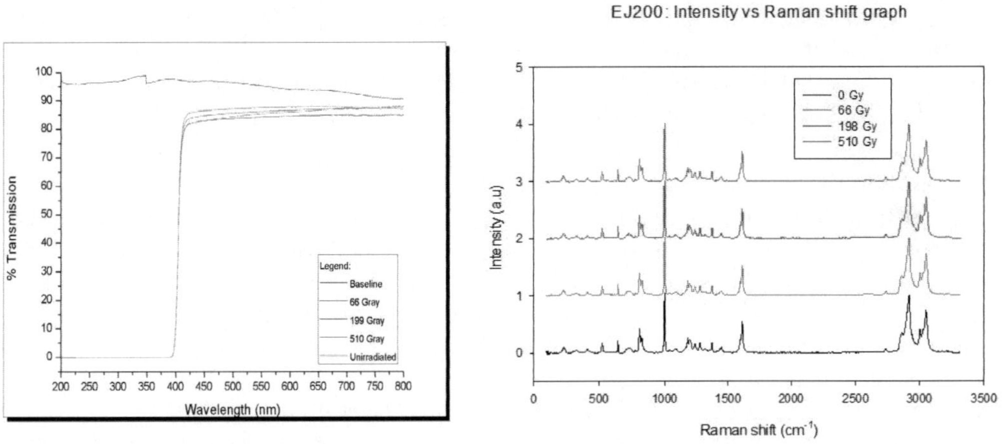

Figure 2. Left: Light transmission spectra for the EJ200 samples. Right: Raman spectra for the EJ200 samples.

Raman spectra were obtained for the various irradiated samples as well as for un-irradiated control samples. The irradiated samples maintained their structure with no additional peaks being formed and no damage was observed. Figure 2 also shows the Raman spectra of the EJ200 samples for the irradiated and un-irradiated samples.

Figure 3 shows the fluorescence spectra of the EJ200 samples.Several fluorescence peak features are observed in the spectra. Peaks in the wavelength region of 300-375 nm correlate to fluorescence of the PVT/PS base. The "two-peak" feature is observed since the fluorescence is predominantly from the benzene ring structure. In the wavelength region of 375-500 nm, the fluorescence correlates to that of the fluor dopants. No change in fluorescence is observed in the spectra for the irradiated and un-irradiated samples.

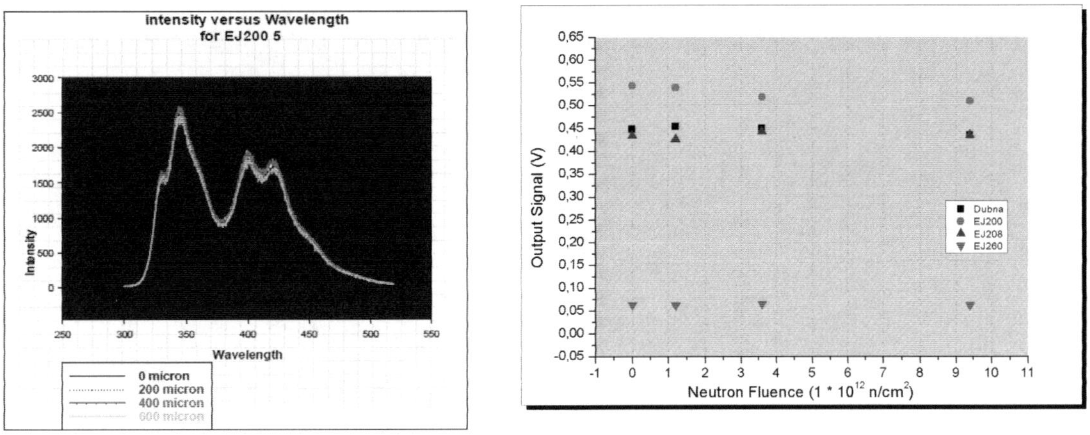

Figure 3. Left: Fluorescence spectra for the EJ200 samples. Right: Light yield output signal versus neutron fluence.

The light yield of the scintillators is also shown in figure 3. The light yield of the samples was measured by testing their response to a Sr-90 source that undergoes beta decay. As the sample flourescence to emit light, the light is passed through four WLS optical fibres and detected by

IOP Conf. Series: Journal of Physics: Conf. Series **878** (2017) 012008 doi:10.1088/1742-6596/878/1/012008

a PMT. The signal from the PMT is further processed through electronics and is digitized in the computer. This setup was aimed at mimicking what happens within the detector when the scintillators are measuring the energy of the particles. From the graph below, we observe no change for all the scintillator grades with irradiation.

5. Further work

During the autumn 2016 run of the IBR-2 reactor, composite scintillators were irradiated at higher neutron fluences than previously done. Composite scintillators were irradiated at higher neutron fluences range of $3.8 \times 10^{-12} - 1.8 \times 10^{14}\ n/cm^2$. Raman Spectroscopy, Light yield measurements and Light transmission testing were performed on the irradiated samples. Light yield measurements were done at the Dzhelepov Laboratory for Nuclear Problems at the Joint Institute for Nuclear Research.

The light yield decreased by $\sim 28\%$ after the neutron irradiation of $1.8 \times 10^{14}\ n/cm^2$. A decrease in the light transmission is also observed from the light transmission spectrum. In figure 4 on the left, we observe the change in the light transmission spectrum on the inset clearly. On the right of figure 4 we observe the light yield graph.

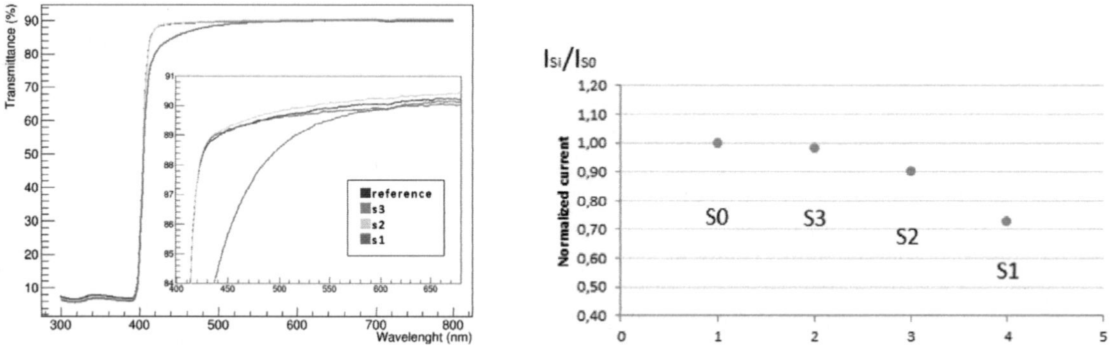

Figure 4. Left: Light transmission of the composite scintillator. Right: Light yield output of the composite scintillator.

No evident structural changes were observed after the irradiation from the Raman spectroscopy measurements. No additional peaks were formed after irradiation. Figure 5 shows the Raman spectra of the composite scintillator, the un-irradiated spectrum with the irradiated spectra above it.

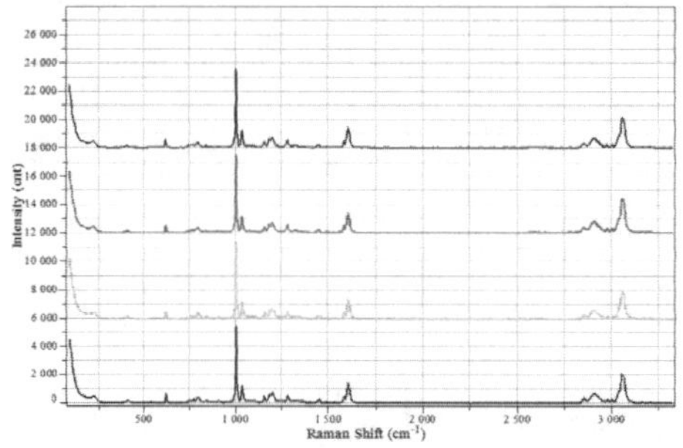

Figure 5. Raman spectrum of the composite scintillators.

6. Conclusion

The radiation damage the plastics were exposed to is very low relative to that within the ATLAS detector. Structural and optical measurements performed to study the damage of the plastics for the first set of samples that were irradiated during the April 2016 run yielded no change. The study was further extended to higher fluences during the autumn 2016 run of the reactor. Changes in the transmission of the scintillators was observed at fluences $1.8 \times 10^{14}\ n/cm^2$. This is a good indication of what range of neutron flux the plastics need to be subjected to in order to start observing changes. Further analysis on the plastics is done and plans to work at even higher fluences are underway.

7. Acknowledgments

We would like to acknowledge the Frank Laboratory of Neutron Physics (FLNP) for making the IBR-2 reactor available for experiments and the staff involved. We also acknowledge SA-CERN consortium, NRF and SA-JINR for the financial support.

References

[1] ATLAS:CERN [Online] Available: *http://home.cern/* [Accessed 24 08 2016]
[2] Jivan H, Erasmus R, Mellado B, Peters G, Sekonya K, Sideras-Haddad E 2014 Radiation hardness of plastic scintillators for the Tile Calorimeter of the ATLAS detector (Submitted to the SAIP conference)
[3] Succurro A 2012 *Physcis Procedia* **37** 229-237
[4] Jivan H, Mellado B, Sideras-Haddad E, Erasmus R, Liao S, Madhuku M, Peters G, Solvyanov O 2015 *Journal of Physics conference series* **645** 012019
[5] Pelwan C, Jivan H, Joubert D, Keartland J, Liao S, Peters G, Sideras-Haddad E 2015 *Journal of Physics conference series* **10** 645:012023
[6] Liao S, Erasmus R, Jivan H, Pelwan C, Peters G, Sideras-Haddad E 2015 *Journal of Physics conference series* **10** 645:012021
[7] Bisanti P 1986 *Magnetic properties of matter* (Turin:World Scientific) ed Borsa F, Tognetti V p 385-408
[8] ELJEN Technology 2013 Products: Plastic Scintilators [Online] Available: [Accessed 26 08 2016]
[9] Knoll G F 1999 *Radiation Detection and Measurement* Third Edition (Michigan:John Wiley and Sons Inc.) chapter 8 pp 220-222
[10] Bulavin M. 2015 Irradiation facility at the IBR-2 reactor for investigation of material radiation hardness *NIM B343* **26 -29** p26.
[11] Monte Carlo Group: Los Alamos National Laboratory [Online] Available: *https://mcnp.lanl.gov/* [Accessed 24 11 2016]

Results on top-quark physics and top-quark-like signatures by CMS

Eric Chabert on behalf the CMS Collaboration
IPHC, IN2P3-CNRS, Université de Strasbourg
E-mail: Eric.Chabert@cern.ch

Abstract. This report reviews the results obtained by the CMS Collaboration on top quark physics, focusing on the latest ones based on p–p collisions provided by the LHC at $\sqrt{s}$=13 TeV during Run II. It covers measurements of single-top, top quark pairs and associated productions as well as measurements of top quark properties. Finally several beyond the standard model searches involving top quark in the final states are presented, such as searches for supersymmetry in the third generation, heavy resonances decaying into a top quark pair, or dark matter produced in association to a single-top or a top quark pair.

1. Introduction

The study of the top quark is a domain in itself within the field of the particle physics. Latest quark being discovered by the Tevatron experiments at Fermilab in 1995, the top quark has been intensively studied at the LHC. In this note we will mainly focus on the results obtained by the CMS collaboration [1]. However, when available, we will present as well results obtained by combining measurements performed by ATLAS [2] and CMS collaborations.

It is the heaviest elementary particle ever discovered; its mass has been measured with an excellent accuracy to 173.34 ± 0.27 (stat.) ± 0.71 (syst.) GeV [3] (world combination) . The existence of the Brout-Englert-Higgs mechanism revealed by the discovery of the Higgs boson [4, 5] induces a strong interplay between Higgs an top quark physics as the Yukawa coupling of the latter is large. Quadratic divergences appear in the computation of the Higgs boson mass and are dominated by top quark loop contributions. This induces a fine-tuning in the renormalisation procedure often referred as the naturalness problem. The top quark has a very short lifetime, about 10^{-25} seconds, which is lower by one order of magnitude than the typical hadronization timescale and by four orders of magnitude than the characteristic spin decorrelation time. This has two important consequences: firstly no strong bound states made of top quark exists and secondly the top quark will decay before hadronizing, transmitting its spin information to its product decays.

Top quarks decay via the electroweak interaction into a W boson and a bottom quark with a branching ratio (BR) of about 1. Tiny other top quark contributions in the CKM matrix are studied by dedicated analysis. The study of the top quark decays allows to test the V-A structure of the electroweak interaction leading to studies such as the W-helicity measurements [7]. As W bosons can decay either leptonically or hadronically, top quark physics can be studied via few final states, three for top quark pair production.

2. Production of top quarks

2.1. Top quark pair production

Many standard model (SM) processes can produce top quarks in the final state. The dominant one is the top quark pair production, $t\bar{t}$, which is dominated by gluon-gluon fusion at LHC (about 90% at 13 TeV). The cross-section of this strong interaction process is proportional to $(\alpha_S/m_{top})^2$ at the leading order of the pertubative theory where α_S is the strong coupling constant and m_{top} is the top quark mass. By consequence its measurement is a multi-purpose test. Comparisons of experimental measurements to the best theoretical calculations performed at next-to-next-to-leading order allow to test the perturvative QCD. Both experimental and theoretical uncertainties have reached a threshold of 2-3% [8, 9]. Moreover cross-section measurements can be reinterpreted in term of constraints either on α_S or on the top quark mass.

Measurement have been performed at all center-of-mass energies delivered by the LHC in several final states by both ATLAS and CMS experiments. These results are summarized in the left plot of the Figure 2.1. The right plot reports the latest measurements done at 13 TeV with their combination performed by the LHC Top Working Group. All reported measurements are compatible with the theoretical predictions. With the high integrated luminosity recorded, those measurements are not dominated by statistic uncertainties but by the systematics ones. The channel with the best accuracy is thus the dileptonic final state with one electron and one muon. Although the branching ratio is small, about 2%, the channel is pure and do not suffer from backgrounds uncertainties, Drell-Yann contributions being heavily suppressed by the requirement of opposite lepton flavor. While the first cross-section measurements were based on counting experiments, the application of profile likelihood ratio methods relying on a categorization of the events have considerably helped to reduce the systematic uncertainties. Indeed by describing the main uncertainties such as the jet energy scale in term of nuisance parameters into the fit, the choice of appropriate categories or distributions led to a reduction of the sensitivity to those parameters. A categorization based on the jet multiplicity and the jet p_T spectrum was used in the 8 TeV measurement [8].

The cross-section measurements were used for several reinterpretations such as an accurate measurement of α_S leading to a measured value of $\alpha_S(m_Z) = 0.1151\pm^{0.0028}_{0.0027}$ [10]. The dependence of the cross-section as function of the top quark mass has been parametrized from the 7 and 8 TeV measurements as following [8]:

$$\sigma_{t\bar{t}}(7\mathrm{TeV}, m_t^{\mathrm{MC}}) = \exp\left[-0.1718\,(m_t^{\mathrm{MC}}/\mathrm{GeV} - 178.5)\right] + 170.9\,\mathrm{pb} \tag{1}$$

$$\sigma_{t\bar{t}}(8\mathrm{TeV}, m_t^{\mathrm{MC}}) = \exp\left[-0.1603\,(m_t^{\mathrm{MC}}/\mathrm{GeV} - 185.4)\right] + 237.0\,\mathrm{pb} \tag{2}$$

Similar results have been obtained and reported with 13 TeV mesurements [11].

2.2. Single top quark production

Top quarks can also be produced from weak interactions through three processes: t-channel, Wt-channel and s-channel. These processes have smaller cross-sections than the pair production mechanism as they involve the weak coupling constant and probe different parton density functions (pdf), in particular the anti-quark or b-quark ones. The cross-sections of these processes are proportional to $|\mathrm{V_{tb}}|^2$ where $\mathrm{V_{tb}}$ is one of the element of the CKM matrix. The study of the single top production allows to probe the tWb vertex through studies of the top quark polarisation. Cross-sections have been measured at 7, 8 and 13 TeV and the measurements and their combinations performed by the LHC Top WG are summarized in Fig. 2.2. Beyond cross-section measurements several reinterpretations have been made. Thereby the ratio of the measurement in the t-channel for top and anti-top, identified from the charge of the lepton in

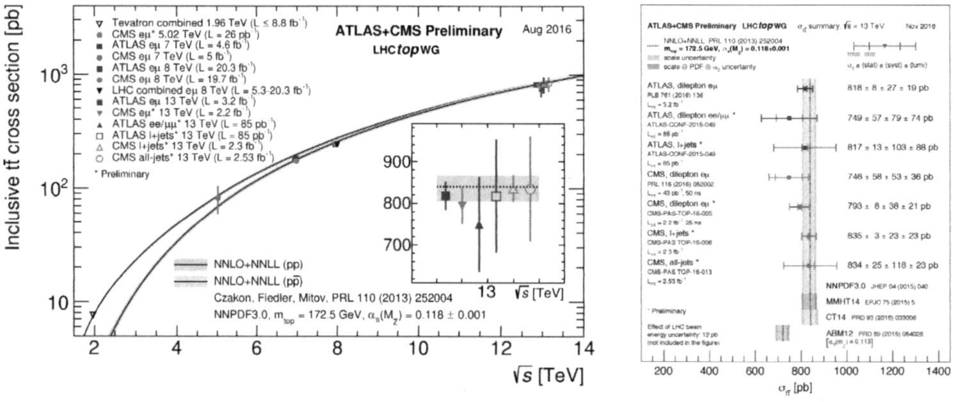

Figure 1. Left plot: Summary of LHC and Tevatron measurements of the top-pair production cross-section as a function of the centre-of-mass energy compared to the NNLO+NLL calculation [6]. The theory band represents uncertainties due to renormalisation and factorisation scale, parton density functions and the strong coupling. **Right plot:** Summary of measurements of the top-pair production cross-section at 13 TeV compared to the NNLO+NLL calculation [6].

the final state, has been used to constrain the relative pdf of up and down quarks in the proton [12]. As presented in Fig. 2.2, the measurement is in agreement with all sets of pdf used in this analysis. Moreover the single top cross-section measurements have been reinterpreted in term of value of $|f_{LV}V_{tb}|$ where f_{LV} is a parameter which encapsulates possible contributions from new physics and is equal to 1 for the SM. The results do not rely on any assumption on unitarity of the CKM matrix. Reported in Fig. 2.2, the results show no deviation to the SM prediction. As a comparison, another constraint on the CKM matrix element V_{tb} have been obtained by fitting the b-jet multiplicity in di-lepton events resulting in a bound of V_{tb} >0.955 at 95% CL [13].

3. Differential cross-section measurements

Although inclusive cross-section measurements of the single top and $t\bar{t}$ productions already offer numerous possibilities for reinterpretation, the study of differential cross-sections provide new opportunities. They are based on unfolding techniques which correct the observed distributions for efficiency, acceptance and resolution effect. The unfolded distributions can then be directly compared to any predicted model. Thus the study of differential cross-sections allow to test and constrain the generator predictions. The Fig. 4 shows differential cross-sections at parton level [14] as function of $p_T(t\bar{t})$, $|y(t\bar{t})|$, $M(t\bar{t})$, and cross sections as a function of the number of additional jets compared to several generator chains. These studies are important for BSM searches where the top quark contribution is an important background as it increases the confidence into the generator chain used or define a domain of validity in case of discrepancies with the data. The study of the jet multiplicity allows to probe the whole simulation chain. The lower multiplicities are mainly sensitive to the Matrix Element description while the higher multiplicities are mainly sensitive to the parton shower description. As seen in Fig. 4, the predictions of $t\bar{t}$ + $\geq$ 4 jets could vary to up-to 50% depending on the generator chain while they agree at lower multiplicity. The measurement in data helps to decide which generators are able to correctly describe the physics of interest.

Measurement of different cross-sections can also be used to measure physical parameters or to constrain them. Thereby the measurement of the double differential cross-section of $m(t\bar{t})$ and $y(t\bar{t})$ [16] has been used to constrain the pdfs and especially to improve the gluon pdf at

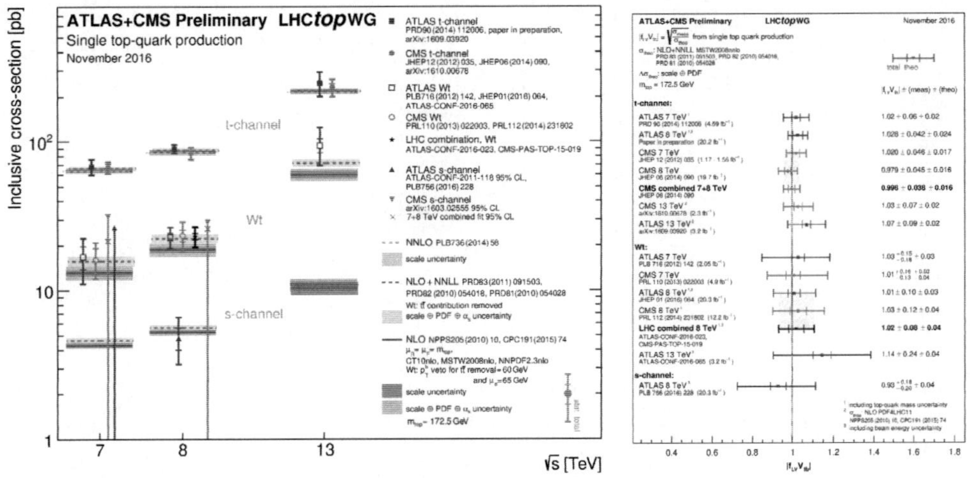

Figure 2. Left plot: Summary of ATLAS and CMS measurements of the single top production cross-sections in various channels as a function of the center of mass energy. The measurements are compared to theoretical calculations (NLO+NLL except for the t-channel: NNLO+NLL) [6]. **Right plot:** Summary of the ATLAS and CMS extractions of the CKM matrix element V_{tb} from single top quark measurements. The measurements below the line were made after the LHC combination that is shown in the upper part of the figure [6].

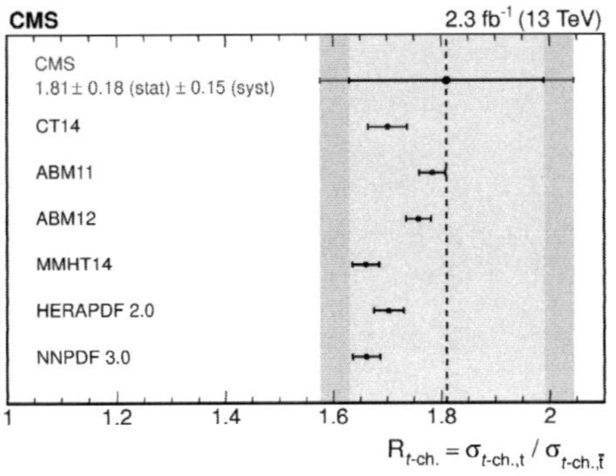

Figure 3. Comparison of the measured t and $\bar{t}$ cross-section ratio in the t-channel (dotted line) with the prediction from different PDF sets. The error bars for the different PDF sets include the statistical uncertainty, the uncertainty due to the factorization and renormalization scales and the uncertainty in the top quark mass [12].

high x as presented in Fig. 3.

3.1. Associated production

Top quark pairs can be produced in association with electroweak bosons. The $t\bar{t}W$ and $t\bar{t}Z$ processes have been observed for the first time at LHC during run I [17] and are now measured at run II [18] with uncertainties equally shared between statistic and systematic uncertainties. The multi-leptons final states are studied in order to reduced the major backgrounds. While the requirement of two leptons compatible with the Z boson significantly improves the purity of the sample for the measurement of $t\bar{t}Z$, the selection of $t\bar{t}W$ rely on a multivariate approach based on kinematic variables such as the transverse missing momenta ($\not{E}_T$), the leading lepton or jet p_T . A categorization of the selected events based on (b-tagged) jets helps to reduce the systematic uncertainties. The latest measurements are:

$$\sigma_{t\bar{t}W} = 0.98 \pm^{0.23}_{0.22}\,(\text{stat}) \pm^{0.22}_{0.18}\,(\text{syst})\,\text{pb} \tag{3}$$

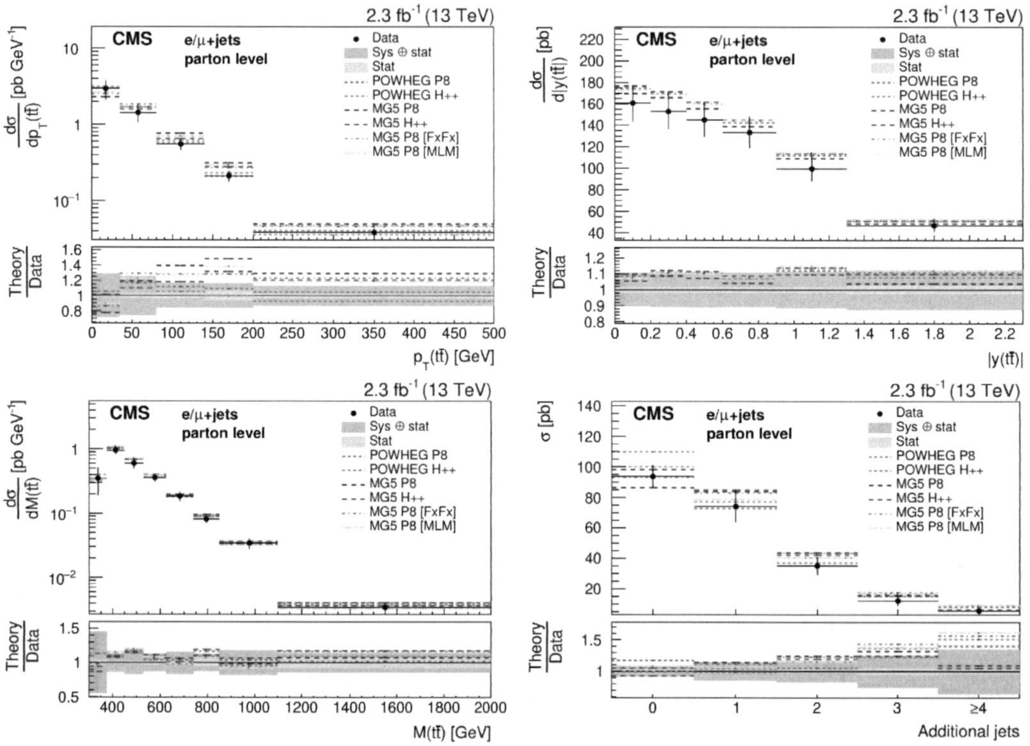

Figure 4. Differential cross sections at parton level as a function of $p_T(t\bar{t})$, $|y(t\bar{t})|$, $M(t\bar{t})$, and cross sections as a function of the number of additional jets compared to the predictions of POWHEG and MG5-MC@NLO (MG5) combined with PYTHIA8 (P8) or HERWIG++ (H++) and the multiparton simulations MG5-MC@NLO +PYTHIA8 MLM and MG5-MC@NLO +PYTHIA8 FxFx. The ratios of the various predictions to the measured cross sections are shown at the bottom of each panel together with the statistical and systematic uncertainties of the measurement [14].

$$\sigma_{t\bar{t}Z} = 0.70 \pm^{0.16}_{0.15} \text{ (stat)} \pm^{0.14}_{0.12} \text{ (syst) pb} \tag{4}$$

The production of top quark pairs in association with a photon have also been studied. The physical process is different as the photon can be emitted by any charged particle including initial states radiations or decay products of the W bosons. Thus one should define a p_T threshold for the studied photons. The analysis strategy is based the measurement of the fraction $t\bar{t}$ events where an additional photon is observed. This ratio has been measured [19] to be $(5.2 \pm 1.1) \times 10^{-4}$ for photon with $p_T > 25$ GeV leading to a measurement of the fiducial $t\bar{t}\gamma$ cross-section of 127 ± 27 fb.

3.2. Rare processes

Top decays with Flavour Changing Neutral Current (FCNC) are suppressed by the GIM mechanism [15] within the SM and lead to rare processes currently not accessible at LHC with the available luminosity. Thus any observation of such decays would be a sign of new phenomena. Many BSM models predict an increase of the cross-section of these processes by many orders of magnitude. All possible decays have been studied: $t \to u/c + (\gamma/Z/g/H)$. The SM branching ratios (BR) range from 10^{-12} to 10^{-17} while they could be as large as $10^{-9} - 10^{-6}$

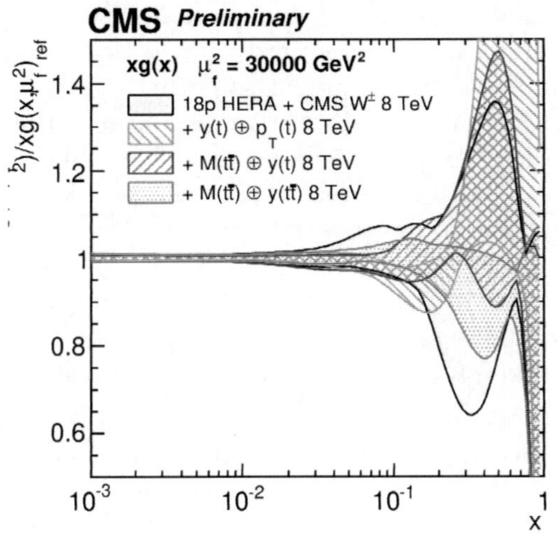

Figure 5. The gluon distributions at μ_f^2= 30 000 GeV^2, as obtained in all variants of the PDF fit, normalized to the results from the fit using the DIS and W$^{\pm}$ boson charge asymmetry data only. The total uncertainty of each distribution is shown by a shaded (or hatched) band [16].

in some BSM models. The studies have been made in single top production as well as in $t\bar{t}$. The analyses often rely on multivariate techniques to enhance the observability of tiny signals. A summary of the state of the art is reported in the Fig. 3.2. No excess has been found so far and the results were interpreted in term of limits which dominated by statistics and thus large improvements are expected in coming years.

Four top quark production is predicted by the SM but the cross-section is so tiny than it is outside the reach of the LHC at run II. However, many BSM processes predict an enhancement of the inclusive production of four top quarks in the final state such as in supersymmetric models. A generic search has been performed and has constrained the cross-section of this process. Using final states with one or two leptons, the analysis reported in [20] makes use of a Boosted Decision Tree (BDT) to reconstruct hadronic top quarks, a global event BDT to separate SM backgrounds from the signal as well as a categorization as function of the (b-tagged) jets multiplicity. The current limit is 10.2 times greater than the cross-section predicted within the SM.

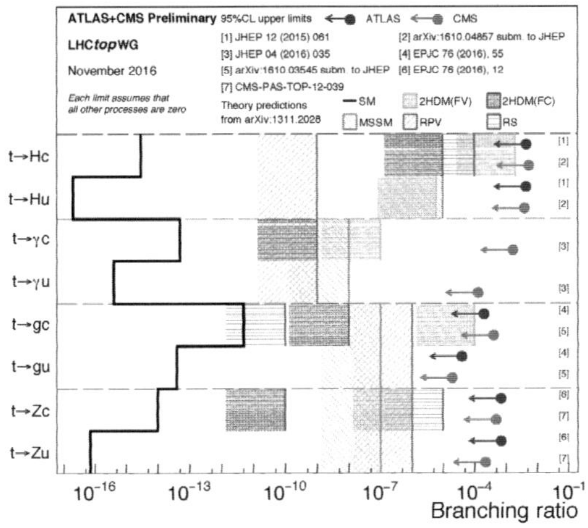

Figure 6. Summary of the current 95% confidence level observed limits on the branching ratios of the top quark decays via flavour changing neutral currents to a quark and a neutral boson t→Xq (X=g, Z, γ or H; q=u or c) by the ATLAS and CMS Collaborations compared to several new physics models.

4. Properties of the top quark

The most well-known and studied property of the top quark is its mass. As highlighted in the introduction, its importance is related to the interplay with the Higgs mechanism. The accuracy reached at the LHC is beyond the original projections thanks to the developments of advanced analysis techniques and a good control of the main systematics. The largest ones such as the jet energy correction were reduced by the use of in-situ measurements in some analyses. A broad variety of techniques have been developed. Their combination does not only improve the combined results but also gives confidence in the measured value as they are all in agreement. CMS has released few month ago a note [21] where alternative methods have been presented and combined. Combined with previous published CMS measurements, the measured value is 172.43 ± 0.13 (stat) $\pm$ 0.46 (syst) GeV.

Less studies, the top quark width has been directly or indirectly measured by few dedicated analyses. A recent paper [22] has studied the invariant mass of the lepton plus closest b-tagged jet. This observable should have a kinematic endpoint with a maximal value defined by $M_{lb}^{max} = \sqrt{m_t^2 - m_W^2}$. Contributions beyond that value are due both the top quark width as well as the experimental resolution and combinatoric effects. Thus this observable is sensitive to the top quark width. A measurement has been performed in the dileptonic channel, setting the following bounds: $0.6 \leq \Gamma_t \leq 2.4$ GeV.

A large program of the top quark physics is dedicated of the measurements of the charge asymmetries and the spin correlation but will not be mentioned here.

5. Search of new physics with top quarks in the final states

5.1. $t\bar{t}$ resonances

Many BSM models predict new particles that could decay into a top quark pair. The nature and spin of the resonance varies accross the models. Spin 1 bosons such as Z' should appear as a peak above the $t\bar{t}$ continuum in the invariant mass spectrum of the $t\bar{t}$ system. The searches are often decoupled into two regimes depending on the invariant mass of the resonances and thus on the kinematic regime of the top quark pair. While at low mass the events will appear in resolved topology where all jets are visible, at high mass the boost is such that then angles are reduced and the jets would merge. The search of Z' [24] with 13 TeV data [24] has reveal no excess and limits have been set on topcolor Z' with several width hypotheses and on Kaluza-Klein excitations of a gluon. With a luminosity of 2.3 $fb^{-1]}$, the sensitivity was equivalent to the one obtained at Run I. The 95% CL limits can reach 4 TeV in the case of Z' with $\frac{\Gamma}{M} = 0.3$.

5.2. Supersymmetric processes with top in the final states

Many supersymmetric (SUSY) processes involve top quark in the final state. The most obvious case is the production of supersymmetric top quark partners often referred as stop. Among the possible decay chain for the stop pair production, the decay into a top quark and the lightest neutralino has been intensively investigated. Top quarks could also be produced in the decay chain of gluino production. In the case of hadronic decay of the top quark, dedicated techniques can be applied. There are a variety of top-tagging techniques on the market but the basic principle remains almost the same. It uses a jet algorithm with a large parameter, sufficiently large to collect the three-body decay of the top quark. Then to discriminate those fat jets from the other jets, one needs to probe their substructure. Declustering techniques are applied associated with pruning/trimming techniques in order to remove pile-up dependencies. The invariant mass of the jet is used to select the jet as well as consideration on the sub-jets. Such techniques have been used for SUSY searches leading to large improvement in term of sensitivity at high sparticle masses. Searches in the hadronic channel [25, 26] using those techniques as set limits on stop mass up to 900 GeV with 12.9 fb^{-1} of p–p collisions at 13 TeV. Reinterpretations have also been done in the context the gluino pair productions with contrain on its mass going up to about 1750 GeV.

5.3. Top quark in association with dark matter candidates

The search of dark matter (DM) at LHC cover many possible final states. The strategy has been established in such a way that it should be possible to more easily reinterpret the constraints or observations made at LHC in term of specific models. Many analyses are looking for final states with DM+X where X could be a jet, a photon which would boost the system and make a clear signature. One of the considered signature is the production of DM in association with a top quark. Two scenarii were considered, one via the resonant production of a colored scalar that couples to a top quark and a fermionic DM candidate, the second one with FCNC production of a top quark and a new vector boson which could decay into a DM candidate pair. The experimental signature is the presence of a top quark in association with a large $\not{E}_T$. Analysis [27] in the hadronic final states has been established with 13 TeV data and put constraints on both considered models.

The production of a pair of DM candidate in association of the top quark pair has also been considered. Such model considered the existence of (pseudo-)scalar mediator between the top quark and the DM sector. A search has been setup looking for signature with 0 or 1 lepton in the final state with 13 TeV data [28]. The selection rely on strong requirement on $\not{E}_T$ as well as a specific reconstruction of the top quark. Limits were obtained by fitting $\not{E}_T$ distributions.

References

[1] S. Chatrchyan *et al.* [CMS Collaboration], JINST **3** (2008) S08004. doi:10.1088/1748-0221/3/08/S08004
[2] G. Aad *et al.* [ATLAS Collaboration], JINST **3**, S08003 (2008). doi:10.1088/1748-0221/3/08/S08003
[3] [ATLAS and CDF and CMS and D0 Collaborations], arXiv:1403.4427 [hep-ex].
[4] S. Chatrchyan *et al.* [CMS Collaboration], Phys. Lett. B **716** (2012) 30 doi:10.1016/j.physletb.2012.08.021 arXiv:1207.7235 [hep-ex].
[5] G. Aad *et al.* [ATLAS Collaboration], Phys. Lett. B **716** (2012) 1 doi:10.1016/j.physletb.2012.08.020 arXiv:1207.7214 [hep-ex].
[6] TOP LHC Working Group, `https://twiki.cern.ch/twiki/bin/view/LHCPhysics/TopLHCWGSummaryPlots`
[7] V. Khachatryan *et al.* [CMS Collaboration], Phys. Lett. B **762** (2016) 512 doi:10.1016/j.physletb.2016.10.007 arXiv:1605.09047 [hep-ex].
[8] V. Khachatryan *et al.* [CMS Collaboration], JHEP **1608** (2016) 029 doi:10.1007/JHEP08(2016)029 arXiv:1603.02303 [hep-ex].
[9] M. Czakon, P. Fiedler and A. Mitov, Phys. Rev. Lett. **110**, 252004 (2013) doi:10.1103/PhysRevLett.110.252004 [arXiv:1303.6254 [hep-ph]].
[10] S. Chatrchyan *et al.* [CMS Collaboration], Phys. Lett. B **728** (2014) 496 Erratum: [Phys. Lett. B **738** (2014) 526] doi:10.1016/j.physletb.2014.08.040, 10.1016/j.physletb.2013.12.009 arXiv:1307.1907 [hep-ex].
[11] A. M. Sirunyan *et al.* [CMS Collaboration], arXiv:1701.06228 [hep-ex].
[12] A. M. Sirunyan *et al.* [CMS Collaboration], arXiv:1610.00678 [hep-ex].
[13] V. Khachatryan *et al.* [CMS Collaboration], Phys. Lett. B **736** (2014) 33 doi:10.1016/j.physletb.2014.06.076 arXiv:1404.2292 [hep-ex].
[14] V. Khachatryan *et al.* [CMS Collaboration], arXiv:1610.04191 [hep-ex].
[15] S. L. Glashow, J. Iliopoulos and L. Maiani, Phys. Rev. D **2**, 1285 (1970). doi:10.1103/PhysRevD.2.1285
[16] CMS Collaboration,arXiv:1703.01630 [hep-ex].
[17] V. Khachatryan *et al.* [CMS Collaboration], JHEP **1601** (2016) 096 doi:10.1007/JHEP01(2016)096 arXiv:1510.01131 [hep-ex].
[18] CMS Collaboration,CMS-PAS-TOP-16-017.
[19] CMS Collaboration,CMS-PAS-TOP-14-008.
[20] A. M. Sirunyan *et al.* [CMS Collaboration], arXiv:1702.06164 [hep-ex].
[21] CMS Collaboration,CMS-PAS-TOP-15-012.
[22] CMS Collaboration,CMS-PAS-TOP-16-019.
[23] V. Khachatryan *et al.* [CMS Collaboration], Phys. Rev. D **93** (2016) no.1, 012001 doi:10.1103/PhysRevD.93.012001 arXiv:1506.03062 [hep-ex].
[24] CMS Collaboration,CMS-PAS-B2G-15-002.
[25] CMS Collaboration,CMS-PAS-SUS-16-029.
[26] CMS Collaboration,CMS-PAS-SUS-16-030.
[27] CMS Collaboration,CMS-PAS-EXO-16-040.
[28] CMS Collaboration,CMS-PAS-EXO-16-005.

IOP Conf. Series: Journal of Physics: Conf. Series **878** (2017) 012010 doi:10.1088/1742-6596/878/1/012010

Results from CMS on Higgs boson physics

Paolo Azzurri on behalf of the CMS Collaboration
INFN sezione di Pisa, Largo Pontecorvo 3, 56127 Pisa, Italy
E-mail: paolo.azzurri@cern.ch

Abstract. A selection of measurements and results of Higgs physics obtained by the CMS experiment are presented, obtained with proton collision data collected in 2015 and 2016 at the center-of-mass energy of 13 TeV.

1. Introduction

After over forty years of search efforts, the Higgs boson has been discovered by the CMS and ATLAS collaborations during the first extended running period of the CERN LHC (Run 1) with proton collisions at 7 and 8 TeV. The Run 1 data also served to measure and set constraints on many accessible properties as the Higgs boson mass, its production and decay rates and its couplings to other standard model (SM) particles.

In this document a minimal summary of Run 1 results from the CMS experiment [1] is given, followed by a review of preliminary results obtained with Run 2 proton collision data at 13 TeV . While the Run 1 data comprised a total integrated luminosity of about 25 fb^{-1}, the preliminary results presented here with Run 2 data are obtained with integrated luminosities of 2.3-2.7 fb^{-1} collected in 2015, and 12-13 fb^{-1} collected in 2016. It is to be noted that given the approximate twofold increase of signal and background cross sections for Higgs boson analyses, between the Run 1 and Run 2 collision energies, the statistical precision of the Run 2 data presented here is roughly equal to the precision obtained with Run 1 data.

2. Summary of CMS results with Run 1 data

The final combined Run 1 results from CMS are published in Ref. [2] making use of proton-proton collisions data collected in 2011 and 2012 and corresponding to integrated luminosities of up to 5.1 fb^{-1} at 7 TeV and up to 19.7 fb^{-1} at 8 TeV. Using the results of the two photon ($\gamma\gamma$) and four lepton (4ℓ) decay channels the mass of the Higgs boson is determined to be $m_{\text{H}} = 125.02^{+0.26}_{-0.27}$ (stat.) $^{+0.14}_{-0.15}$(syst.) GeV.

The event yields obtained in the different analyses production and decay channels are consistent with those expected for a standard model Higgs boson with the measured mass. No significant deviations were found, with the largest one related to the top-quark pair associated production mechanism (ttH), about 2.2 standard deviation higher than expectations, as visible in Fig. 1 (left). The excess in the ttH channel was driven by the measurements in the same-sign (SS) dilepton decay mode, aiming mostly at fully leptonic W-pair decays of the Higgs boson [3]. On the other hand, when including constraints from other channels through gluon and photon loop-induced interactions, the Higgs boson coupling to the top quark does not appear to deviate

IOP Conf. Series: Journal of Physics: Conf. Series **878** (2017) 012010 doi:10.1088/1742-6596/878/1/012010

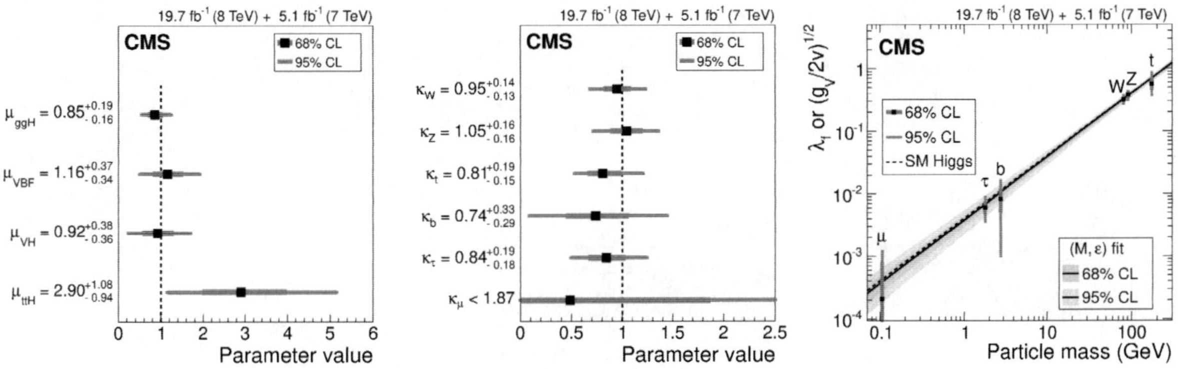

Figure 1. CMS Run 1 combined results [2]. Higgs boson production signal strengths relative to the SM expectations (left). Higgs boson couplings relative to the SM expectations (center), and versus the particle mass (right).

from the SM expectations, as also visible in Fig. 1 (right). It is therefore important to verify the top-Yukawa coupling both directly and indirectly with the new Run 2 data.

The Run 2 data will furthermore be crucial to establish the Higgs boson couplings to the down-type sector b-quarks, and to muons in the second fermion family.

3. Top-associated productions

3.1. With decays to b-quarks

First CMS results of the search for the associated production of a Higgs boson with a top quark-antiquark pair, decaying into a bottom quark-antiquark pair with 2016 Run 2 data have been recently released [4], and are shown in Fig. 2.

Events are split into several categories according to their expected signal purity. In each category the signal presence is measured using a multivariate approach that combines a matrix element method (MEM) with boosted decision trees (BDT). The ttH production cross section relative to the SM expectation is fitted to $\mu = -0.19^{+0.45}_{-0.44}(\text{stat.})^{+0.66}_{-0.68}(\text{syst.})$, that is compatible with SM expectations at the level of 1.5 standard deviations.

Results obtained with 2.7 fb^{-1} of 2015 data, also at 13 TeV, yielded a fitted $\mu = -2.0 \pm 1.8$ [5]. Previous results with Run 1 data yielded $\mu = 0.7 \pm 1.8$ with a BDT method [3], and $\mu = 1.2^{+1.6}_{-1.5}$ with a MEM method [6].

3.2. With decays to leptons

The ttH search with leptons [7] selects events with two same charge sign (SS) leptons or at least three charged (3L) leptons, produced together with b jets. After the selection of events based on the expected signal topology, two BDT classifiers are trained to separate the signal from the main ttW and ttZ backgrounds. The outputs of the two BDT classifiers in each event category are combined in bins of different signal purity, shown in Fig. 3, and the signal contribution is extracted with a simultaneous fit of the all bins contents. The results are combined with the 2015 dataset and yield a ttH signal strength of $\mu = 2.0^{+0.8}_{-0.7}$ relative to the SM prediction. Results separated by event category and lepton flavour are shown in Fig. 3.

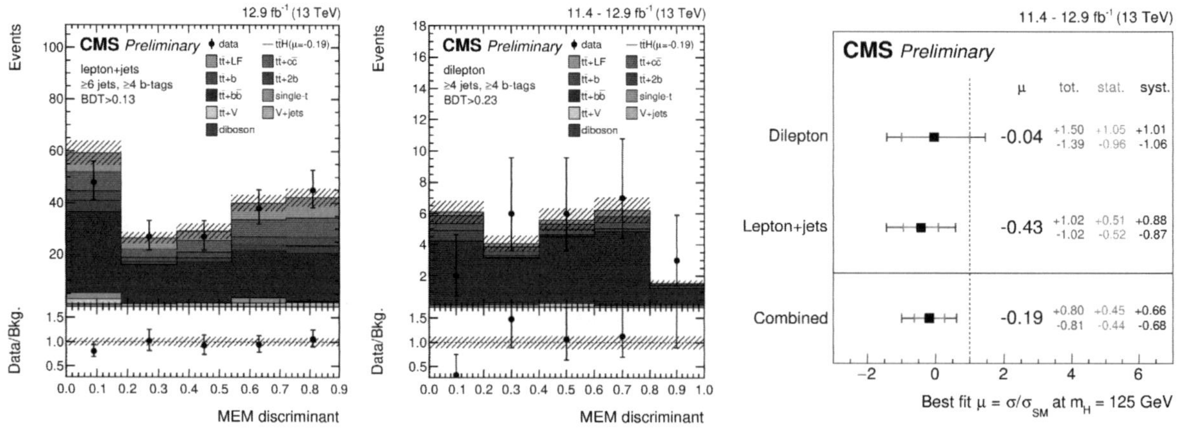

Figure 2. Preliminary Run 2 results in the top-pair associated production channel, with Higgs boson decays to b-quarks [4]. Final discriminant shapes (MEM) in the most signal-sensitive analysis categories for lepton+jets (left), and dilepton (center) events. Fitted values of the signal strength split per channel and into their statistical and systematic components (right).

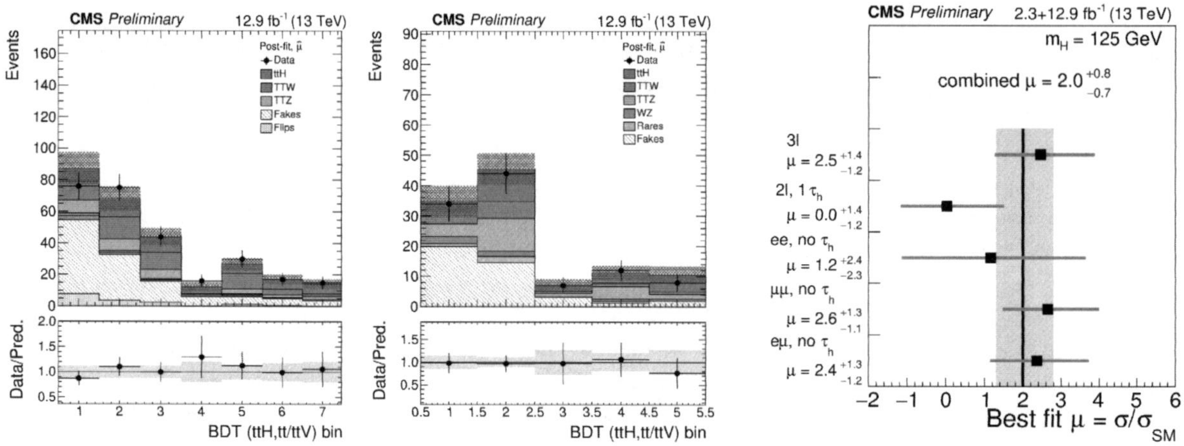

Figure 3. Preliminary Run 2 results in the top-pair associated production channel, with Higgs boson decays to leptons [7]. Combination of the BDT classifier outputs in the bins used for signal extraction, for the same-sign dilepton (left) and three-lepton (center) channels. Fitted signal strengths for the combined 2015 and 2016 data, for trileptons, dileptons and split into flavor categories (right).

3.3. With decays to photons

The search for ttH with photons [8] selects events with two photons, at least one b-jet, and additional jets or isolated leptons, defining hadronic and leptonic categories. The combined fit to the diphoton mass spectrum with 2016 data yields a signal strength of $\mu = 1.9^{+1.5}_{-1.2}$ relative to the SM prediction.

4. Two photon decays

The preliminary measurements of Higgs boson diphoton decays in different production processes with Run 2 data [8] follows closely the Run 1 strategy, with events divided into eight classes. The fitted signal strength with $m_{\mathrm{H}} = 125.09$ GeV,is reported to be $\mu = 0.91 \pm 0.20$ relative to the SM prediction, corresponding to a significance of 5.6 standard deviations. A fiducial region is defined with generator-level requirements, and the fiducial cross section is measured to be 69^{+18}_{-22} fb compared to the 73.8 ± 3.8 fb theoretical prediction.

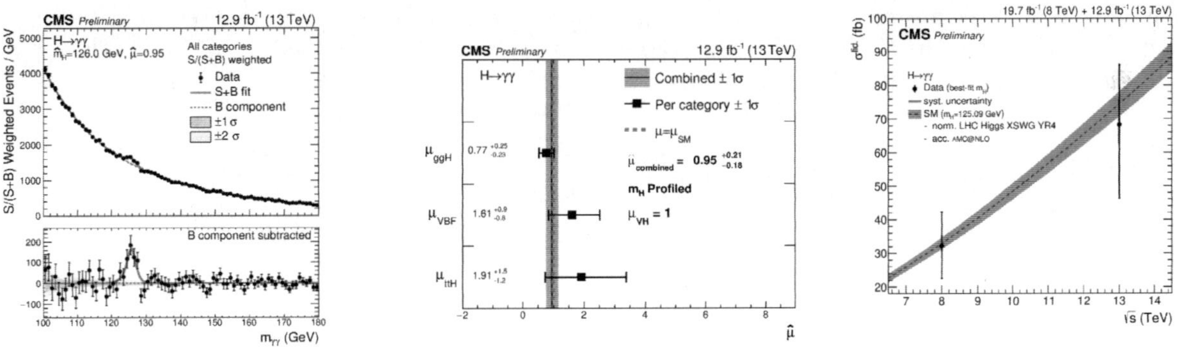

Figure 4. Preliminary Run 2 results with Higgs boson decays to photons [8]. Data points and signal plus background model fits of the diphoton invariant mass for all event categories summed and weighted by their sensitivity (left). Fitted signal strengths for each production process (center). Measurements of the fiducial cross section at 8 TeV and at 13 TeV (right).

5. Four lepton decays

Higgs boson decays to four-leptons (4ℓ) are selected with two pairs of opposite-sign, same-flavor isolated leptons. The analysis with 2016 data [9], makes use of new vector boson fusion (VBF), V-associated (VH) and ttH event categories with additional jets and leptons, and kinematic discriminants D are built in each event category.

The inclusive distribution of the 4ℓ invariant mass is shown if Fig. 5, where about 20 signal events are expected. Signal strengths are obtained with likelihood fits of $m_{4\ell}$ and D in 18 event categories. Results for different production channels are also shown in Fig. 5 for $m_{\mathrm{H}} = 125.09$ GeV, with the final combined signal strength of $0.99^{+0.33}_{-0.26}$ relative to the SM prediction, corresponding to a significance of 6.2 standard deviations. A fiducial region is defined with generator-level requirements, and the cross section is measured to be $2.29^{+0.74}_{-0.64}(\mathrm{stat.})^{+0.30}_{-23}(\mathrm{syst.})^{+0.01}_{-0.05}(\mathrm{model})$ fb compared to the theoretical prediction of 2.53 ± 0.13 fb. The energy dependence of the $\mathrm{H} \rightarrow 4\ell$ fiducial cross section is shown in Fig. 5.

The Higgs boson mass in the 4ℓ channel is also extracted with likelihood scans shown in Fig. 6, leading to the nominal result $m_{\mathrm{H}} = 124.50^{+0.47}_{-0.45}(\mathrm{stat.})^{+0.13}_{-0.11}(\mathrm{syst.})$ GeV. The Higgs boson width can be constrained by the off-shell production, fitting the 4ℓ mass distribution in the

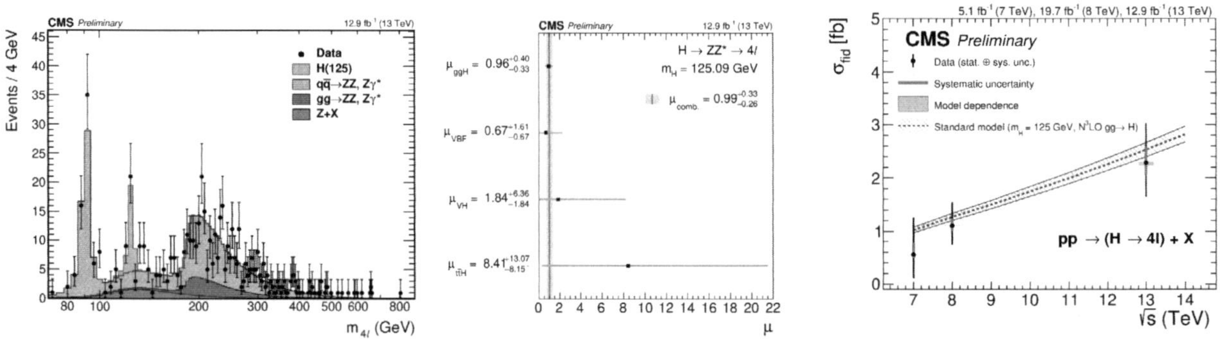

Figure 5. Preliminary Run 2 results with Higgs boson decays to four leptons [9]. Distribution of the four-lepton reconstructed invariant mass. Points represent the data and stacked histograms represent expected distributions with a 125 GeV Higgs boson (left). Fitted signal strengths for different production processes (center). The measured fiducial cross section as a function of $\sqrt{s}$ (right).

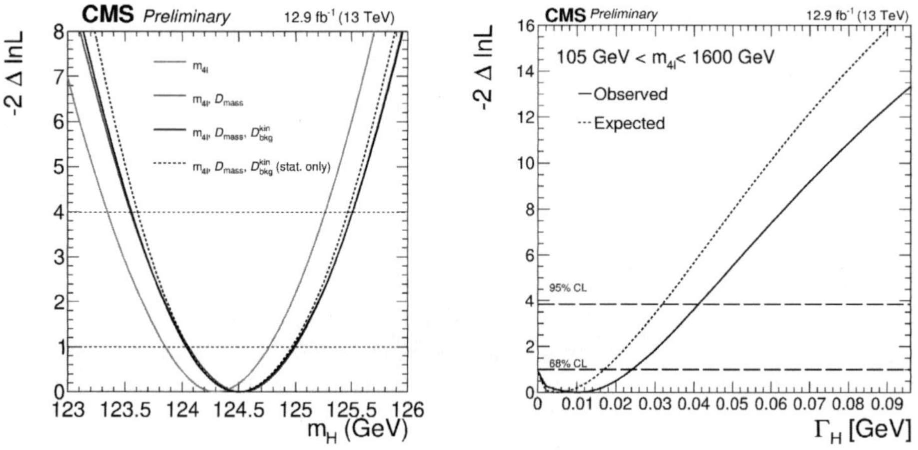

Figure 6. Preliminary Run 2 results with Higgs boson decays to four leptons [9]. Likelihood scan of the Higgs boson mass (left). Observed and expected likelihood scan of the Higgs boson width using the full four lepton mass range (right).

100-1500 GeV range, the resulting observed (expected) 95% confidence level (CL) upper limits are 41(32) MeV, as also shown in Fig. 6.

6. Higgs pair searches

6.1. Non-resonant HH

The production of Higgs bosons pairs allows to determine the self-coupling and the details of the SM Higgs potential. The Higgs boson pair production cross section predicted by the SM in 13 TeV proton collisions and for $m_{\mathrm{H}} = 125$ GeV is about 33 fb. A search for non-resonant Higgs boson pair production with both Higgs bosons decaying into b-quark pairs has been performed with 2.3 fb^{-1} of 13 TeV collision data collected in 2015 [10]. Events are collected by an hadronic multijet trigger with online b-tagging, and selected offline with a BDT classifier. The two-dimensional distribution of reconstructed dijet masses is fitted to search for the signal presence, as shown in Fig 7, and no excess over the data-driven predicted background is found. The resulting observed (expected) 95% CL upper limits are 3880(3490) fb, corresponding to 342(308) times the SM prediction.

The non-resonant Higgs boson pair production has been also searched for in the $bb\tau\tau$ final state with 12.9 fb^{-1} of 13 TeV collision data collected in 2016 [11]. No excess is found over the background expectations and he resulting observed (expected) 95% CL upper limits are 508(420) fb, corresponding to 200(170) times the SM prediction.

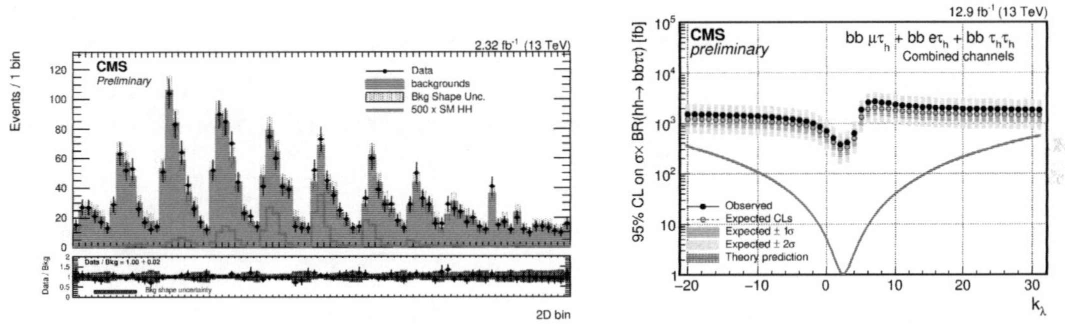

Figure 7. Preliminary Run 2 results of non-resonant Higgs boson pair productions [10, 11]. Background-only fit of the 2D dijet mass plane for the search in the four b-jet channel (left). Observed and expected 95% CL upper limits on HH cross section times branching ratio in the $bb\tau\tau$ channel, as a function of the trilinear coupling κ_λ relative to the SM prediction (right).

6.2. Resonant HH

Searches for narrow-width resonances decaying into two Higgs bosons have been performed in the bbbb [12], $bb\tau\tau$ [13] $bb\gamma\gamma$ [14], and bbWW [15] final states, using 2015 and 2016 proton collision data. No evidence for a signal is observed in any of the searched final states and upper limits at a 95% CL are set on the resonance production cross section, in the mass range from 260 to beyond 1 TeV are set, as shown in Fig. 8.

7. Future prospects

Updates on projections of physics reach with the upgraded CMS detector for high luminosity LHC have been recently released [16], taking into account the effects of high pileup conditions and expected detector performances. The extrapolated results are given for dataset of 300 fb^{-1} and 3000 fb^{-1} in two more or less conservative scenarios for systematic uncertainties. The effects of higher pileup conditions and detector upgrades are taken into account for the 3000 fb^{-1} projections.

The updated projections for the diphoton and four lepton decay channels are shown in Fig. 9. Most sensitivity projections are statistically limited, and scale roughly with $1/\sqrt{\mathcal{L}}$ up to 300 fb^{-1} with the potential of reaching a precision three to four times better than what was obtained with the Run1 data. Systematic limitations become important in many sensitivity projections with 3000 fb^{-1}, when the precision could reach 5-10 times the Run 1 sensitivity levels.

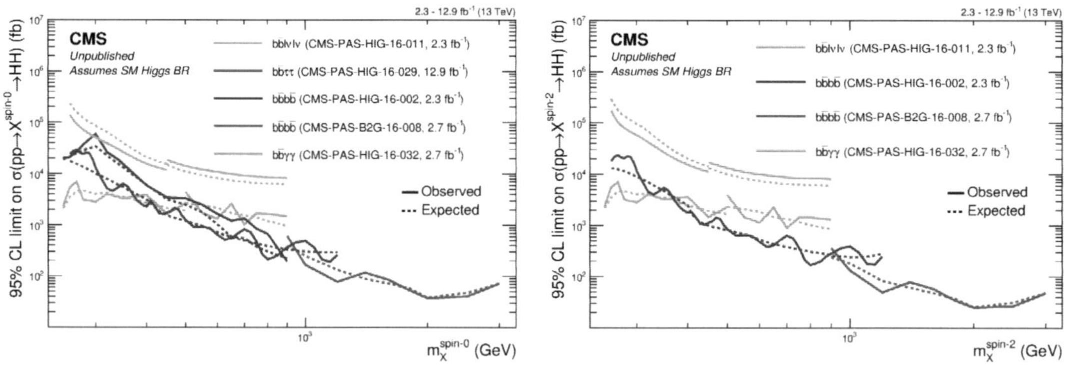

Figure 8. Preliminary Run 2 results for resonant pair productions of Higgs bosons [12, 13, 14, 15]. Observed and expected 95% CL upper limits on the resonant HH cross section as a function of the resonance mass in the spin-0 (left) and spin-2 (right) hypotheses.

8. Summary

Results with the first 2.7+12.9 fb^{-1} of 13 TeV proton collisions data have been produced by the CMS experiment, leading to the rediscovery of the Higgs boson with significances in excess of five standard deviations both in the diphoton and four-lepton channels. The new data did not reveal any significant deviations from the SM, any additional Higgs bosons, nor di-higgs productions, yet. Higgs boson physics at CMS is gradually evolving from discovery to higher precision measurements, looking forward to establish the coupling to down-type quarks, through b-jets decays, and then the coupling to the second generation, through dimuon decays. Exciting prospects are ahead of us with the full Run 2 data and beyond.

References

[1] S. Chatrchyan *et al.* [CMS Collaboration], "The CMS Experiment at the CERN LHC," JINST **3** (2008) S08004. doi:10.1088/1748-0221/3/08/S08004

[2] V. Khachatryan *et al.* [CMS Collaboration], "Precise determination of the mass of the Higgs boson and tests of compatibility of its couplings with the standard model predictions using proton collisions at 7 and 8 TeV," Eur. Phys. J. C **75** (2015) no.5, 212 doi:10.1140/epjc/s10052-015-3351-7 [arXiv:1412.8662 [hep-ex]].

[3] V. Khachatryan *et al.* [CMS Collaboration], "Search for the associated production of the Higgs boson with a top-quark pair," JHEP **1409** (2014) 087 Erratum: [JHEP **1410** (2014) 106] doi:10.1007/JHEP09(2014)087, 10.1007/JHEP10(2014)106 [arXiv:1408.1682 [hep-ex]].

[4] CMS Collaboration [CMS Collaboration], "Search for $\mathrm{t\bar{t}H}$ production in the $\mathrm{H} \rightarrow \mathrm{b\bar{b}}$ decay channel with 2016 pp collision data at $\sqrt{s} = 13$ TeV," CMS-PAS-HIG-16-038.

[5] CMS Collaboration [CMS Collaboration], "Search for $\mathrm{t\bar{t}H}$ production in the $\mathrm{H} \rightarrow \mathrm{b\bar{b}}$ decay channel with $\sqrt{s} = 13$ TeV pp collisions at the CMS experiment," CMS-PAS-HIG-16-004.

[6] V. Khachatryan *et al.* [CMS Collaboration], "Search for a Standard Model Higgs Boson Produced in Association with a Top-Quark Pair and Decaying to Bottom Quarks Using a Matrix Element Method," Eur. Phys. J. C **75** (2015) no.6, 251 doi:10.1140/epjc/s10052-015-3454-1 [arXiv:1502.02485 [hep-ex]].

[7] CMS Collaboration [CMS Collaboration], "Search for associated production of Higgs bosons and top quarks in multilepton final states at $\sqrt{s} = 13$ TeV," CMS-PAS-HIG-16-022.

[8] CMS Collaboration [CMS Collaboration], "Updated measurements of Higgs boson production in the diphoton decay channel at $\sqrt{s} = 13$ TeV in pp collisions at CMS.," CMS-PAS-HIG-16-020.

[9] CMS Collaboration [CMS Collaboration], "Measurements of properties of the Higgs boson and search for an additional resonance in the four-lepton final state at sqrt(s) = 13 TeV," CMS-PAS-HIG-16-033.

[10] CMS Collaboration [CMS Collaboration], "Search for non-resonant pair production of Higgs bosons in the $\mathrm{b\bar{b}b\bar{b}}$ final state with 13 TeV CMS data," CMS-PAS-HIG-16-026.

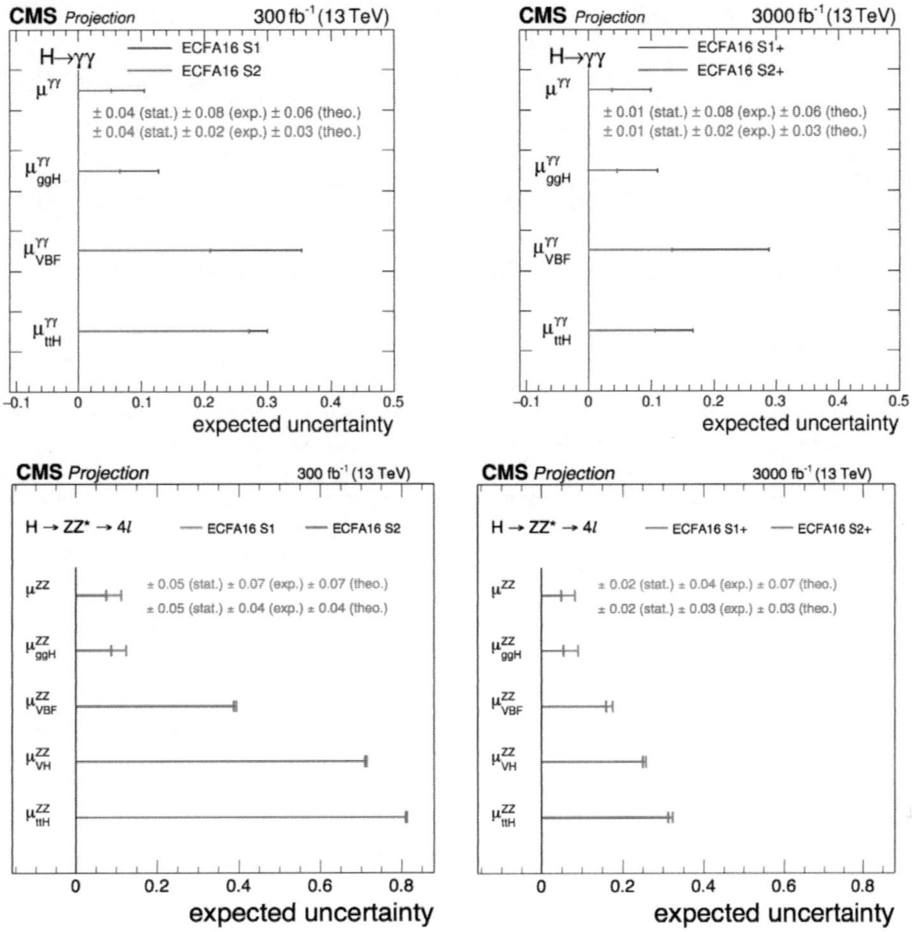

Figure 9. Projected 68% CL uncertainties on the Higgs boson productions [16] with decays to diphotons (top) and four leptons (bottom), with 300 fb^{-1} (left) and 3000 fb^{-1} (right) under the two different scenarios described in the text.

[11] CMS Collaboration [CMS Collaboration], "Search for non-resonant Higgs boson pair production in the bbtautau final state using 2016 data," CMS-PAS-HIG-16-028.
[12] CMS Collaboration [CMS Collaboration], "Search for resonant pair production of Higgs bosons decaying to two bottom quark-antiquark pairs in proton-proton collisions at 13 TeV," CMS-PAS-HIG-16-002.
[13] CMS Collaboration [CMS Collaboration], "Search for resonant Higgs boson pair production in the $b\bar{b}\tau^+\tau^-$ final state using 2016 data," CMS-PAS-HIG-16-029.,
[14] CMS Collaboration [CMS Collaboration], "Search for H(bb)H($\gamma\gamma$) decays at 13 TeV," CMS-PAS-HIG-16-032.
[15] CMS Collaboration [CMS Collaboration], "Search for resonant Higgs boson pair production in the $b\bar{b}l\nu l\nu$ final state at $\sqrt{s} = 13$ TeV," CMS-PAS-HIG-16-011.
[16] CMS Collaboration [CMS Collaboration], "Updates on Projections of Physics Reach with the Upgraded CMS Detector for High Luminosity LHC," CMS-DP-2016-064.

Heavy Flavour and Quarkonia production at LHCb

Katharina Müller on behalf of the LHCb collaboration
Winterthurerstr. 190, CH-8057 Zurich, Switzerland
E-mail: kmueller@physik.uzh.ch

Abstract. The LHCb detector, with its excellent momentum resolution and flexible trigger strategy, is ideally suited for measuring heavy quark and quarkonia production properties. Recent LHCb measurements of inclusive and differential cross-sections of the production of J/ψ and Υ resonances, as well as charm, bottom and top quarks, in pp collisions at different centre-of-mass energies are presented. Finally, results on the associated production of Υ and open charm hadrons and the exclusive production of charmonium are discussed.

1. Introduction

Heavy-flavour hadrons, containing open or hidden charm and beauty flavour, are among the most important tools for the study of Quantum Chromodynamics (QCD) in high-energy hadronic collisions. The heavy-quark mass acts as a long distance cut-off so that the partonic hard-scattering process can be calculated in the framework of perturbative QCD (pQCD) down to low transverse momenta (p_{T}). The (differential) cross section for open-heavy-flavour production is sensitive to the gluon and the heavy-quark content in the proton. Due to the forward geometry of the LHCb detector, the LHCb measurements provide valuable constraints on these parton-distribution functions (PDFs) at high and low values of Bjorken-x. When the heavy-quark pair forms a quarkonium bound state, this process is non-perturbative as it involves long distances and soft momentum scales. Therefore, the detailed study of heavy-flavour production and the comparison to predictions provide an important testing ground for both perturbative and non-perturbative aspects of QCD calculations. The dominant uncertainties for the theoretical predictions originate from higher order contributions. However, by measuring ratios of cross-sections at different centre-of-mass-energies, $\sqrt{s}$, the scale uncertainties almost cancel, while some sensitivity to the PDFs remains.

2. LHCb detector

The LHCb detector [1] is a single-arm forward spectrometer, fully instrumented in the pseudorapidity range $2 < \eta < 5$. The detector includes a high-precision tracking system consisting of a silicon-strip vertex detector (VELO) surrounding the pp interaction region and tracking stations in front and behind of the dipole magnet. To increase the pseudorapidity coverage, forward shower counters consisting of five planes of scintillators (HeRSCheL) at -114, -19.7, -7.5, +20 and +114 m from the interaction point have been installed for Run II, which started in 2015. The combination of VELO and HeRSCheL has sensitivity to particles in the regions $-10 < \eta < -5$, $-3.5 < \eta < -1.5$, $1.5 < \eta < 10$.

The LHCb experiment is operated at a low number of interactions per beam crossing. The data used in the analyses presented below use data collected in pp collisions at $\sqrt{s} = 5$, 7, 8 and 13 TeV.

3. Heavy quark production

3.1. Open charm production

Predictions of charm meson cross-sections are available in pQCD at next-to-leading order (NLO) using the general-mass variable-flavour-number scheme [2] (GMVFNS) and at fixed order with next-to-leading-log resummation [3] (FONLL). The cross-sections are calculated as a convolution of the PDFs, the partonic hard scattering cross-section, and a fragmentation function that parametrises the hadronisation of the charm quark into a given type of charm hadron. Since open charm production at LHCb is dominated by gluon-gluon scattering, the results have a sensitivity to the gluon distribution in the proton at high and low x (down to 10^{-4}) [4]. In pp collisions, charmed mesons can be produced directly from hard collisions of partons, through the feed-down of excited states, or via decays of b-flavoured hadrons. The first two sources are referred to as prompt production, while the third process is referred to in the following as from-b. For this analysis the latter is treated as background.

Double differential cross-sections measurements for the prompt production of D^0, D^+, D_s^+, and $D^{\star+}$ mesons at LHCb have been performed for $\sqrt{s} = 5$ [5], 7 [6] and 13 TeV [7] with respect to p_T and y of the charmed meson. Figure 1 shows a comparison of the double differential prompt D^0 cross section at $\sqrt{s} = 13$ TeV [7] and the ratio of the cross sections between $\sqrt{s} = 13$ and 5 TeV [5] to the predictions. The individual cross-sections and the ratio of the cross-sections at 13 and 5 TeV are described by the predictions reasonably well, also for the other charm mesons. As expected, the uncertainties of the predictions are significantly reduced for the ratio.

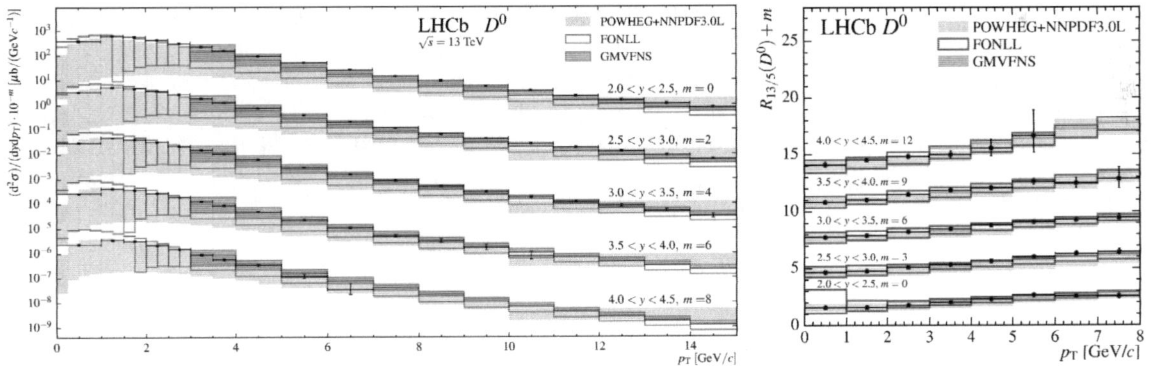

Figure 1. Measurements and predictions for (left) the double differential prompt D^0 cross-section at $\sqrt{s} = 13$ TeV (from [7]) and (right) the prompt D^0 cross-section ratios between $\sqrt{s} = 13$ and 5 TeV (from [5]).

3.2. Bottom production

The uncertainty in the knowledge of the b quark production cross section limits the sensitivity to searches for physics beyond the Standard Model (SM) as decays of hadrons containing a b quark are often dominant background proceses. In addition, knowledge of the b-quark yield is essential for calculating the sensitivity of experiments testing the SM by measuring CP-violating and rare decay processes [8].

The b-quark production cross-section is measured using semileptonic decays of b hadrons to charm hadrons (D^0, D^+, D_s, Λ_c), where the vertex formed by the muon and the charm hadron is detached from the PV. Prompt production of charm hadrons is treated as background. Figure 2 shows the differential cross-section as a function of η for $\sigma(pp \to H_b X)$, where H_b is a hadron that contains either a b or a $\bar{b}$ quark, at $\sqrt{s} = 7$ TeV and 13 TeV along with the predictions from FONLL. The predictions describe the measurement at both centre-of-mass energies within uncertainties.

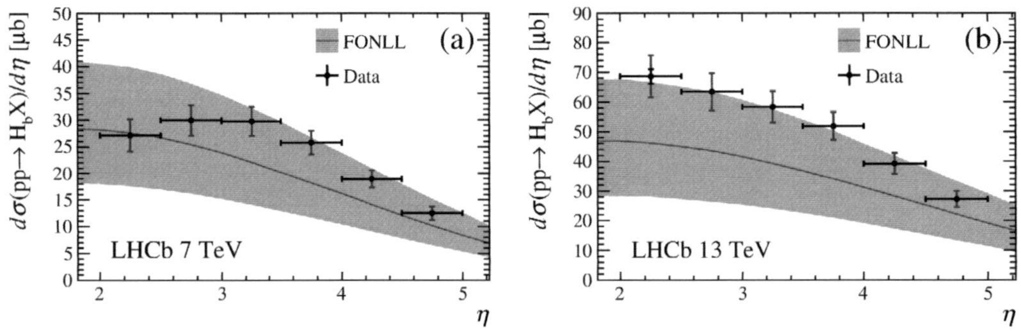

Figure 2. Differential cross-section as a function of η for $\sigma(pp \to H_b X)$, where H_b is a hadron that contains either a b or a $\bar{b}$ quark, at $\sqrt{s} = 7$ TeV (a) and 13 TeV (b) (from [9]). The measurements are compared to the predictions from FONLL.

3.3. Top production in the forward region

The production of top quarks (t) in the forward region is of considerable experimental and theoretical interest. Top quarks decay almost entirely via $t \to Wb$. The Standard Model (SM) predicts that about 75% of top production in the forward region is due to $t\bar{t}$ pair production. The remaining 25% is mostly due to t-channel single-top production. In the forward region $t\bar{t}$ production via $q\bar{q}$ and qg scattering is enhanced relative to gg fusion. This can result in larger charge asymmetries, which may be sensitive to physics beyond the SM. Forward $t\bar{t}$ events can be used to constrain the gluon PDF at large and low x.

Events with a $t \to Wb$ candidate are selected by requiring one isolated muon and a b-tagged jet in the final state. The $W+$ jet yield is determined by performing a fit to the distribution of the isolation of the muon. Two boosted decision trees, trained on the characteristics of the secondary vertex and the jet, are used to separate heavy-flavour jets from light-parton jets, and to separate b jets from c jets [10]. The excess of the observed yield relative to the direct $W + b$ prediction is attributed to top quark production as shown in Fig. 3. The resulting inclusive top production cross-sections in the fiducial region of the measurement[1] are

$$\sigma(\text{top})[7\ \text{TeV}] = 239 \pm 53(\text{stat}) \pm 33(\text{syst}) \pm 24(\text{theory})\ \text{fb},$$
$$\sigma(\text{top})[8\ \text{TeV}] = 289 \pm 43(\text{stat}) \pm 40(\text{syst}) \pm 29(\text{theory})\ \text{fb},$$

in general agreement with the SM expectations at NLO from MCFM [11] of $180^{+51}_{-41}(312^{+83}_{-68})$ fb at 7(8) TeV [12].

Events with an isolated lepton plus two tagged jets have been used to measure $W+c\bar{c}$, $W+b\bar{b}$ and $t\bar{t}$ production in the forward region. The yields are extracted from a multi-dimensional fit to the BDTs to separate c from b jets the di-jet mass and a multivariate discriminant to separate

[1] defined by $p_{\rm T}(\mu) > 25$ GeV, $2.0 < \eta(\mu) < 4.5$, $50 < p_{\rm T}(b) < 100$ GeV, $2.2 < \eta(b) < 4.2$, $\Delta R(\mu, b) > 0.5$, and $p_{\rm T}(\mu + b) > 20$ GeV

between $W + c\bar{c}$ and $W + b\bar{b}$ from $t\bar{t}$. The measured cross-sections are in agreement with the SM predictions calculated at NLO using MCFM with large uncertainties due to the limited sample size [13] as shown in Fig. 3. With the full statistics at the end of Run II differential measurements can be performed and also other decay channels will be accessible.

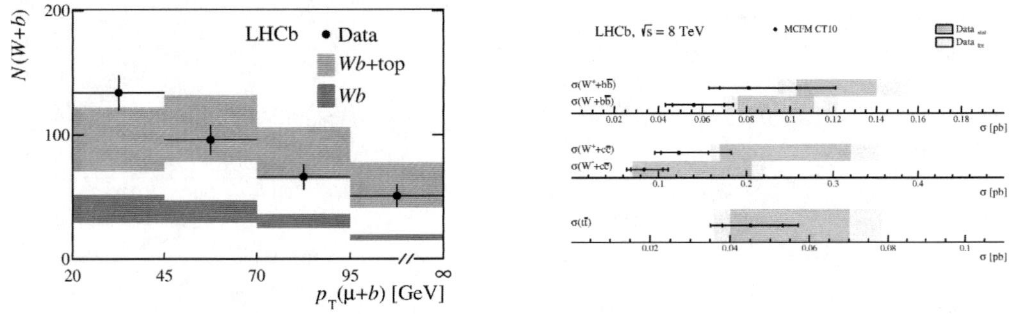

Figure 3. Left: Distribution of $p_{\rm T}(\mu + b)$ for the W plus b-jet compared to SM predictions at NLO for $W + b$ and $W + t$ production (from [12]). Right: cross-sections for $W + c\bar{c}$, $W + b\bar{b}$ and $t\bar{t}$ production in the forward region (from [13]).

4. Quarkonia production

4.1. J/ψ production at $\sqrt{s}$=13 TeV

J/ψ mesons are reconstructed in the di-muon final state with $p_{\rm T}$ of the di-muon system restricted to $p_{\rm T} < 14$ GeV/c. The excellent vertexing capability of LHCb allows a separation of prompt J/ψ mesons and J/ψ-from-b. The number of prompt J/ψ and J/ψ-from-b candidates is determined by a combined fit to the di-muon invariant mass and the pseudo-proper time distribution, where the pseudo-proper time is defined as

$$t_z = \frac{(z_{J/\psi} - z_{PV}) \times M_{J/\psi}}{p_z}.$$

Here, $z_{J/\psi}$ is the z position of the J/ψ decay vertex, z_{PV} that of the primary vertex, p_z the z component of the measured J/ψ momentum, and $M_{J/\psi}$ the reconstructed mass of the J/ψ candidate.

Figure 4 shows the single differential production cross-sections for prompt J/ψ and J/ψ-from-b as a function of $p_{\rm T}$ in the forward region at $\sqrt{s} = 13$ TeV [14]. The cross-section for J/ψ-from-b is about a factor of ten smaller than for prompt J/ψ with a $p_{\rm T}$ and y dependence similar to what was observed by LHCb at lower $\sqrt{s}$. The measurements are described by non-relativistic QCD (NRQCD) [15] for prompt J/ψ and FONLL for J/ψ-from-b.

4.2. Υ production at $\sqrt{s} = 7$ and 8 TeV

The production of Υ mesons in pp collisions can occur either directly or via feed down from the decay of heavier bottomonium states. The cross-section for Υ is measured differentially in $p_{\rm T}$ and y in the di-muon final state in the fiducial volume $1 < p < 30$ GeV/c, $2 < \eta < 4.5$ at $\sqrt{s} = 7$ and 8 TeV. The signal yields in each $(p_{\rm T}, y)$ bin are determined from an unbinned extended maximum likelihood fit to the di-muon mass spectrum of the selected candidates. Fig. 5 a) shows the differential cross-section for $\Upsilon(1S)$, $\Upsilon(2S)$ and $\Upsilon(3S)$ at $\sqrt{s} = 8$ TeV [16]. The shape of all three states are well described by predictions based on the CO model [17].

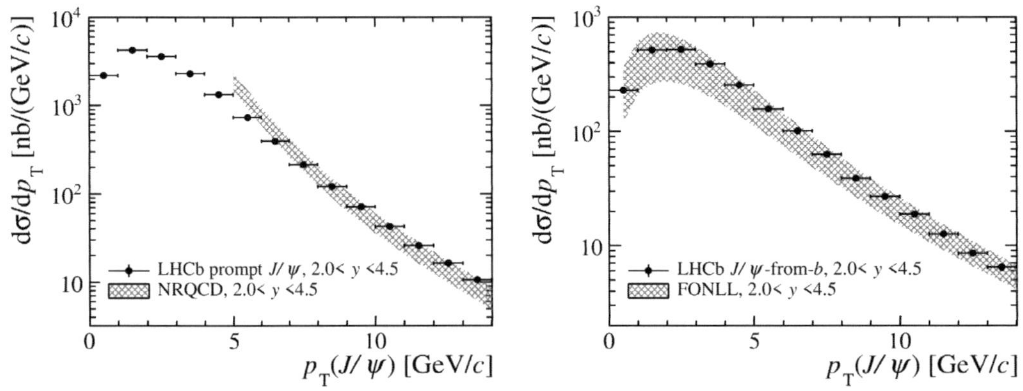

Figure 4. Differential cross-sections as a function of p_T, (left) compared with the NRQCD calculation for prompt J/ψ and (right) with the FONLL calculation for J/ψ-from-b meson (from [14]).

The ratio of the cross-sections at $\sqrt{s} = 8$ and 7 TeV as a function of rapidity, integrated over the region $p_T < 30$ GeV/c is shown in Fig. 5b). They are compared with the expectations from the CO mechanism with normalisation factors fixed from the fits to the differential cross-sections. The shape observed in data does not agree with the pure CO model.

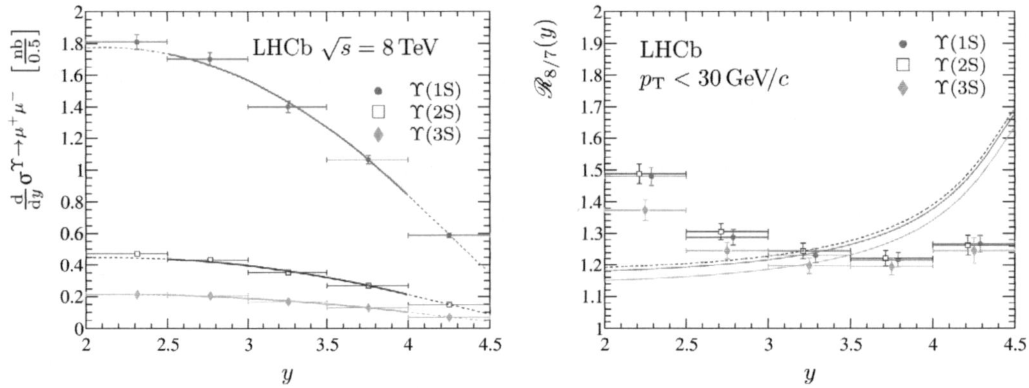

Figure 5. Left: differential cross-sections in bins of y for $\Upsilon(1S)$(red), $\Upsilon(2S)$(blue) and $\Upsilon(3S)$(green) at $\sqrt{s} = 8$ TeV. Right: ratio of differential cross-sections at $\sqrt{s} = 8$ and 7 TeV (from [16]). Lines in both plots show fit results with the CO model predictions.

4.3. Υ plus open charm

Production of multiple heavy quark pairs in pp collisions receives contributions from single parton (SPS) as well as double parton scattering (DPS). Within the DPS mechanism, the Υ meson and $c\bar{c}$ (C hadron) pair are produced independently in different partonic interactions. Neglecting the parton correlations in the proton, the contribution is estimated according to the formula

$$\sigma^{\Upsilon \times c\bar{c}} = \frac{\sigma^{\Upsilon} \times \sigma^{c\bar{c}}}{\sigma_{\text{eff}}},$$

where σ^{Υ} and $\sigma^{c\bar{c}}$ are the inclusive Υ and charm cross-sections, and σ_{eff} is an effective cross-section.

The associated production of Υ mesons with open charm hadrons (D^0, D^+, D_s^+ and Λ_c^+) was measured by LHCb using the data collected at $\sqrt{s} = 7$ and 8 TeV [18]. The selected Υ and C candidates are paired to form ΥC candidates. A global fit to the ΥC candidates is performed, which requires both hadrons to be consistent with originating from a common vertex. A cut on the quality of the fit reduces the background from the pile-up of two independent pp interactions producing separately a Υ and C hadron to a negligible level. The event yields are determined using unbinned extended maximum likelihood fits to the two-dimensional ΥC mass distributions of the selected candidates[2]. For $\Upsilon(1S)D^0$, $\Upsilon(2S)D^0$, $\Upsilon(1S)D^+$, $\Upsilon(2S)D^+$ and $\Upsilon(1S)D_s^+$ a signal is observed with significances exceeding five standard deviations. Cross-sections are measured for $\Upsilon(1S)D^0$ and $\Upsilon(1S)D^+$. They are in agreement with DPS expectations and significantly exceed the expectations from SPS. Assuming only DPS production, the measured cross-sections are used to determine σ_{eff} to be $\sigma_{\text{eff}} = 18.0 \pm 1.8$ mb, in agreement with most previous measurements as shown in Fig 6. This supports the assumption of σ_{eff} being process and energy independent.

The differential kinematic distributions for ΥD^0 and ΥD^+ are studied and all are found to be in good agreement with the DPS expectations as the main production mechanism. As an example Fig. 6 shows the azimuthal angular separation of $\Upsilon(1S)$ and D^0 which is consistent with a flat distribution as expected for uncorrelated scattering processes (DPS).

4.4. Central exclusive J/ψ and $\psi(2S)$ production at $\sqrt{s} = 13$ TeV

Central exclusive production (CEP), $pp \to pXp$, in which the protons remain intact and the system X is produced with a rapidity gap on either side, proceeds via the exchange of colourless, neutral particles, either photons or combinations of gluons, for example pomerons. Experimentally, this leads to a unique signature with a small number of particles in the detector, either produced directly or as decay products, and two rapidity gaps that extend to the outgoing protons. The outgoing protons are not detected but escape through the beam-pipe. CEP is attractive to study QCD and the role of the pomeron, particularly when the mass of the central system is high enough to allow perturbative calculations. Furthermore, it allows searches for exotic states in a low-background experimental environment and probes of the gluon distribution of the proton.

CEP charmonia candidates are selected through their characteristic signature, which is only the charmonium meson that is reconstructed from its decay to two muons. The addition of HeRSCheL extends the pseudorapidity region in which charged particles can be vetoed and roughly halves the inelastic background contribution compared to the previous measurements at lower energies [21]. The largest background is due to inelastic production of J/ψ and ψ(2S) mesons with additional gluon radiation or proton dissociation where the additional particles are not detected in LHCb. Since the inelastic background has a higher p_{T} than the signal the background contribution is determined by a fit to the p_{T}^2 distribution of the meson. The inelastic background accounts for approximately 20% of the selected candidates. The differential cross-sections for J/ψ and ψ(2S) mesons at $\sqrt{s} = 13$ TeV as functions of the rapidity of the mesons are shown in Fig. 7 [22]. Both measurements are better described by NLO than by LO predictions [23, 24].

[2] The fit model is a sum of several components, each of which is the product of a di-muon mass distribution, corresponding to an individual Υ state or combinatorial background, and a C candidate mass distribution,

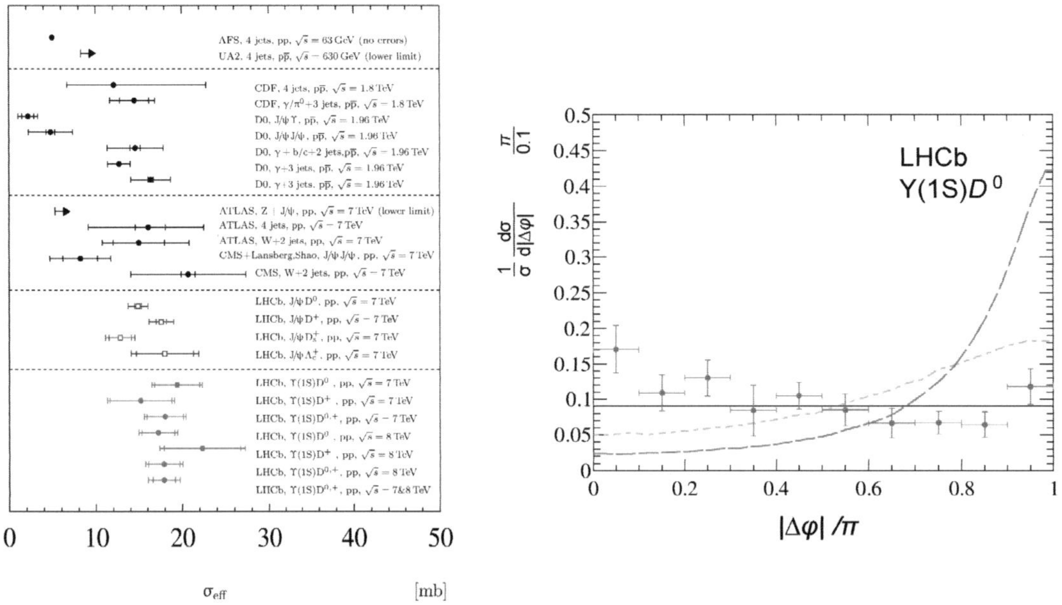

Figure 6. Left: effective cross-section as determined from the associated production of Υ and open charm together with results from other experiments; right: corrected distribution of $|\Delta\phi|/\pi$ for $\Upsilon(1S)D^0$ events (from [18]). The blue line shows the result of the fit with a constant function as a model for uncorrelated scattering processes (DPS). The SPS predictions are shown with dashed (orange) and longdash-dashed (magenta) curves for calculations based on the k_T-factorization [19] and the collinear approximation [20], respectively.

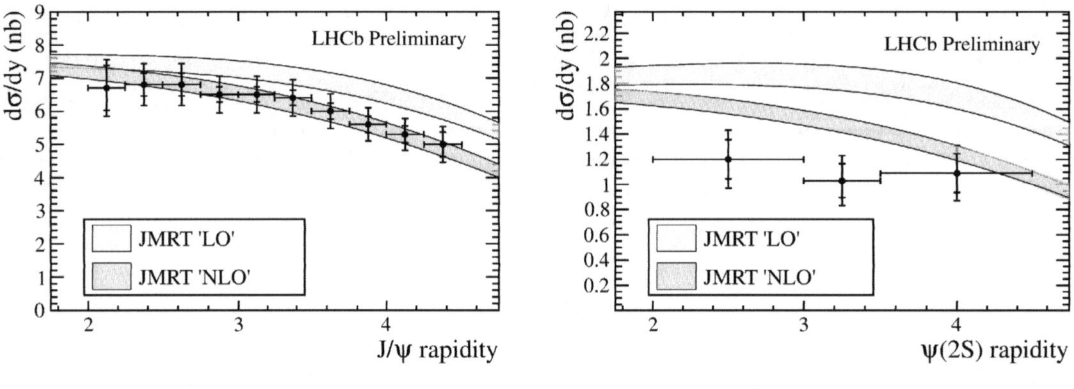

Figure 7. Differential cross-section for central exclusive J/ψ (left) and $\psi(2S)$ (right) production compared to LO and NLO predictions (from [22]).

5. Conclusions

Differential cross section measurements and ratios of cross sections at different centre-of-mass energies have been presented for heavy flavour (charm and beauty) and quarkonia (J/ψ and Υ) production. In general the differential cross-sections and the ratio of cross-sections at different

corresponding to a C signal or combinatorial background component.

$\sqrt{s}$ are described by the predictions within uncertainties. Uncertainties for the predictions are significantly reduced by taking the ratios and also many experimental uncertainties cancel.

Top production has been observed in agreement with NLO predictions in the forward region with low statistics. The full Run II dataset will allow to perform differential measurements and to access different decay channels.

Measurements of the associated production of Υ and open charm have been presented. The cross-sections and the differential distributions indicate the dominance of double parton scattering as the main production mechanism.

Measurements of differential cross-sections for central exclusive J/ψ and $\psi(2S)$ have been discussed. They agree with the shapes predicted by NLO calculations. The additional instrumentation of the LHCb detector with forward shower counters for Run II significantly improved the background evaluation for central exclusive processes.

References

[1] A. A. Alves Jr. et al. The LHCb detector at the LHC. *JINST*, 3:S08005, 2008.
[2] I. Schienbein B. A. Kniehl, G. Kramer and H. Spiesberger. Open charm hadroproduction and the charm content of the proton. *Phys. Rev.*, D79:094009, 2009.
[3] M. Cacciari et al. Theoretical predictions for charm and bottom production at the LHC. *JHEP*, 10:137, 2012.
[4] O. Zenaiev et al. Impact of heavy-flavour production cross sections measured by the LHCb experiment on parton distribution functions at low x. *Eur. Phys. J.*, C75(8):396, 2015.
[5] R. Aaij et al. Measurements of prompt charm production cross-sections in pp collisions at $\sqrt{s} = 5$ TeV. *arXiv:1610.02230*, 2016.
[6] R. Aaij et al. Prompt charm production in pp collisions at $\sqrt{s}$=7 TeV. *Nucl. Phys.*, B871:1–20, 2013.
[7] R. Aaij et al. Measurements of prompt charm production cross-sections in pp collisions at $\sqrt{s} = 13$ TeV. *JHEP*, 03:159, 2016. [Erratum: JHEP09,013(2016)].
[8] R. Aaij et al. Implications of LHCb measurements and future prospects. *Eur. Phys. J.*, C73(4):2373, 2013.
[9] R. Aaij et al. Measurement of the b-quark production cross-section in 7 and 13 TeV pp collisions. *Phys. Rev. Lett.*, 118(5):052002, 2017.
[10] R. Aaij et al. Identification of beauty and charm quark jets at LHCb. *JINST*, 10(06):P06013, 2015.
[11] J. M. Campbell and R. K. Ellis. Radiative corrections to Z b anti-b production. *Phys. Rev.*, D62:114012, 2000.
[12] R. Aaij et al. First observation of top quark production in the forward region. *Phys. Rev. Lett.*, 115(11):112001, 2015.
[13] Roel Aaij et al. Measurement of forward $t\bar{t}$, $W + b\bar{b}$ and $W + c\bar{c}$ production in pp collisions at $\sqrt{s} = 8$ TeV. *Phys. Lett.*, B767:110–120, 2017.
[14] R. Aaij et al. Measurement of forward J/ψ production cross-sections in pp collisions at $\sqrt{s} = 13$ TeV. *JHEP*, 10:172, 2015.
[15] Peter L. Cho and Adam K. Leibovich. Color octet quarkonia production II. *Phys. Rev.*, D53:6203–6217, 1996.
[16] R. Aaij et al. Forward production of Υ mesons in pp collisions at $\sqrt{s} = 7$ and 8 TeV. *JHEP*, 11:103, 2015.
[17] L. S. Kisslinger and D. Das. ψ and Υ production in pp collisions at 8.0 TeV. *Mod. Phys. Lett.*, A29:1450082, 2014.
[18] R. Aaij et al. Production of associated Υ and open charm hadrons in pp collisions at $\sqrt{s}$=7 and 8 TeV via double parton scattering. *JHEP*, 07:052, 2016.
[19] S. P. Baranov. Pair production of J/ψ mesons in the k_T-factorization approach. *Phys. Rev.*, D84:054012, 2011.
[20] A. V. Berezhnoy, A. K. Likhoded, A. V. Luchinsky, and A. A. Novoselov. Double J/ψ-meson Production at LHC and 4c-tetraquark state. *Phys. Rev.*, D84:094023, 2011.
[21] R. Aaij et al. Updated measurements of exclusive J/ψ and $\psi(2S)$ production cross-sections in pp collisions at $\sqrt{s} = 7$ TeV. *J. Phys.*, G41:055002, 2014.
[22] The LHCb Collaboration. Central exclusive production of J/ψ and $\psi(2S)$ mesons in pp collisions at $\sqrt{s} = 13$ TeV. *LHCb-CONF-2016-007*, 2016.
[23] S. P. Jones, A. D. Martin, M. G. Ryskin, and T. Teubner. Exclusive J/ψ and Υ photoproduction and the low x gluon. *J. Phys.*, G43(3):035002, 2016.
[24] S. P. Jones, A. D. Martin, M. G. Ryskin, and T. Teubner. Predictions of exclusive $\psi(2S)$ production at the LHC. *J. Phys.*, G41:055009, 2014.

IOP Conf. Series: Journal of Physics: Conf. Series **878** (2017) 012012 doi:10.1088/1742-6596/878/1/012012

The LHCb Upgrade

Alessio Piucci

Physikalisches Institut, Im Neuenheimer Feld 226, 69120 Heidelberg, DE

E-mail: alessio.piucci@cern.ch

Abstract. During the LHC Run 1 the LHCb experiment has successfully performed a large number of high precision measurements in heavy flavour physics using 3 fb^{-1} of data collected at centre-of-mass energies of 7 TeV and 8 TeV. In LHC Run 2 the LHCb is expected to integrate an additional 5 fb^{-1} data, however many of the measurements will remain limited by statistics. For this reason LHCb will undergo in 2020 a major upgrade during the Long Shutdown 2 of LHC, with the aim to collect 50 fb^{-1} of data by 2028. To achieve this goal the LHCb detector readout rate will be upgraded from the current 1 MHz to the LHC bunch crossing rate of 40 MHz. The luminosity delivered to the experiment will increase by a factor five, up to $2 \cdot 10^{33}$ cm^{-2} s^{-1}. The online selection of events will be uniquely performed by a pure software trigger, improving the trigger efficiencies. In order to sustain the increased luminosity and readout rate, all the sub-detectors will be upgraded. The architecture of the upgraded DAQ system and trigger strategy will be presented, as well an overview of the sub-detector upgrades.

1. Introduction

The LHCb experiment apparatus [1] is a single-arm forward spectrometer dedicated to the study of heavy flavour physics, installed on the Large Hadron Collider (LHC). Its physics program includes precise measurement of CP violation and rare decays of B and charm hadrons, study of electroweak and QCD processes, hadron spectroscopy and heavy ion collisions.

During the Run 1 of the LHC, the LHCb experiment collected 3 fb^{-1} of data and performed many important measurement in the heavy flavour sector, such as the first observation of the $B_s^0 \to \mu^+\mu^-$ [2] joint with the CMS collaboration, the measurement of a flavour changing neutral current anomaly in the $B^0 \to K^{*0}\mu^+\mu^-$ decay [3], and the observation of pentaquark states [4] in the $\Lambda_b^0 \to J/\psi\ p\ K^-$ decay.

However, many crucial measurements are statistically limited [5], with experimental precisions not directly comparable to the uncertanties of the theoretical predictions. The scenario will only partially improve with the LHC Run 2, during which LHCb is expected to reach an integrated luminosity of 8 fb^{-1}. For this reason LHCb will undergo in 2020, during the LHC Long Shutdown 2, a major upgrade [6] that will make it feasible to collect at least 50 fb^{-1} of data by 2028. With such a large dataset, the measurements will finally have comparable precision as the theoretical predictions.

2. The LHCb Upgrade

To be able to record at least 50 fb^{-1} of data, during the LHC Run 3 the luminosity delivered to the experiment will be increased by a factor five with respect to the current conditions, to $2 \cdot 10^{33}$ cm^{-2} s^{-1}. Moreover, the online selection of events will be revolutionised, fully-reconstructing

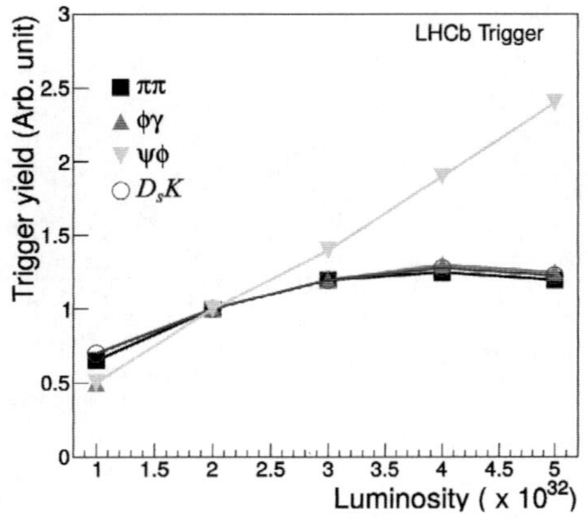

Figure 1: Trigger yield dependance on the instantaneous luminosity.

all the events at the LHC inelastic event rate exploiting a fully software-based trigger. To sustain the new experimental conditions, all the LHCb sub-detectors will be upgraded as well the detector readout system.

2.1. The LHCb trigger and readout upgrade

The current trigger strategy at LHCb is based on a first hardware-based trigger step, called Level 0 (L0), followed by a selection performed by a software-based High Level Trigger (HLT). The L0 reduces the event rate from the LHC bunch crossing rate of 40 MHz up to 1 MHz, basing on measurements performed by the calorimeters and the muon chambers. The HLT then performs a first partial event reconstruction at the rate of 1 MHz, that is the LHCb detector readout rate. Then, a full reconstruction of events with offline-like quality is performed at the rate of 150 KHz. Finally, the selected events are stored at a rate of 12.5 KHz.

However, the current readout and triggering scheme is a limiting factor for the trigger effectiveness at Upgrade conditions. Figure 1 shows that the trigger yield on many hadronic channels already saturates at the current luminosity of LHCb, not increasing for higher values of the luminosity.

To be able to efficiently run at increased luminosity, the L0 hardware-based trigger will be removed, and events will be selected by the software-based HLT alone. The HLT will perform a full reconstruction of events at the LHC inelastic event rate of 30 MHz. Offline precision particle identification and track quality information will be exploited to reduce the rate down to 20-100 KHz. The new trigger strategy will increase the triggering efficiency on the hadronic channels by a factor 2 to 4 with respect to Run 1 [7], corresponding to an increase by a factor 10 to 20 of the hadronic yields.

The detector readout scheme will be upgraded, in order to process events at 40 MHz rate. The readout signals will be sent to the surface by 300 m long optical links to an event builder farm. An uniform infrastructure based on PCIe40 cards will be used for data readout, slow and fast control. Events will be distributed by data-centre technology networking to the triggering farm, composed by standard dual-socket x86 servers.

2.2. The LHCb detector upgrade

The LHCb detector needs to be upgraded to sustain the renewed experimental challenges due to the increased luminosity. Higher granularity and radiation tolerance are required for the sub-detectors, in particular for the tracking sub-detectors, as well as new front-end and readout electronics. Figure 2 shows a sketch of the upgraded LHCb detector.

IOP Conf. Series: Journal of Physics: Conf. Series **878** (2017) 012012 doi:10.1088/1742-6596/878/1/012012

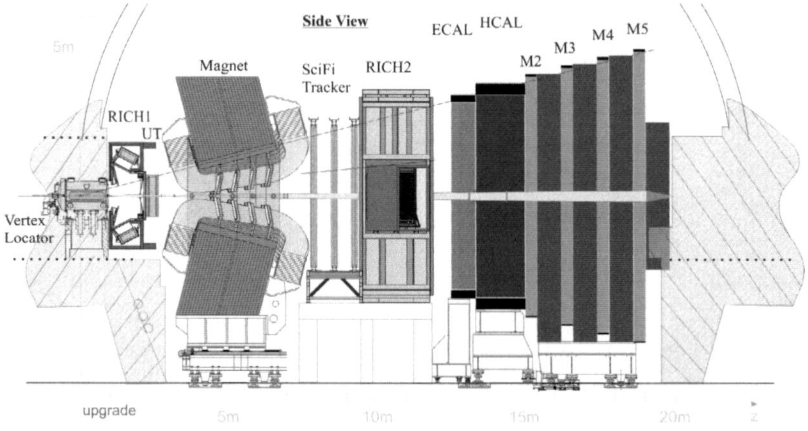

Figure 2: Side view of the LHCb upgraded detector.

The Vertex Locator (VELO) [8] is the tracking detector devoted to the precise measurement of primary vertices and displaced vertices of short living particles. The current VELO, based on silicon microstrips technology, will be replaced by 26 tracking layers based on 50x50 μm^2 pixel technology that will ensure a better hit resolution and simpler track reconstruction. Figure 3.a shows a sketch of the tracking detector. The upgraded VELO will be closer to the beam axis, from the current 8.4 mm up to 5.1 mm from it, and the particles will see substantially less detector material, from 4.6% to 1.7% radiation lengths, before the intersection of the first tracking layers which are the crucial ones to determine the resolution on the measurement of the impact parameter of the particles. These improvements will improve the impact parameter resolution by a factor of about 40%, increase the VELO tracking efficiency especially for low momentum tracks, and provide a better decay time resolution. In order to reduce the radiation damage to the sensors, the detector layers will be opened while not taking data and closed when stable beam condition is reached. The sensors will be cooled at the temperature of -20 °C exploiting an innovative microchannel CO_2 cooling technology.

The Upstream Tracker (UT) [9] will be used for downstream reconstruction of long lived particles decaying after the VELO. It will be also essential to improve the trigger timing, and the momentum resolution. The UT will be composed by 4 tracking layers based on silicon strip technology. A UT layer is sketched in Figure 3.b. The inner sensors will be closer to the beam pipe with respect to the current tracker, in order to increase the geometrical acceptance. The segmentation and technology of the sensors is driven by the expected particle occupancy and radiation dose. In the outer region, the strips will be 99.5 mm long with 180 μm pitch, and based on p^+-in-n technology. The strips in the central region will have same length but 95 μm pitch, and they will be based on n^+-in-p technology in order to better sustain the higher radiation dose. Finally, the strips closest to the beam will be 51.5 mm long, with 95 μm pith and based on n^+-in-p technology.

The Scintillating Fiber tracker (SciFi) [9] will be structured in 12 detector layers, and used for track reconstruction after the magnet region thus providing measurement of the particle momentum. Figure 3.c shows a sketch of a SciFi layer. The SciFi will be based on 2.4 m long plastic scintillating fibres with 250 μm diameter, arranged on vertical direction. Each of the detector layers will be composed of 6 layers of fibres, with a total layer thickness of 1.35 mm and transversal dimensions of about 6 m x 5 m. On vertical direction the layers will be made of two series of fibres separated by mirrors. The fibres will be readout by SiPMs placed on the top and on the bottom of the detector layers, reading arrays of 50x50 μm^2 sized pixels grouped in arrays of 128 channels. SiPMs will be cooled at the temperature of -40 °C in order to decrease

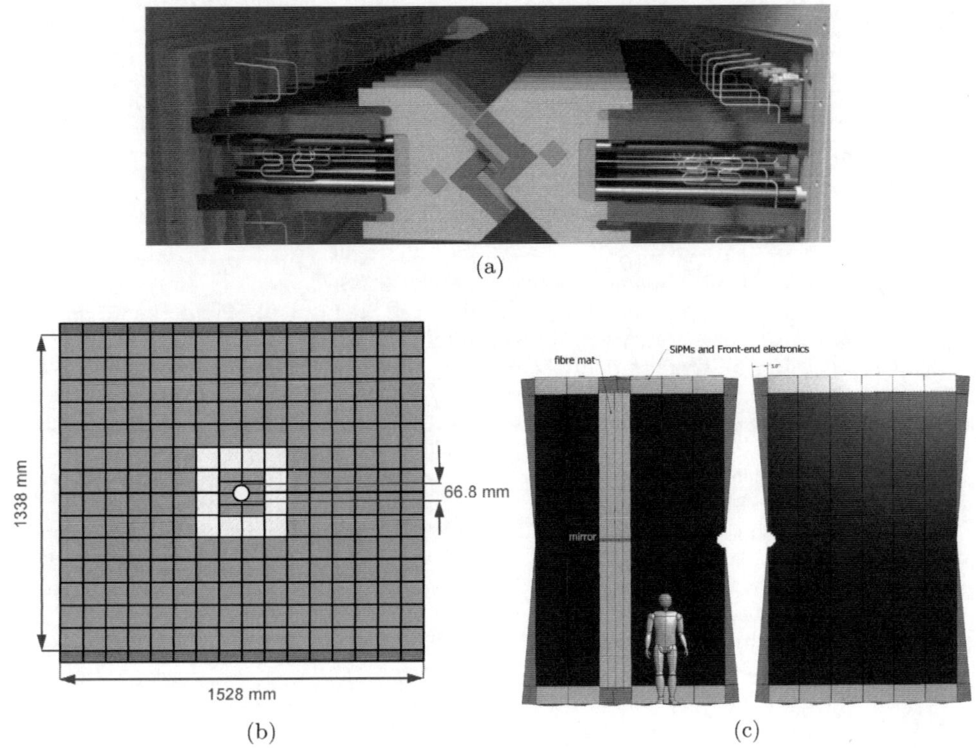

Figure 3: Sketches of the upgraded LHCb trackers: VELO (a), UT (b) and SciFi (c).

the radiation damage and the dark noise.

The usage of UT hits in the track extrapolation from the VELO to the SciFi detectors will allow to reduce the number of fake tracks reconstructed by the tracking algorithms, by a factor 50-70%. Consequently, the trigger timing will be largely improved. The SciFi will improve the tracking efficiency, reducing in the meanwhile the amount of reconstructed fake tracks, with respect using the current detector in the Upgrade.

The two RICH detectors of LHCb are used for p, π and K particle identification. The optical layout of the RICH1, which is more near to the interaction point, will be modified to handle the much higher particle occupancy of the upgrade conditions [10]. In particular, the focal length of the mirrors will be increased by a factor $\sqrt{2}$ thus halving the occupancy. The current readout of both RICH detectors, performed by HPDs at 1 MHz rate, will be replaced by Multianode PMTs working at 40 MHz.

The LHCb calorimeters and muon chambers, used for p, e, γ, μ particle identification, will not undergo substantial upgrades, apart by the replacement of the front-end electronics and the removal of some stations currently used by the L0 trigger only [10].

The installation of the LHCb upgraded detector will take place during the LHC Long Shutdown 2. In 2018 the old sub-detectors will be removed, and in 2019 the Upgrade ones will be installed in order to get ready to run for the LHC Run 3, on 2020-2021.

3. Conclusion

The LHCb experiment experienced a great success in the heavy flavour sector, during the LHC Run 1. Several crucial measurements are currently statistically limited, and will not significantly improve with the integrated luminosity of 8 fb^{-1} expected to be collected until the end of the LHC Run 2.

IOP Conf. Series: Journal of Physics: Conf. Series **878** (2017) 012012 doi:10.1088/1742-6596/878/1/012012

The LHCb will undergo in 2020 a major upgrade to collect 50 fb^{-1} of data by the end of LHC Run 4. The Upgrade will consists of new trackers based on completely new technologies, as well of updates of the RICH, calorimeters and muon chambers. The online selection of events will be performed by a software-based trigger only, reconstructing the full events at the LHC inelastic interaction rate of 30 MHz. The Upgrade project is fully approved and funded in its completeness, and the construction of all sub-detectors are on the way.

References

[1] Alves Jr A A *et al.* (LHCb collaboration) 2008 *JINST* **3** S08005
[2] Aaij R *et al.* (LHCb collaboration) 2013 *Phys. Rev. Lett.* **111** 101805 (*Preprint* 1307.5024)
[3] Aaij R *et al.* (LHCb collaboration) 2016 *JHEP* **02** 104 (*Preprint* 1512.04442)
[4] Aaij R *et al.* (LHCb collaboration) 2015 *Phys. Rev. Lett.* **115** 072001 (*Preprint* 1507.03414)
[5] Aaij R *et al.* (LHCb collaboration) 2013 *Eur.Phys.J.* **C73: 2373** (*Preprint* hep-ph/0506026)
[6] Aaij R *et al.* (LHCb collaboration) 2012 Framework TDR for the LHCb Upgrade: Technical Design Report Tech. Rep. CERN-LHCC-2012-007
[7] Aaij R *et al.* (LHCb collaboration) 2014 LHCb Trigger and Online Technical Design Report Tech. Rep. CERN-LHCC-2014-016
[8] Aaij R *et al.* (LHCb collaboration) 2013 LHCb VELO Upgrade Technical Design Report Tech. Rep. CERN-LHCC-2013-021
[9] Aaij R *et al.* (LHCb collaboration) 2014 LHCb Tracker Upgrade Technical Design Report Tech. Rep. CERN-LHCC-2014-001
[10] Aaij R *et al.* (LHCb collaboration) 2013 LHCb PID Upgrade Technical Design Report Tech. Rep. CERN-LHCC-2013-022

IOP Conf. Series: Journal of Physics: Conf. Series **878** (2017) 012013 doi:10.1088/1742-6596/878/1/012013

CP Violation in *b*- and *c*-hadron decays at LHCb

Olaf Steinkamp ***on behalf of the LHCb Collaboration***
Physik-Institut der Universität Zürich, Winterthurerstrasse 190, 8057 Zürich, Switzerland
E-mail: olafs@physik.uzh.ch

Abstract. Testing the Standard Model of particle physics by precision measurements of *CP* violating observables in the decays of b and c hadrons has been one of the design goals of the LHCb experiment. World-leading measurements have been performed of the semileptonic asymmetry, a_s^{sl}, and of the mixing-induced *CP*-violating phase ϕ_s in the $B_s^0\overline{B}_s^0$ system. The CKM angle γ is still the least known angle of the Unitarity Triangle, and the only one easily accessible using tree-level decays. A recent combination of LHCb measurements in various $B \to DK$ decay modes has yielded the most precise determination of γ from a single experiment to date. The LHCb experiment is collecting unprecedented samples of beauty baryons, allowing for the first time to study *CP* violating observables in their decays. A recent analysis provided the first evidence for *CP* violation in the beauty baryon sector. Finally, LHCb has the largest samples of charmed hadron decays collected by any experiment to date. These samples yield some of the world's most sensitive searches for direct and indirect *CP* violation in the charm sector.

1. Introduction

The main goal of the LHCb experiment [1] at CERN's Large Hadron Collider (LHC) is to perform precision measurements in *CP* violating observables and rare decays of hadrons containing beauty or charm quarks. This contribution to the proceedings is focused on LHCb measurements of *CP* violating observables. The comparison of these measurements with predictions from the Standard Model (SM) of particle physics tests the SM and can reveal the presence of contributions from physics Beyond the Standard Model (BSM).

The single source of *CP* violation in the SM is a complex phase in the Cabbibo-Kobayashi-Maskawa (CKM) quark mixing matrix [2, 3]. The CKM matrix parametrises transitions between the three quark families in charged current interactions. It is fully described by four independent real parameters, namely three rotation angles and one complex phase. A non-zero value of this complex phase can induce *CP* violating asymmetries in processes in which two or more amplitudes with different weak phases interfere. Usually, one of the interfering amplitudes involves loop diagrams with internal quark lines, such as box diagrams or penguin diagrams.

Most BSM physics models predict the existence of hitherto unobserved heavy particles. Such particles can appear in the internal lines of the loop diagrams and modify observed *CP* asymmetries with respect to SM predictions. Precision measurements of such asymmetries therefore test the SM and can reveal the presence of BSM physics. The pattern of observed deviations from SM predictions can hint at the underlying dynamics of the BSM physics at play. Since the deviations that these measurements search for are caused by the appearance of virtual particles in internal lines of loop diagrams, the sensitivity of these searches extends to mass scales that exceed the centre of mass energy of the accelerator at which the measurement is performed.

The B^+B^-, $B^0\overline{B}^0$ and $B_s^0\overline{B}_s^0$ meson systems give rise to a wide range of *CP* violating observables for which precise SM predictions can be made. *CP* violating observables can also be measured in the decays of b baryons and in the charm sector, although predictions are less precise here.

The CKM mechanism gives rise to three possible sources of *CP* violating asymmetries:

CP **violation in mixing** in the $B^0\overline{B}^0$ and $B_s^0\overline{B}_s^0$ systems is caused by the interference of box diagrams with different internal quark lines. The mixing phase in the $B_s^0\overline{B}_s^0$ system is predicted with good precision to be close to zero in the Standard Model and is therefore particularly sensitive to possible contributions from New Physics.

CP **violation in decay** is caused by the interference of decay diagrams with different weak and strong phases leading to the same final state. This is the only possible source of *CP* violation for in the $B^+ - B^-$ system. In general, poorly known strong phases limit the precision with which the weak phase can be extracted from the measured asymmetry. There are, however, cases in which the weak phase can be cleanly extracted using a combination of several related decay modes and exploiting symmetries of the strong interaction between these modes.

CP **violation in the interference of mixing and decay** to a final state that is accessible to both the B^0 or B_s^0 meson and its respective antiparticle is caused by the interference of the direct decay amplitude of a B^0 or B_s^0 meson to the given final state and the process where the B^0 or B_s^0 meson first mixes to a $\overline{B}^0$ or $\overline{B}_s^0$, which then decays to the same final state. A prominent example is the "golden" decay mode $B^0 \to J/\psi K_S^0$, which allowed the precise measurement of the mixing phase $\sin 2\beta$ in the $B^0\overline{B}^0$ system by the B factories BaBar [4] and Belle [5].

All three sources of *CP* violation will play a role in the measurements that are discussed in the remainder of this contribution, which is organized as follows. Section 2 is dedicated to a recent LHCb measurement of the *CP* violating phase in $B_s^0 - \overline{B}_s^0$ mixing using semileptonic B_s^0 decays, Sec. 3 to the measurement of the *CP* violating phase ϕ_s in the decay channel $B_s^0 \to J/\psi K^+K^-$, and Sec. 4 to measurements of the CKM phase γ from $B^\pm \to DK^\pm$ tree decays; Sec. 5 and Sec. 6 briefly touch on recent measurements of *CP* violating asymmetries in the decays of Λ_b^0 baryons and of charmed mesons, respectively. Unless mentioned otherwise, all measurements discussed here have been performed using the full LHCb Run-I data set, corresponding to an integrated luminosity of 3.0 fb^{-1} of pp collisions collected at pp centre of mass energies of 7 TeV in 2011 and 8 TeV in 2012.

2. *CP* violation in $B_s^0 - \overline{B}_s^0$ mixing from $B^0s \to D_s^-\mu^+X$ decays

Second-order weak transitions (box diagrams) induce particle-antiparticle mixing in the $K^0\overline{K}^0$, $D^0\overline{D}^0$, $B^0\overline{B}^0$ and $B_s^0\overline{B}_s^0$ systems. The time evolution of the coupled $B_s^0\overline{B}_s^0$ system can be described by an effective Schrödinger equation of the form

$$-i\frac{\partial}{\partial t}\begin{pmatrix} B_s^0 \\ \overline{B}_s^0 \end{pmatrix} = H \cdot \begin{pmatrix} B_s^0 \\ \overline{B}_s^0 \end{pmatrix} = \begin{pmatrix} M_{11} - \frac{i}{2}\Gamma_{11} & M_{12} - \frac{i}{2}\Gamma_{12} \\ M_{12}^\star - \frac{i}{2}\Gamma_{12}^\star & M_{22} - \frac{i}{2}\Gamma_{22} \end{pmatrix} \cdot \begin{pmatrix} B_s^0 \\ \overline{B}_s^0 \end{pmatrix}$$

(and similar for the other neutral meson–antimeson systems). The eigenstates of the effective Hamiltonian, H, are the heavy and light mass eigenstates, B_H and B_L, which have well defined masses, m_H and m_L, and lifetimes, Γ_H and Γ_L. Relevant observables are the mass difference $\Delta m_s \equiv m_H - m_L$ and the lifetime difference $\Delta\Gamma_s \equiv \Gamma_H - \Gamma_L$. Particle–antiparticle mixing is described by the off-diagonal elements of H, where M_{12} describes the dispersive part of

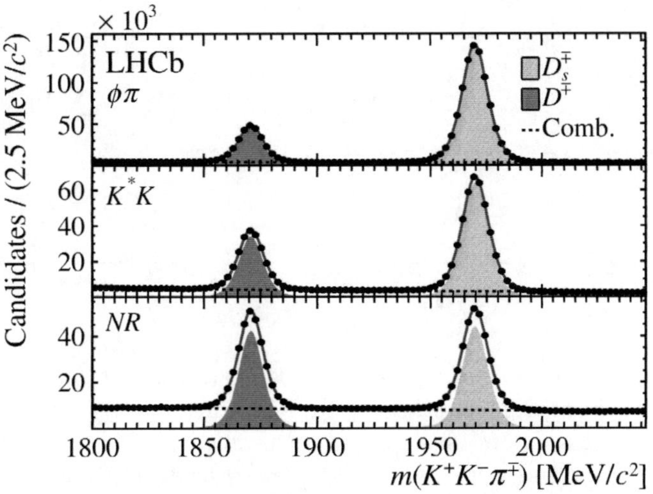

Figure 1. Combined invariant mass spectra of $K^+K^-\pi^-$ and $K^+K^-\pi^+$ candidates for each of the three kinematical regions used in the measurement of $a^s_{\rm sl}$. Top: $\phi\pi^-$ with $\phi \to K^+K^-$, middle: $K^{0*}K^-$ with $K^{0*} \to K^+\pi^-$, bottom: non-resonant (NR) $K^+K^-\pi^-$. In all three samples a clear $D_s^\mp$ signal is visible as well as a $D^\mp$ signal at lower invariant mass.

the transition amplitude, i.e. transitions via off-shell intermediate states, and Γ_{12} describes its absorptive part, i.e. transitions via on-shell intermediate states. A relative phase ϕ_M between M_{12} and Γ_{12} causes *CP* violation in the mixing process, causing the probability for the transition $B_s^0 \to \overline{B}{}_s^0$ to be different from that for the transition $\overline{B}{}_s^0 \to B_s^0$.

This *CP* asymmetry can be measured in the time-integrated semileptonic asymmetry

$$a^s_{\rm sl} \equiv \frac{\Gamma(\overline{B}{}_s^0 \to \ell^+ X) - \Gamma(B_s^0 \to \ell^- X)}{\Gamma(\overline{B}{}_s^0 \to \ell^+ X) + \Gamma(B_s^0 \to \ell^- X)} = \frac{\Delta\Gamma_s}{\Delta m_s} \tan\phi_M,$$

where $\ell^\pm$ stands for a positively or negatively charged lepton. In the SM, the phase ϕ_M in the $B_s^0\overline{B}{}_s^0$ system is small and the semileptonic asymmetry is predicted to be $a^s_{\rm sl} = (1.9 \pm 0.3) \times 10^{-5}$ [6]. This makes measurements of $a^s_{\rm sl}$ sensitive to possible contributions from BSM physics in $B_s^0 - \overline{B}{}_s^0$ mixing.

LHCb has published [7] a measurement of $a^s_{\rm sl}$ using the decay mode $B_s^0 \to D_s^-\mu^+X$ with $D_s^- \to K^+K^-\pi^-$. The measured raw asymmetry was defined as

$$A_{\rm raw} \equiv \frac{N(D_s^-\mu^+X) - N(D_s^+\mu^-X)}{N(D_s^-\mu^+X) + N(D_s^+\mu^-X)},$$

where the numbers of candidates were extracted from a fit to the $K^+K^-\pi^-$ and $K^+K^-\pi^+$ invariant mass spectra, as illustrated in Fig. 1. A total of about 1.6 million $D_s^\mp\mu^\pm$ candidates was reconstructed in the 3 fb^{-1} sample from run I. No attempt was made to determine the initial flavour of the B_s^0 or $\overline{B}{}_s^0$ meson at the time of its production. This reduces the measured asymmetry by a factor of two but avoids the significantly larger loss in statistical precision due to flavour-tagging algorithms. The $B_s^0 - \overline{B}{}_s^0$ production asymmetry in *pp* collisions is expected to be of the order of a few percent, but is diluted to a negligible level by the rapid $B_s^0 - \overline{B}{}_s^0$ oscillations. Taking into account trigger and detection asymmetries, $A_{\rm D}$, as well as the fraction of backgrounds peaking at the $D_s^\pm$ mass in the $K^+K^-\pi^\pm$ mass spectra, $f_{\rm bkg}$, and

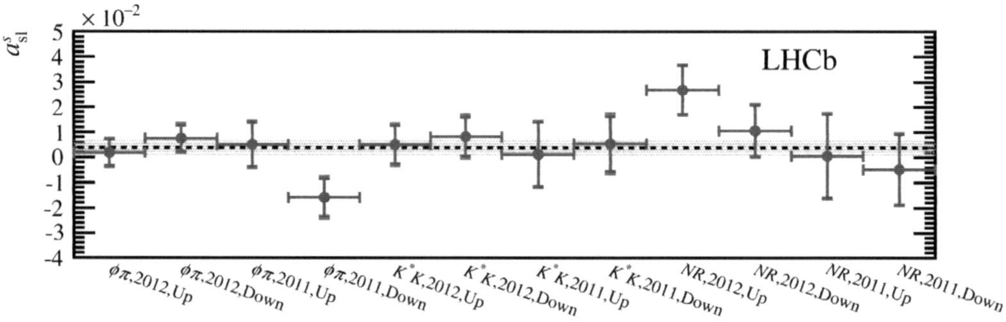

Figure 2. Results for $a_{\rm sl}^s$ from each of the twelve disjoint data samples used in the analysis. The inner (blue) vertical error bars indicate statistical uncertainties, while the outer (red) error bars indicate the combined statistical and systematic uncertainties. The dashed (black) horizontal line indicates the combined result and the shaded (green) horizontal band its uncertainty.

possible charge asymmetries in these backgrounds, $A_{\rm bkg}$,

$$\frac{a_{\rm sl}^s}{2} = \frac{1}{1 - f_{\rm bkg}} \left(A_{\rm raw} - A_{\rm D} - f_{\rm bkg} A_{\rm bkg} \right) .$$

Detection asymmetries cancel to first order between the two oppositely charged final-state kaons on the one hand and between the oppositely charged pion and muon on the other hand. Small remaining asymmetries due to different kinematical properties of the four final-state particles, particle-identification criteria and trigger requirements on the muon were estimated and corrected for using data-driven techniques in a variety of control channels. The fractions of peaking backgrounds and charge asymmetries were derived from known branching fractions and production asymmetries. The analysis was performed separately for three disjoint kinematic regions of the D_s^- decay products, namely $\phi\pi^-$ with $\phi \to K^+K^-$, $K^{0*}K^-$ with $K^{0*} \to K^+\pi^-$ and non-resonant $K^+K^-\pi^-$. The polarity of the LHCb spectrometer magnet was reversed in regular intervals throughout the data taking periods and the analysis was performed separately for the data collected with the magnetic field pointing up and down. Finally, the analysis was performed separately for the data collected in 2011 and 2012. The results obtained for the in total twelve disjoint data samples are in good agreement with each other, as is illustrated in Fig. 2. The twelve individual measurements were combined to yield the final result,

$$a_{\rm sl}^s = (0.39 \pm 0.26({\rm stat}) \pm 0.20({\rm syst})) \times 10^{-2} \,,$$

which is the most precise measurement of the *CP* violating asymmetry in $B_s^0 - \overline{B}_s^0$ mixing to date and is in good agreement with the SM prediction. The result is still limited by statistical uncertainties and a more precise measurement can be expected from extending the analysis to include run-II data.

3. The *CP*-violating phase ϕ_s in $B_s^0 \to J/\psi K^+K^-$

The final state $J/\psi K^+K^-$ is accessible to decays of both B_s^0 and $\overline{B}_s^0$. The interference between the decay amplitudes and the $B_s^0 - \overline{B}_s^0$ mixing amplitude can give rise to a *CP* violating phase

$$\phi_s = \phi_M - 2\phi_D,$$

where ϕ_M is the $B_s^0 - \overline{B}_s^0$ mixing phase discussed in the previous section and ϕ_D is the weak phase between the B_s^0 and $\overline{B}_s^0$ decay amplitudes. In the SM, both ϕ_M and ϕ_D are very small,

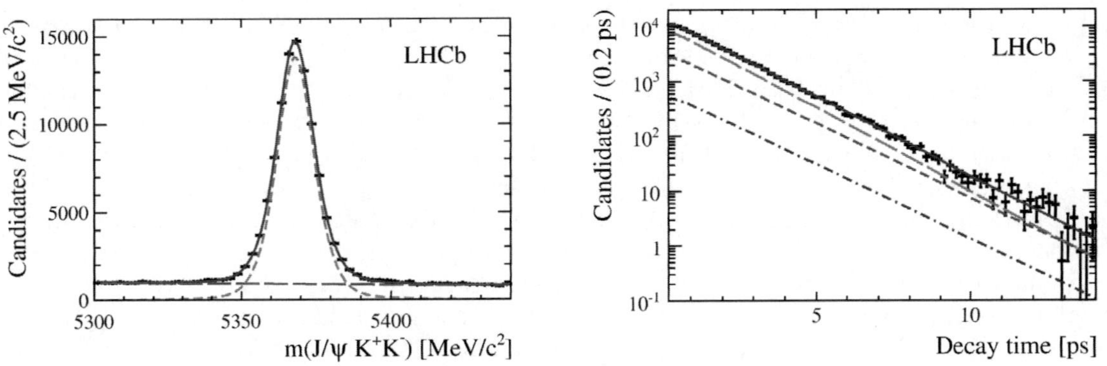

Figure 3. Left: $J/\psi K^+K^-$ invariant mass distribution of the event sample employed for the ϕ_s measurement. Overlayed is the result of the invariant-mass fit. Right: decay time distribution with the projection of the time-dependent angular fit overlayed. The long-dashed (red) line indicates the *CP* even contribution, while the short-dashed (green) and dash-dotted (purple) lines indicate *CP* odd contributions.

resulting in a prediction of $\phi_s = -0.038 \pm 0.001$ rad [8]. This makes measurements of ϕ_s sensitive to possible contributions from BSM physics in $B_s^0 - \overline{B}_s^0$ mixing.

Measurements of ϕ_s proceed through the time-dependent *CP* asymmetry

$$A_{CP}(t) \equiv \frac{\frac{\partial\Gamma}{\partial t}(\overline{B}_s^0 \to J/\psi K^+K^-) - \frac{\partial\Gamma}{\partial t}(B_s^0 \to J/\psi K^+K^-)}{\frac{\partial\Gamma}{\partial t}(\overline{B}_s^0 \to J/\psi K^+K^-) + \frac{\partial\Gamma}{\partial t}(B_s^0 \to J/\psi K^+K^-)} = \frac{-\eta_f \sin\phi_s \sin(\Delta m_s t)}{\cosh(\Delta\Gamma_s \frac{t}{2}) - \eta_f \cos\phi_s \sinh(\Delta\Gamma_s \frac{t}{2})},$$

where B_s^0 and $\overline{B}_s^0$ depict the initial flavour of the decaying meson at the time of its production, η_f is the *CP* eigenvalue of the final state, and Δm_s and $\Delta\Gamma_s$ are the mass and lifetime differences between the heavy and light mass eigenstates in the $B_s^0\overline{B}_s^0$ system.

Reconstructing the J/ψ in its decay to two muons provides for a clear trigger signature and clean event samples. Several factors, however, make this a challenging measurement:

- The initial flavour of the beauty meson at the time of its production has to be derived using information from the remainder of the event. This flavour tagging is not perfect and the associated mis-tag fraction dilutes the oscillation amplitude. The dilution factor has to be known precisely in order to extract $\sin\phi_s$ from the measured oscillation amplitude.
- An excellent decay time resolution is required to resolve the rapid $B_s^0 - \overline{B}_s^0$ oscillation. The decay-time resolution causes a dilution of the oscillation amplitude and has to be known precisely.
- The final state in the decay $B_s^0 \to J/\psi K^+K^-$ is not a *CP* eigenstate. A time-dependent angular analysis of the four final state particles is required to statistically separate the *CP* odd and *CP* even contributions.
- The value of the decay-width difference $\Delta\Gamma_s$ has to be determined simultaneously with that of ϕ_s.

The latest LHCb measurement of ϕ_s in $B_s^0 \to J/\psi K^+K^-$ [9] is based on 96'000 signal candidates reconstructed from the 3 fb^{-1} collected in run I. The $J/\psi K^+K^-$ invariant mass spectrum and decay-time distribution are shown in Fig. 3.

A combination of flavour tagging algorithms using opposite-side lepton, kaon and vertex charge and same-side kaon charge was employed to derive the initial flavour of the beauty meson. Opposite-side tagging algorithms exploit the fact that b and $\overline{b}$ quarks are predominantly produced in quark-antiquark pairs. The algorithms search for flavour-specific signatures from the decay of the second beauty hadron in the event, which must have been produced with flavour opposite to that of the signal B_s^0 or $\overline{B}_s^0$ candidate. The same-side kaon charge algorithm exploits the fact that the strange-type valence quark in the signal B_s^0 or $\overline{B}_s^0$ meson must have been produced via a $s\overline{s}$ pair during the hadronisation process. The other strange-type quark can give rise to a charged kaon, which should be close in phase space to the signal meson. The charge of that kaon can then be employed to tag the initial flavour of the signal B_s^0 or $\overline{B}_s^0$ candidate. Mis-tag fractions, i.e. the fractions of events in which the tagging algorithms give a wrong decision, were determined from data, employing control samples of flavour-specific decays $B^+ \to J/\psi K^+$ for the opposite-side algorithms and $B_s^0 \to D_S^- \pi^+$ for the same-side kaon algorithm. The decay-time resolution was modeled from data, using random combinations of prompt J/ψ candidates with two oppositely charged kaons from the same primary vertex. The angular acceptance, which needs to be known to correctly estimate the *CP* even and *CP* odd contributions in the final state, was extracted from samples of simulated events, which were reweighted to reproduce kinematic distributions observed in collision data.

The projection of the time-dependent angular fit onto the decay time distribution is shown in the right panel in Fig. 3. The different lifetimes of the *CP* even and *CP* odd components are clearly visible, indicating a non-zero value of $\Delta\Gamma_s$.

The combination of this measurement with an earlier, less precise, LHCb measurement of ϕ_s using the decay mode $B_s^0 \to J/\psi\pi^+\pi^-$ [10], yielded $\phi_s = -0.010 \pm 0.039$ rad. This is the most precise determination of ϕ_s from a single experiment to date and it is in excellent agreement with the SM prediction. The uncertainty of the measurement is still large compared to the quoted precision of the SM prediction. The precision of the measurement is limited by statistical uncertainties and is therefore expected to improve significantly as the size of the analysed data sample increases.

4. The CKM angle γ from $B^\pm \to DK^\pm$ decays

The unitarity of the CKM matrix gives rise to six orthogonality relations, which can be represented as triangles in the complex plane. The triangle derived from multiplying the first and third columns of the CKM matrix is of particular interest since its three sides are of comparable length. Its three angles are closely related to *CP* violating observables and the lengths of its sides can be measured from *CP* conserving observables. The overconstrained determination of the sides and angles of this Unitarity Triangle provides a powerful consistency test of the SM picture of *CP* violation.

The least well determined parameter of the Unitarity Triangle to date is the angle $\gamma \equiv \arg\left(-(V_{ud}V_{ub}^*)/(V_{cd}V_{cb}^*)\right)$. The CKMfitter group [11] quotes $\gamma = (73.2^{+6.3}_{-7.0})^\circ$ from a global fit of CKM parameters, while the UTfit group [12] quotes $\gamma = (68.3 \pm 7.5)^\circ$.

A theoretically clean determination of γ can be derived from measurements of time integrated decay rates for different $B^\pm \to DK^\pm$ tree decays. If the decay is observed in a final state $[f]_D$ that is accessible to both D^0 and $\overline{D}^0$ decays, the interference of the decay amplitudes

$$B^\mp \to D^0 K^\mp \to [f]_D K^\mp \quad \text{and} \quad B^\mp \to \overline{D}^0 K^\mp \to [f]_D K^\mp$$

leads to a *CP* asymmetry that is sensitive to γ. Since the decays involve no loop diagrams, the extracted value for γ should be unaffected by possible contributions from BSM Physics. However, the measurements are experimentally challenging since they involve purely hadronic final states, which are difficult to trigger, and require excellent K/π separation. Moreover, branching fractions for the interesting decays are small.

Several methods, using different combinations of final states $[f]_D$, have been proposed. The GLW (Gronau-London-Wyler) method [13, 14] uses *CP* eigenstates such as $[K^+K^-]_D$. The disadvantage of this method is that the decay amplitude for $B^- \to \overline{D}^0 K^-$ is suppressed with respect to $B^- \to D^0K^-$ by a factor of $r_B \approx 0.1$, limiting possible interference effects and the sensitivity to γ. The ADS (Atwood-Dunietz-Soni) method [15, 16] uses quasi flavour-specific final states such as $[\pi^-K^+]_D$, where the decay $\overline{D}^0 \to \pi^-K^+$ is Cabibbo favoured while the decay $D^0 \to \pi^-K^+$ is doubly Cabibbo suppressed. The ratio of magnitudes of the D^0 and $\overline{D}^0$ decay amplitudes, $r_D \approx 0.05$, compensates for the small value of r_B and leads to similar overall magnitudes for the two $B^- \to [f]_DK^-$ decay paths. Large interference effects are therefore possible, giving better sensitivity to γ. Disadvantages of the ADS approach are the very small total branching fraction for the doubly Cabibbo suppressed final states and the introduction of two additional parameters that have to be determined experimentally, namely r_D and the strong phase δ_D between the D^0 and $\overline{D}^0$ decay amplitudes. The GGSZ or Dalitz-plot method [17, 18] exploits interference patterns across the phase space for self-conjugate three-body decays $D^0 \to K^0_S\pi^+\pi^-$. A rich resonance structure is observed in these decays, giving rise to potentially large interference effects and good sensitivity to γ. However, variations of the strong phase δ_D across the phase space can be large and have to be taken into account in the extraction of γ. The associated systematic uncertainties ultimately limit the achievable precision of the measurement.

All three methods are pursued in LHCb. Measurements of GLW and ADS observables in $B^\mp \to DK^\mp$ decays, based on the 3 fb^{-1} collected in Run I, have been published in Ref. [19]. Invariant mass distributions for the GLW modes and for the suppressed ADS modes are shown in Figs. 4 and 5, respectively. In both cases, the effect of the *CP* asymmetry is visible as different event yields for the respective B^- and B^+ decay modes. As expected, the suppressed ADS modes give smaller event yields but the observed asymmetry is significantly larger than in the GLW modes. The deviation from a vanishing asymmetry in the suppressed ADS modes corresponds to eight Gaussian standard deviations, making this the first observation of *CP* violation in a single $B \to DK$ decay mode. The lower panels in the two figures show invariant mass spectra for $B^\mp \to D\pi^\mp$ decays. In these decays, *CP* asymmetries are expected to be significantly smaller than in $B \to DK$, due to a smaller value of r_B. In addition to the classic GLW and ADS decay modes discussed above, Ref. [19] also includes measurements using four-body D-meson decays to $\pi^\mp K^\pm\pi^+\pi^-$ and $\pi^+\pi^-\pi^+\pi^-$. A recent analysis employing GLW and ADS approaches to $B^\pm \to DK^{*\pm}$ decay modes [20] is one of the first LHCb measurements to include data from Run II, adding another 1 fb^{-1} to the 3 fb^{-1} from Run I. Dalitz-plot analyses have been performed using $B^\mp \to DK^\mp$ decays with D mesons decaying to $K^0_S\pi^+\pi^-$ and $K^0_SK^+K^-$ [22] and, more recently, using $B^0 \to DK^{*0}$ decays with D mesons decaying to $K^0_S\pi^+\pi^-$ [23] and to $K^0_S\pi+\pi-$ or $K^0_SK^+K^-$ [24].

A different approach to measuring γ is via the time-dependent *CP* asymmetry in the decay $B^0_s \to D^+_sK^-$. The final state is accessible for B^0_s and $\overline{B}^0_s$ decays and *CP* violation can be caused by interference of mixing and decay. The two decay amplitudes are of similar magnitude and the mixing is strong, such that large asymmetries can be expected. The *CP*-violating weak phase is given by $\gamma - 2\beta_s$, where $-2\beta_s$ is the $B^0_s - \overline{B}^0_s$ mixing phase. Independent measurements of this mixing phase can be employed to extract γ. LHCb has presented a preliminary measurement [21] of γ using 6'000 $B^0_s \to D^\pm_sK^\mp$ candidates extracted from the 3 fb^{-1} collected in Run I.

A combined fit to the results of several recent LHCb analyses using $B \to DK$-like decay modes, including several of those mentioned above, has yielded [25]

$$\gamma = (72.2^{+6.6}_{-7.3})^\circ ,$$

which is the most precise determination of γ from a single experiment to date. The result is in good agreement with SM expectations.

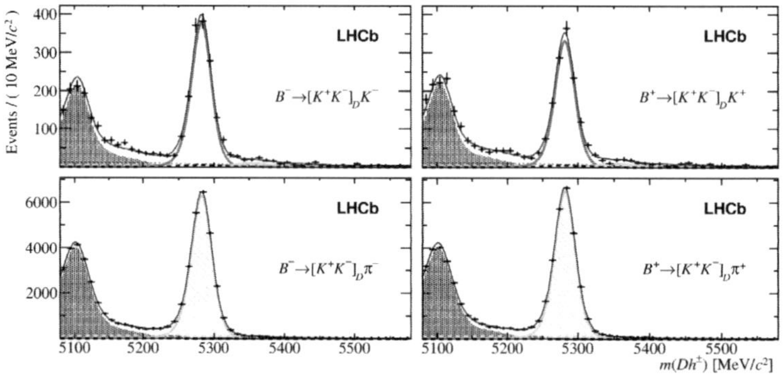

Figure 4. Invariant mass distributions for GLW modes, upper left: $B^- \to [K^+K^-]_D K^-$ and upper right: $B^+ \to [K^+K^-]_D K^+$. The lower panels show the corresponding distributions for $B^\mp \to D\pi^\mp$ decays.

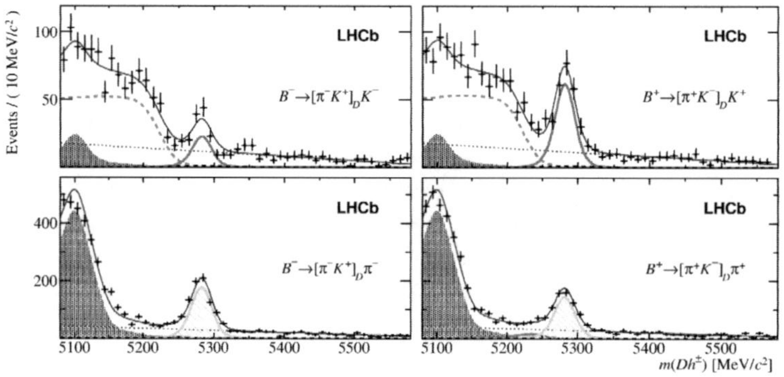

Figure 5. Invariant mass distributions for the suppressed ADS modes, upper left: $B^- \to [\pi^- K^+]_D K^-$, upper right: $B^+ \to [\pi^+ K^-]_D K^+$. The lower panels show the corresponding distributions for $B^\mp \to D\pi^\mp$ decays.

5. *CP* violation in $\Lambda_b^0 \to p\pi^-\pi^+\pi^-$ decays

In the SM, sizeable *CP* asymmetries are predicted to occur in decays of beauty baryons to final states that contain no charmed hadrons. Asymmetries as large as 20% can be expected in some three-body decay modes. The LHCb experiment has collected an unprecedented number of beauty baryon decays, which allows for the first time to perform sensitive studies of *CP* violating observables. A systematic study of *CP* violation in beauty baryon decays can test the validity of the CKM mechanism in the baryon sector, for example by comparing measured asymmetries with those measured in B meson decays with identical underlying quark transitions.

LHCb has published a study of *CP* violating observables in the decay $\Lambda_b^0 \to p\pi^-\pi^+\pi^-$, based on 6'650 signal candidates reconstructed in the 3 fb^{-1} collected in Run I [26]. In this decay, direct *CP* violation can arise from the interference of a tree amplitude and a penguin amplitude. The two decay amplitudes are of similar magnitude and large interference effects are therefore possible. The weak phase between the two amplitudes corresponds to the angle α of the CKM Triangle. However, the poorly kown relative strong phase between the two decay amplitudes makes predictions of the expected *CP* asymmetry challenging.

The search for a *CP* violating asymmetry in this decay cannot rely on a comparison of Λ_b^0 and $\overline{\Lambda}_b^0$ decay rates, due to the poorly known $\Lambda_b^0 - \overline{\Lambda}_b^0$ production asymmetry in *pp* collisions. Instead, the four-body topology of the decay is exploited to combine momenta of final-state particles into a triple products, $C_{\hat{T}}$ and $\overline{C}_{\hat{T}}$, which are odd under parity and time reversal. For

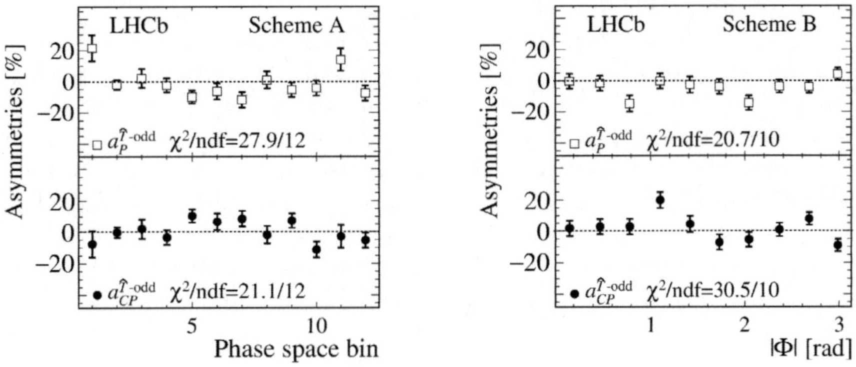

Figure 6. The asymmetry $a_P^{\hat{T}-\mathrm{odd}}$ (upper panels) and $a_{CP}^{\hat{T}-\mathrm{odd}}$ (lower panels) in bins of left: phase space and right: Φ.

Λ_b^0 decays,

$$C_{\hat{T}} \equiv \vec{p}_p \cdot \left(\vec{p}_{\pi^-_{\mathrm{slow}}} \times \vec{p}_{\pi^+}\right) \propto \sin\Phi\,,$$

where $\vec{p}_p$, $\vec{p}_{\pi^+}$ and $\vec{p}_{\pi^-_{\mathrm{slow}}}$ are the three-momenta of the final-state proton, the π^+ and the lower-momentum π^-, respectively, and Φ is the angle between the two decay planes defined by the flight directions of the proton and the higher-momentum π^- on the one hand and by the flight directions of the final-state π^+ and the lower-momentum π^- on the other hand. For $\overline{\Lambda}_b^0$ decays, correspondingly,

$$\overline{C}_{\hat{T}} \equiv \vec{p}_{\overline{p}} \cdot \left(\vec{p}_{\pi^+_{\mathrm{slow}}} \times \vec{p}_{\pi^-}\right) \propto \sin\overline{\Phi}\,.$$

Defining furthermore asymmetries in event yields,

$$A_{\hat{T}} \equiv \frac{N(C_{\hat{T}} > 0) - N(C_{\hat{T}} < 0)}{N(C_{\hat{T}} > 0) + -N(C_{\hat{T}} < 0)}$$

for Λ_b^0 decays and

$$\overline{A}_{\hat{T}} \equiv \frac{N(\overline{C}_{\hat{T}} > 0) - N(\overline{C}_{\hat{T}} < 0)}{N(\overline{C}_{\hat{T}} > 0) + -N(\overline{C}_{\hat{T}} < 0)}$$

for $\overline{\Lambda}_b^0$ decays, parity is violated if

$$a_P^{\hat{T}-\mathrm{odd}} \equiv \frac{1}{2}\left(A_{\hat{T}} + \overline{A}_{\hat{T}}\right) \neq 0$$

and CP is violated if

$$a_{CP}^{\hat{T}-\mathrm{odd}} \equiv \frac{1}{2}\left(A_{\hat{T}} - \overline{A}_{\hat{T}}\right) \neq 0\,.$$

The analysis was performed in ten bins of $|\Phi|$ and in twelve bins of phase-space, as illustrated in Fig. 6. Combining the two binning schemes, and including systematic uncertainties, a p-value of 9.8×10^{-4} was found for the hypothesis of no CP violation. This corresponds to 3.3 Gaussian standard deviations and makes this result the first evidence for CP violation in b baryon decays. No deviation from parity conservation was found.

6. *CP* violation in the $D^0\overline{D}^0$ system

The $D^0\overline{D}^0$ system is the only neutral meson system that consists of up-type quarks. Measurements therefore probe different quark dynamics from those in the $K^0\overline{K}^0$, $B^0\overline{B}^0$, and $B_s^0\overline{B}_s^0$ systems. In the SM, *CP* violating effects in the $D^0\overline{D}^0$ system are expected to be small, with asymmetries generally predicted to be no larger than of the order of 10^{-3}. In principle, measurements of *CP* violating observables therefore provide good sensitivity to possible contributions from BSM physics. A caveat is that predictions suffer from sizeable uncertainties on hadronic form factors, caused by the relatively small mass of the charm quark.

The LHCb experiment is well suited for studies in the charm sector due to its excellent vertex resolution, its excellent kaon/pion identification capabilities, and its flexible trigger scheme, which allows to trigger on purely hadronic states down to small transverse momenta. The $c\bar{c}$ production cross section is about 20 times larger than the $b\bar{b}$ production cross section [27], allowing to collect large event samples. The LHCb collaboration has published various measurements of *CP* violating observables using 2-body, 3-body and 4-body decays of D^0 and $\overline{D}^0$ mesons. For a relatively recent overview, see Ref. [28]. A very recent measurement compared the ratios of doubly Cabibbo suppressed "wrong-sign" decays to Cabibbo favoured "right-sign" decays for D^0 and $\overline{D}^0$ candidates,

$$R^+ \equiv \frac{N(D^0 \to K^+\pi^-)}{N(D^0 \to K^-\pi^+)} \quad \text{and} \quad R^- \equiv \frac{N(\overline{D}^0 \to K^-\pi^+)}{N(\overline{D}^0 \to K^+\pi^-)}.$$

A non-vanishing difference between R^+ and R^- would imply *CP* violation. The measurement made use of flavour-tagged samples of D^0 and $\overline{D}^0$ candidates from $D^{*+} \to D^0\pi^+$ and $D^{*-} \to \overline{D}^0\pi^-$ decays as well as samples of doubly-tagged candidates, in which the $D^{*\pm}$ was produced in a semileptonic B-meson decay, $B \to D^{*\pm}\mu^{\mp}$. The doubly-tagged samples have lower yields, but better purity and better trigger acceptance at small D^0 and $\overline{D}^0$ decay times. A total of 6'700 wrong-sign and 1'730'000 right-sign candidates was extracted from the 3 fb^{-1} collected in Run 1 and used to determine $R^+ - R^-$ as a function of the reconstructed D^0 or $\overline{D}^0$ decay time. No significant deviations from zero were found, compatible with the hypothesis of no *CP* violation.

References

[1] A. A. Alves, Jr. *et al.* [LHCb Collaboration], JINST **3** (2008) S08005.
[2] N. Cabibbo, Phys. Rev. Lett. **10** (1963) 531.
[3] M. Kobayashi and T. Maskawa, Prog. Theor. Phys. **49** (1973) 652.
[4] B. Aubert *et al.* [BaBar Collaboration], Phys. Rev. D **79** (2009) 072009 [arXiv:0902.1708 [hep-ex]].
[5] I. Adachi, H. Aihara, D. M. Asner, V. Aulchenko, T. Aushev, T. Aziz, A. M. Bakich and A. Bay *et al.*, Phys. Rev. Lett. **108** (2012) 171802 [arXiv:1201.4643 [hep-ex]].
[6] A. Lenz, arXiv:1205.1444 [hep-ph].
[7] R. Aaij *et al.* [LHCb Collaboration], "Measurement of the *CP* asymmetry in $B_s^0 - \overline{B}_s^0$ mixing," Phys. Rev. Lett. **117** (2016) 061803 [arXiv:1605.09768].
[8] J. Charles, O. Deschamps, S. Descotes-Genon, R. Itoh, H. Lacker, A. Menzel, S. Monteil and V. Niess *et al.*, Phys. Rev. D **84** (2011) 033005 [arXiv:1106.4041 [hep-ph]].
[9] R. Aaij *et al.* [LHCb Collaboration], Phys. Rev. Lett. **114** (2015) 041801 [arXiv:1411.3104 [hep-ex]].
[10] R. Aaij *et al.* [LHCb Collaboration], Phys. Lett. B **736** (2014) 186 [arXiv:1405.4140 [hep-ex]].
[11] J. Charles *et al.* [CKMfitter Group Collaboration], Eur. Phys. J. C **41** (2005) 1 [hep-ph/0406184], updated results and plots available at: http://ckmfitter.in2p3.fr.
[12] D. Derkach [UTfit Collaboration], arXiv:1301.3300 [hep-ph].
[13] M. Gronau and D. London, Phys. Lett. B **253** (1991) 483.
[14] M. Gronau and D. Wyler, Phys. Lett. B **265** (1991) 172.
[15] D. Atwood, I. Dunietz and A. Soni, Phys. Rev. Lett. **78** (1997) 3257 [hep-ph/9612433].
[16] D. Atwood, I. Dunietz and A. Soni, Phys. Rev. D **63** (2001) 036005 [hep-ph/0008090].
[17] A. Giri, Y. Grossman, A. Soffer and J. Zupan, Phys. Rev. D **68** (2003) 054018 [hep-ph/0303187].
[18] A. Poluektov *et al.* [Belle Collaboration], Phys. Rev. D **70** (2004) 072003 [hep-ex/0406067].

[19] R. Aaij *et al.* [LHCb Collaboration], Phys. Lett. B **760** (2016) 117 doi:10.1016/j.physletb.2016.06.022 [arXiv:1603.08993 [hep-ex]].
[20] The LHCb Collaboration [LHCb Collaboration], LHCb-CONF-2016-014.
[21] The LHCb Collaboration [LHCb Collaboration], LHCb-CONF-2016-015
[22] R. Aaij *et al.* [LHCb Collaboration], JHEP **1410** (2014) 097 [arXiv:1408.2748 [hep-ex]].
[23] R. Aaij *et al.* [LHCb Collaboration], JHEP **1608** (2016) 137 [arXiv:1605.01082 [hep-ex]].
[24] R. Aaij *et al.* [LHCb Collaboration], JHEP **1606** (2016) 131 doi:10.1007/JHEP06(2016)131 [arXiv:1604.01525 [hep-ex]].
[25] R. Aaij *et al.* [LHCb Collaboration], JHEP **1612** (2016) 087 [arXiv:1611.03076 [hep-ex]].
[26] R. Aaij *et al.* [LHCb Collaboration], Nature Phys. (2017) [arXiv:1609.05216 [hep-ex]].
[27] R. Aaij *et al.* [LHCb Collaboration], JHEP **1603** (2016) 159 Erratum: [JHEP **1609** (2016) 013] [arXiv:1510.01707 [hep-ex]].
[28] M. J. Morello [LHCb Collaboration], PoS BEAUTY **2016** (2016) 010.
[29] R. Aaij *et al.* [LHCb Collaboration], arXiv:1611.06143 [hep-ex].

International Workshop on Discovery Physics at the LHC IOP Publishing
IOP Conf. Series: Journal of Physics: Conf. Series **878** (2017) 012014 doi:10.1088/1742-6596/878/1/012014

The SHiP experiment at CERN

Walter M. Bonivento

Istituto Nazionale di Fisica Nucleare, Sezione di Cagliari, Cittadella Universitaria di Monserrato, Monserrato (CA),Italy

E-mail: walter.bonivento@cern.ch

Abstract. The discovery of the Higgs boson has fully confirmed the Standard Model of particles and fields. Nevertheless, there are still fundamental phenomena, like the existence of dark matter and the baryon asymmetry of the Universe, deserving an explanation that could come from the discovery of new particles. Searches for new physics with accelerators are performed at the LHC, looking for high massive particles coupled to matter with ordinary strength. A new experiment at CERN meant to search for very weakly coupled particles in the few GeV mass domain has been recently proposed. The existence of such particles, foreseen in different theoretical models beyond the Standard Model, is largely unexplored. A beam dump facility using high intensity 400 GeV protons is a copious source of such unknown particles in the GeV mass range. The beam dump is also a copious source of neutrinos and in particular it is an ideal source of tau neutrinos, the less known particle in the Standard Model. The neutrino detector can also search for dark matter through its scattering off the electrons. We report the physics potential of the SHiP experiment.

1. Introduction

The discovery of the Higgs boson is certainly a big triumph of the Standard Model. Nevertheless, there are several phenomena deserving an explanation that the Standard Model is unable to provide: the existence of dark matter and its nature, the baryonic asymmetry of the Universe and neutrino masses. It is therefore clear that new physics is there and presumably several new particles have still to be discovered. Searches for new physics with accelerators are being carried out at the LHC, especially suited to look for high mass particles with ordinary couplings to matter. Complementary searches for very weakly coupled and therefore long lived particles require a beam dump facility. Such a facility is made of a high density proton target, followed by a hadron stopper and a muon shield. Apart from residual muons, the only remaining particles are electron, muon and tau neutrinos on top of hidden, long lived particles produced either in proton interactions or in secondary particle decays.

A new experiment, Search for Hidden Particles (SHiP), has been proposed [1], designed to operate at a beam dump facility to be built at CERN and to search for weakly coupled particles in the few GeV mass range. Since a high intensity tau neutrino flux is produced by such a facility from D_s decays, the experimental apparatus foresees a neutrino detector to study the tau neutrino cross-section and discover the tau anti-neutrino. This detector would also be suited to detect dark matter through the scattering off the atomic electrons of its target. The physics case for such an experiment is widely discussed in Ref. [2]. We will review here the physics potential of this experiment for a selection of physics channels.

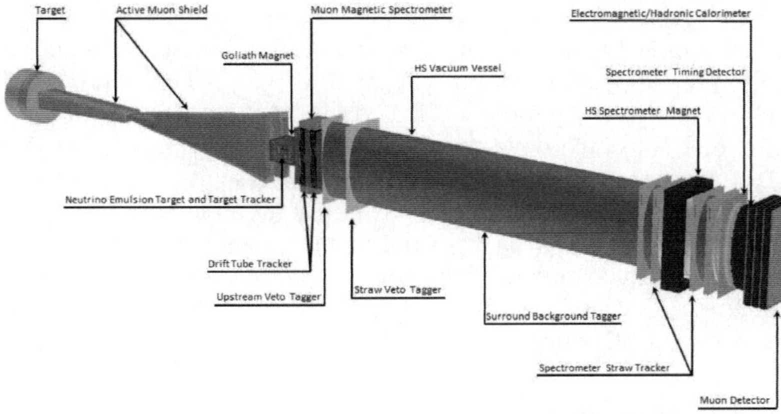

Figure 1. The beam dump facility and the SHiP detector.

2. The SHiP facility and apparatus

In five years, the facility will integrate 2×10^{20} 400 GeV protons, produced by the SPS accelerator complex, impinging on a 12 interaction length (λ_{int}) target made of Molybdenum and Tungsten, followed by a 30 λ_{int} iron hadron absorber. Hidden particles in the GeV mass range would be produced mostly by the decay of charmed hadrons produced in proton interactions. D_s mesons, copiously produced among charmed hadrons, are a source of tau neutrinos through their fully leptonic decay. Therefore, the SHiP facility is ideal also to study the physics of tau neutrinos, the less known particle in the Standard Model.

Figure 1 shows the SHiP facility to be placed in the North Area: downstream of the target, the hadron absorber filters out all hadrons, therefore only muons and neutrinos are left. An active muon shield is designed with two sections with opposite polarity to maximize the muon flux reduction: it reduces the muon flux from $\sim 10^{10}$ down to $\sim 10^{5}$ muons per spill. 4×10^{13} protons are extracted in each spill, designed to be 1s long to reduce the detector occupancy. A first successful test of the SPS cycle with a 1s long spill was performed in April 2015.

An emulsion spectrometer to detect tau neutrinos and search for dark matter is located downstream of the muon shield, followed by the decay vessel and the detector for hidden particles. The Collaboration has been requested by the SPSC Committee to prepare a Comprehensive Design Report by 2018, in the framework of the Physics Beyond Colliders working group, launched in March 2016 at CERN. This working group is meant to prepare proposals as input for the High Energy panel of the European Strategy to be held in 2019-2020; approval of the experiment will follow the recommendation of this panel. The construction and installation will last until the third long shutdown of the LHC and the data taking is assumed to start in 2026.

The emulsion spectrometer detector is made of a magnetised target region, followed by a muon spectrometer. The neutrino target is based on the emulsion cloud chamber technology employed by the OPERA experiment [3], with a compact emulsion spectrometer, made of a sequence of very low density material and emulsion films to measure the charge and momentum of hadrons in magnetic field. Indeed, this feature would allow to discriminate between tau neutrinos and anti-neutrinos also in the hadronic decay channels of the tau lepton. The emulsion target is complemented by high resolution tracking chambers to provide the time stamp to the event and connect muon tracks from the target to the muon spectrometer. The muon spectrometer is based on the concept developed for the OPERA apparatus: a dipolar iron magnet where drift tubes provide the momentum and resistive plate chambers provide the tracking within the iron slabs. The atoms of the emulsion target would also act as the target of dark matter coming from the decays of the dark photons produced from the proton interaction in the target.

IOP Conf. Series: Journal of Physics: Conf. Series **878** (2017) 012014 doi:10.1088/1742-6596/878/1/012014

The detector for hidden particles is located in the downstream part of a 60 m long evacuated decay vessel, with an elliptical transverse section of 5×10 m^2, with the longer axis vertically. The hidden particles are supposed to decay within the vessel. The requirement to have less than 1 neutrino interaction in the vessel over five years sets the pressure to about 10^{-3} mbar. A magnetic spectrometer is located downstream of the vessel: it is made of straw tubes with a material budget of 0.5% X_0 per station, achieving a position resolution of 120 μm per straw, with 8 hits per station on average. This gives a momentum resolution of about 1%. The vessel would be surrounded by a liquid scintillator layer to tag particles coming from outside. Downstream of the spectrometer, an hadronic and electromagnetic calorimeter and a muon filter are meant for particle identification. A timing detector with a resolution better than 100 ps makes the combinatorial background negligible.

3. Search for hidden particles

Extensions of the Standard Model in the low mass region foresee the existence of particles as singlets with respect to the Standard Model gauge group. These particles couple to different singlet composite operators (so-called Portals) of the Standard Model. The neutrino can be considered as the first example of a hidden particle: it was postulated in 1930 to explain the puzzle of the beta decay, it is very light and extremely weakly interacting. The SHiP detector located immediately downstream of a beam dump has the potentiality to discover all of them, since it is sensitive to very weakly interacting and long lived particles in a wide unexplored range of their masses and couplings. We describe in the following its sensitivity to dark photons, heavy neutral leptons and dark matter. For the sensitivity to the other portals, we refer to Ref. [1].

3.1. Search for hidden photons

Dark photons are vectorial particles coupled to the photon with a lagrangian term $L = -kB_{\mu\nu}V^{\mu\nu}/2$. In a beam dump facility, they can be produced directly from the proton interaction with the target and in the decay of particles like π^0, η, ω and η'. The dark photon may decay into dark matter particles or, if no lighter hidden particles exist, into final states like e^+e^-, $\mu^+\mu^-$ or $q\bar{q}$. Recently, data from old experiments have been reanalysed and exclusion plots have been derived together with cosmological limits [4, 5]. The left plot of figure 2 shows the exclusion limits at 90% C.L. set by SHiP in case no signal is found. The parameters displayed are the mass and ε^2 squared coupling of the dark photon. It is worth noting that the region investigated by SHiP covers a new region, unexplored by previous searches.

3.2. Search for heavy neutral leptons

Neutrino masses are explained by the see-saw mechanism with a lagrangian term given by

$$L = \imath\bar{N}_I\partial_\mu\gamma^\mu N_I - Y_{I\alpha}\bar{N}_I^C\tilde{H}L_\alpha - M_I\bar{N}_I^C N_I + h.c. \tag{1}$$

where 3 right-handed heavy neutrinos N_I are introduced. The mass of active neutrinos is given by $m_\nu \simeq m_D^2/M_I$ where $m_D \simeq Y_{I\alpha}v$ and $v \simeq 246$ GeV is the Higgs vacuum expectation value.

If several conditions are satisfied, among which that CP is not conserved in the leptonic sector, these right-handed neutrinos could also explain the leptogenesis and threfore the baryogenesis [6, 7]. Such models do not claim to explain also dark matter: a dark matter candidate could be provided by a fourth right-handed neutrino or an axion. Moreover they do not require two of the three heavy neutrinos to be quasi-degenerate because CP violation does not need to be enhanced by mass degeneracy.

In the attempt of explaining also dark matter with three heavy neutrinos, the so-called νMSM [8, 9] was first introduced in 2005, followed by more refined versions [10] later on. The role of the three right-handed neutrinos is different as well as their mass: M_1 could be a dark

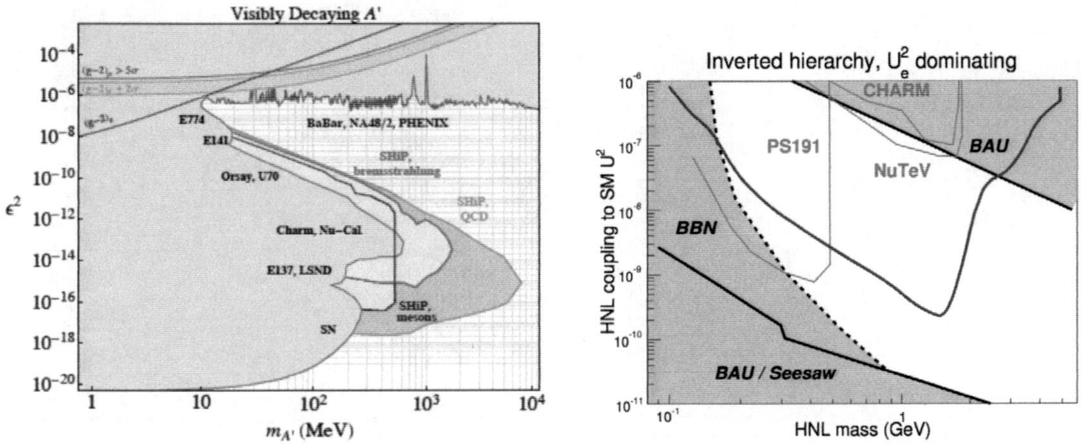

Figure 2. Left: exclusion limit at 90% C.L. in the dark photon search, in the plane defined by its mass and ϵ^2 squared coupling. Right: exclusion limits at 90% C.L. in the heavy neutral lepton search, for inverted hierarchy with U_e^2 dominating, according to Ref. [11].

matter candidate with the mass of several keV while M_2 and M_3 should be states with a mass of about 1 GeV and quasi-degenerate to provide the baryon asymmetry of the Universe. The states N_2 and N_3 could be produced in the decay of sufficiently massive particles like charmed hadrons. These states would be long lived particles that in turn could decay, for example, into a pion-muon pair.

The right plot of Figure 2 shows the experimental and cosmological bounds on the search for heavy neutral leptons. It also reports superimposed the exclusion limits at 90% C.L. SHiP could set in case no signal is found, assuming the model reported in Ref. [11] with inverted hierarchy and a dominant electron coupling, $U_e^2 : U_\mu^2 : U_\tau^2 = 48 : 1 : 1$. The parameter space explored by SHiP extends in the cosmologically relevant region that is experimentally unexplored.

4. Physics with the neutrino detector

The tau neutrino is the less known particle in the Standard Model. The observation of this particle was finally confirmed in 2008 when 9 candidates events were reported with an estimated background of 1.5 [12]. In the same paper, the tau neutrino cross-section, with a 50% uncertainty due to the small statistical sample and to the rather poor knowledge of the incoming flux, was also reported. The DONUT experiment was unable to separate neutrinos from anti-neutrinos. The OPERA experiment has detected five tau neutrinos [13, 14, 15, 16, 17], discovering the tau neutrino appearance from muon neutrino oscillations. The only leptonic decay observed by OPERA [15] shows negative charge as expected from a ν_τ interaction. Therefore, so far there is no direct evidence for tau anti-neutrinos.

The number of ν_τ and $\bar{\nu}_\tau$ emerging from the molybdenum target can be estimated as:

$$N_{\nu_\tau+\bar{\nu}_\tau} = 4N_p\frac{\sigma_{c\bar{c}}}{\sigma_{pN}}f_{D_s}Br(D_s \to \tau) = 6.6 \times 10^{15} \tag{2}$$

where N_p is the number of interacting protons, $\sigma_{c\bar{c}} = 18.1 \pm 1.7$ μbarn [18] is the associated charm production per nucleon, $\sigma_{pN} = 10.7$ mbarn is the hadronic cross-section per nucleon in a Mo target, $f_{D_s} = (8.8 \pm 0.6^{+0.5}_{-0.9})\%$ [19] is the fraction of D_s mesons produced, $Br(D_s \to \tau) = (5.54 \pm 0.24)\%$ [20] is the D_s branching ratio into τ and the factor 4 accounts for the charm

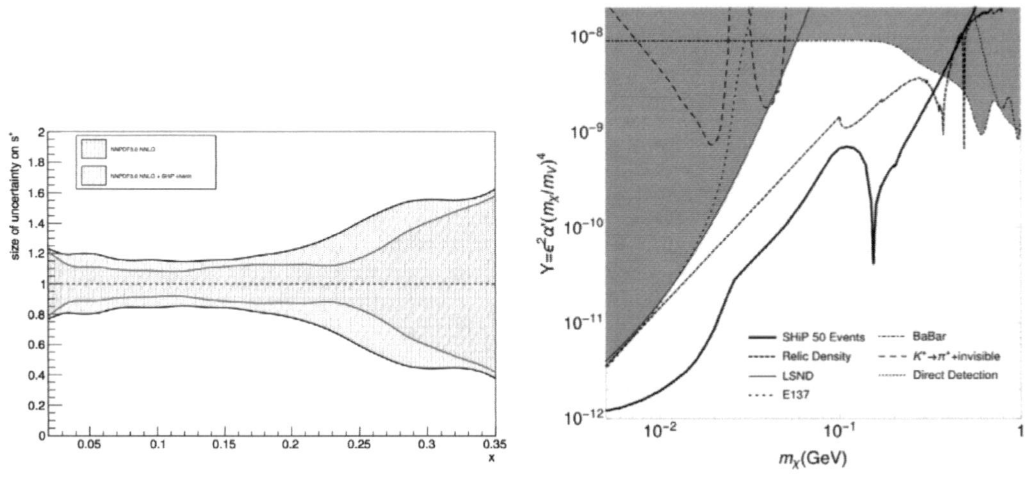

Figure 3. Left: improvement of the accuracy on s^+ with SHiP (red) compared to the present status (blue) in the $0.02 < x < 0.35$ range. Right: sensitivity to the dark matter in the assumption that $m_\chi/m_A = 1/5$.

pair production and for the two ν_τ produced per D_s decay. The SHiP facility is therefore a ν_τ factory.

Given the neutrino target mass of 9.6 tons, the neutrino flux, the geometrical acceptance of the detector and the standard model cross-section [21], one expects about 7000 interactions of tau neutrinos and 3500 of tau anti-neutrinos. The charged-current ν_τ ($\overline{\nu}_\tau$) differential cross-section is given by five structure functions. The contribution to the cross-section of F_4 and F_5 structure functions [22] is negligible in muon and electron neutrino interactions due to the light charged lepton mass. On the contrary, given the non-negligible mass of the τ lepton, tau neutrino scattering is sensitive to their contribution. Indeed, the hypothesis of $F_4 = F_5 = 0$ would significantly increase the ν_τ and $\overline{\nu}_\tau$ charged-current deep-inelastic cross sections and the corresponding number of expected ν_τ and $\overline{\nu}_\tau$ interactions. SHiP is sensitive to a non-null value of F_4 and F_5 [1].

4.1. Strange parton distribution

Charmed hadrons are produced in neutrino and anti-neutrino charged-current interactions at the level of about 5%. Advantages of the nuclear emulsion technology in the detection of neutrino-induced charm production are discussed in Ref. [23]. Despite the fact that 1.280.000 ν_μ and 270.000 $\bar{\nu}_\mu$ charged-current interactions were collected by the NuTeV/CCFR Collaboration, only 5102 ν_μ and 1458 $\bar{\nu}_\mu$ events were identified as charm production induced by neutrino charged-current interactions. The largest number of events in an emulsion experiment was collected by CHORUS with 2013 charm candidates from ν_μ and only 32 from $\bar{\nu}_\mu$ [24]. SHiP will integrate about 10^5 charm candidates, more than one order of magnitude larger than the present statistics, with a large ($\sim 30\%$) contribution from anti-neutrinos. Charm production in neutrino scattering is extremely sensitive to the strange quark content of the nucleon, especially with anti-neutrinos where the s-quark is dominant. SHiP will improve significantly the uncertainty on the strange quark distribution in the nucleon as shown in the left plot of Figure 3 in terms of $s^+ = s(x) + \bar{s}(x)$ in the $0.02 < x < 0.35$ range.

4.2. Dark Matter search

The dark matter at an accelerator can be produced through the decay of a dark photon and it can thus be copiously produced at a beam dump facility. Given the proton energy, a dark matter in the GeV and sub-GeV mass range can be produced at SHiP. Direct searches performed in the underground laboratories rely on the observation of nuclear recoils induced by the scattering of dark matter. Given the non-relativistic nature of dark matter, the recoils induced show a kinetic energy of the keV order if the dark matter mass is around 1 GeV. This makes the search for dark matter at around 1 GeV or below extremely difficult. The sensitivity of current experiments in that range is therefore limited by the capability of observing such low energy recoils. At the accelerator, on the contrary, the dark matter would be ultra-relativistic and it could be observed through its scattering off the electrons of the lead-emulsion target of the neutrino detector. The elastic interaction of dark matter particles with electrons produces one electron in the final state, thus mimicking elastic interaction of neutrinos that constitute the main background for this search. The total background to this search is estimated in Ref. [1] and the right plot of Figure 3 shows the sensitivity in the plane of the coupling Y and the dark matter mass m_χ, assuming $m_\chi/m_A = 1/5$ where m_A is the dark photon mass. In the same plot, the sensitivity of other experimental programs is shown together with the relic dark matter density. SHiP is unique in its sensitivity for masses below 500 MeV/c^2 and it allows to cover completely the predictions made from the relic dark matter density.

References

[1] M. Anelli et al., arXiv:1504.04956.
[2] S. Alekhin et al., Rept. Prog. Phys. **79** (2016) no.12, 124201.
[3] N. Agafonova et al., OPERA Coll., JINST **4** (2009) P04018.
[4] J. Brunner, Phys. Lett. **B731** (2014) 320.
[5] A. Fradette et al., Phys. Rev. **D90** (2014) 035022.
[6] M. Drewes, et al., JHEP **1303** (2013) 096.
[7] M. Drewes, Int. J. Mod. Phys. **E22** (2013) 1330019.
[8] T. Asaka and M. Shaposhnikov, Phys. Lett. **B620** (2005) 17.
[9] M. Shaposhnikov, Nucl. Phys. **B763** (2007) 49.
[10] L. Canetti et al., Phys. Rev. Lett. **110** (2013) 061801.
[11] L. Canetti and M. Shaposhnikov, JCAP **1009** (2010) 001.
[12] K. Kodama et al., Phys. Rev. **D78** (2008) 052002.
[13] N. Agafonova et al., Phys. Lett. **B691** (2010) 138.
[14] N. Agafonova et al., JHEP **1311** (2013) 036.
[15] N. Agafonova et al., Phys. Rev. **D89** (2014) 051102.
[16] N. Agafonova et al., PTEP **2014** (2014) 101C01.
[17] N. Agafonova et al., Phys. Rev. Lett. **115** (2015) 121802.
[18] C. Lourenco, H.K. Wohri, Phys. Rept. **433** (2006) 127.
[19] H. Abramowicz et al., JHEP **1309** (2013) 058.
[20] K.A. Olive et al., Chin. Phys. **C38** (2014) 090001.
[21] M. H. Reno, Phys. Rev. D **74** (2006) 033001.
[22] C. H. Albright and C. Jarlskog, Nucl. Phys. B **84** (1975) 467.
[23] G. De Lellis et al., Physics Reports **399** (2004) 227.
[24] A. Kayis-Topaksu et al., New J. Phys. **13** (2011) 093002.

IOP Conf. Series: Journal of Physics: Conf. Series **878** (2017) 012015 doi:10.1088/1742-6596/878/1/012015

The Beam Energy Scan at the Relativistic Heavy Ion Collider

Declan Keane

Department of Physics, Kent State University, Kent, Ohio 44242, USA

E-mail: keane@kent.edu

Abstract. The Beam Energy Scan (BES) at the Relativistic Heavy Ion Collider (RHIC) is based on Au + Au collision data acquired between 2010 and 2014 at beam energies of $\sqrt{s_{NN}} = 7.7$, 11.5, 14.5, 19.6, 27 and 39 GeV. These measurements constitute Phase-I of BES (also known as BES-I), and along with higher-energy data at 62.4 and 200 GeV, they allow the phase diagram of QCD matter to be probed. BES-I has three physics goals: investigation of a turning-off of the Quark-Gluon Plasma (QGP) signatures that are by now well established at higher energies, the search for a possible first-order phase transition between hadronic and QGP phases, and the search for a possible critical point. Several promising signals have been reported, but since RHIC luminosity decreases steeply as the beam energy is scanned down, statistical errors are excessively large at the lower BES-I energies where potentially novel phenomena are observed. In 2019 and 2020, BES-II will take data with large improvements in both RHIC luminosity and in detector performance.

1. Introduction

During the years 2000 to 2010, corresponding to the first decade of operation of the Relativistic Heavy-Ion Collider at Brookhaven National Laboratory, the physics focus was on gold-gold collisions at $\sqrt{s_{NN}} = 200$ GeV [1, 2, 3, 4]. During that period, there was an accumulation of evidence pointing to the production of a deconfined partonic phase in the early stages of the collisions when the energy density is high. Our current understanding is illustrated by the conceptual phase diagram on the left in Fig. 1. Lattice QCD calculations [5, 6, 7] indicate that the transition is a smooth crossover at top RHIC energies and above. By colliding nuclei at progressively lower energies, it should be possible to probe phase diagram regions of increasingly high baryon chemical potential μ_B (equivalent to increasingly high density of baryons minus antibaryons), exploring the possible region of a first-order phase transition and the vicinity of the associated critical end point. Lattice calculations suggest that the critical point does not lie at or below $\mu_B \sim 250$ MeV [8, 9, 10, 11]. However, our knowledge to date from theory and experiment does not exclude the possibility that a crossover extends to much higher baryon chemical potentials than depicted on the left in Fig. 1.

The first physics goal of the BES-I program is to investigate the expected turning-off of the Quark-Gluon Plasma (QGP) signatures that are by now well established at higher collision energies. A disappearance of these signatures would not necessarily flag the location on the phase diagram where the deconfined phase begins, since factors unrelated to the onset of deconfinement might weaken or even completely obscure individual signatures as the beam energy is scanned down. For example, jet quenching may disappear at lower beam energies simply because all hard scattering effects become rare at lower beam energies, as the high-p_T regions are much less populated. However, if it is observed that a few independent signatures all point to a turn-off at about the same beam energy, a more convincing inference might be reached.

The second BES-I goal is to search for the predicted first-order phase transition [12, 13] between hadronic and QGP phases. Observables like anisotropic flow are promising for this purpose, since flow is a proxy for the pressure in a hydrodynamic picture. As the beam energy increases above the region where the system is always in a hadronic phase, new degrees of freedom begin to open up, and spinodal decomposition and a long-lived mixed phase may play important roles. In some models, there can be a dramatic drop in the pressure, or using equivalent terminology, there is a softening of the Equation of State (EOS) at a bombarding energy where the system first reaches a region where there is a first-order phase transition [14, 15].

The third goal is to search for a possible critical end point, which must be present if the phase diagram includes a first-order phase transition. The characteristic length scale in a macroscopic system (the correlation length) becomes infinite at a critical point, and causes observable effects such as critical opalescence. In a small system such as in a heavy-ion collision, the potential for growth of the correlation length is very limited, but theory still suggests that the local increase in fluctuations caused by a critical point could be observed experimentally [16].

2. BES-I and freeze-out parameters

The STAR detector [17] is well suited for investigating the physics objectives outlined above. In contrast to fixed target experiments, STAR has the advantage that its acceptance changes relatively little as $\sqrt{s_{NN}}$ is scanned from 7.7 to 39 GeV. Furthermore, STAR offers a large and uniform acceptance in azimuthal angle and transverse momentum in the midrapidity region, with excellent particle identification through dE/dx and Time-of-Flight [18], allowing many observables and their correlations to be measured. These detector characteristics make optimum use of the very limited statistics available towards the low end of the BES-I range, where RHIC's luminosity drops very steeply.

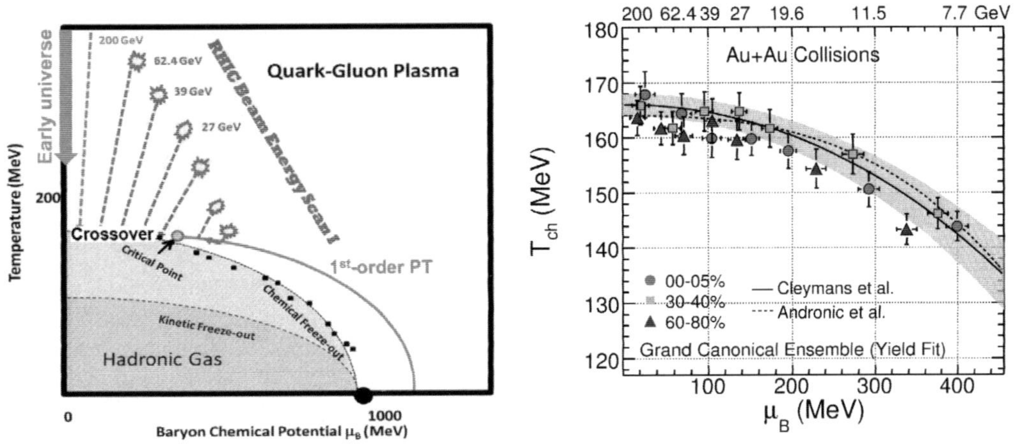

Figure 1. Left panel: A conceptual QCD phase diagram, showing the relevant regions in the plane of T versus μ_B. Various details, especially the position of the critical point, are highly uncertain. Right panel: Inferred coordinates of chemical freeze-out in the plane of T versus μ_B, for various beam energies (see upper horizontal scale) and centralities, based on measured particle spectra [19] and a thermal model [20].

In 2010, 2011 and 2014, the original RHIC measurements at $\sqrt{s_{NN}} = 200$ and 62.4 GeV were extended to lower beam energies in the form of the BES-I data at 39, 27, 19.6, 14.5, 11.5, and 7.7 GeV. These $\sqrt{s_{NN}}$ values correspond to baryon chemical potentials in the range 25 to about 400 MeV [19, 20] (see panel on the right of Fig. 1). This correspondence is determined by combining measured spectra for identified particles with a thermal model to infer the position on the phase diagram where interactions that can change the particle species cease (chemical freeze-out), and any subsequent interactions are elastic only. The panel on the right of Fig. 1 suggests that the existing set of BES-I measurements offers adequate coverage of the phase diagram near and below $\mu_B \sim 400$ MeV.

As of the end of 2016, STAR's BES-I effort has resulted in 18 published journal papers with final results [21, 22, 23, 24, 25], with several more in the pipeline. The final published BES papers account for about one-third of the total physics output from the STAR collaboration since 2012. There have also been many proceedings that cover recent preliminary results from BES-I. The present proceedings have space to touch on only a subset of BES-I findings and on the outlook for BES-II.

3. Directed flow of protons

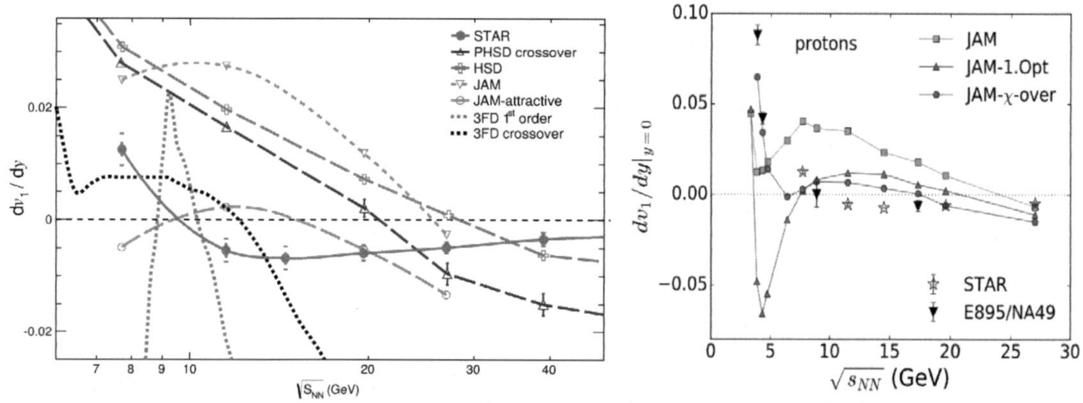

Figure 2. Left panel: Directed flow slope dv_1/dy near midrapidity as a function of beam energy for protons in 10-40% centrality Au+Au collisions from STAR [21], compared with recent models [26, 27, 28], as compiled in Ref. [29]. Other recent model comparisons [30] lie above the experimental data and are off-scale at all plotted beam energies. Right panel: A comparison between the same STAR dv_1/dy data and a new prescription for incorporating the EOS and the possible phase change in the JAM transport model [31]. The red triangle markers correspond to the EOS option with a first-order phase transition.

The first harmonic coefficient in the Fourier expansion of the distribution of azimuths ϕ of final-state particles relative to the event reaction plane azimuth $\Psi_{\rm RP}$, is known as directed flow: $v_1 = \langle \cos(\phi - \Psi_{\rm RP}) \rangle$ [32, 33, 34]. It describes a collective sideward deflection of the emitted particles. Models in the nuclear transport [35, 36] and hydrodynamic [37] categories support the interpretation that directed flow is a good probe of the participant matter during the early stages of the collision [38, 39, 40], e.g., can serve as a signature of the pressure. The slope of the rapidity dependence dv_1/dy close to mid-rapidity is a convenient way to characterize the overall magnitude of the directed flow signal. In particular, STAR's measurements at $\sqrt{s_{NN}} = 7.7$ to 200 GeV reveal that proton directed flow goes through a minimum between 10 and 20 GeV [21, 41]. Three-fluid hydrodynamic calculations [14, 15] qualitatively predicted this minimum, dubbed the "softest point collapse", and reported this model behavior when the EOS has a first-order phase transition, but not when the EOS is hadronic only.

Following the publication of the first set of STAR BES results for directed flow [21], several theorists reported model comparisons incorporating various options for the assumed EOS [26, 27, 28, 30, 31]. Generally, these theorists conclude that an assumption of purely hadronic physics is not favored, whereas they are divided on the issue of whether a crossover or first-order phase transition is closer to the STAR measurements. No EOS option in any of these models gives rise to a minimum in proton directed flow anywhere close in beam energy to the measured minimum [26, 27, 28, 30, 42] with the arguable exception of Nara *et al.* [31], whose most recent paper, in the case of a first-order phase transition, reports a minimum that is about an order of magnitude deeper than the measured one, and lies about a factor of three lower in beam energy (see Fig. 2). Furthermore, for equations of state that are nominally the same (e.g., either crossover or first order), directed flow from different model implementations are often very widely divergent, at the level of an order of magnitude larger than the measured signal and two orders larger than the measured errors [29]. Therefore, in order to reach a definite conclusion about the nature of the transition, further progress in the area of model calculations of $v_1(y)$ is needed. This is an active area for the BEST topical theory collaboration. On the experimental side, new directed flow measurements for more baryon species, and in fine centrality bins, will further constrain models and will help resolve the pending questions [21, 41, 43].

4. The Chiral Magnetic Effect

The non-trivial structure of the QCD vacuum is a topic of high importance, and many open questions remain. There is a family of quantum phenomena that can arise from interactions with the very high vorticity and the exceptionally intense localized magnetic field ($\sim 10^{15}\,\mathrm{T}$) that are briefly generated during non-central collisions of heavy ions. Differences in the number of left-handed and right-handed quarks in the QGP can cause a characteristic signature of charge separation along the magnetic field. These considerations are not just relevant to relativistic heavy ion physics, but have implications for all areas of physics touched by QCD (high energy physics, astrophysics, cosmology, etc.) A number of related anomalous chiral effects are discussed in the literature, but the primary one is known as the Chiral Magnetic Effect (CME). Many articles examine these topics in depth [44, 45, 46, 47, 48, 49, 50, 51] and the interested reader should consult the extensive reference list in the recent review by Kharzeev *et al.* [51].

For the purpose of the present brief summary of BES-I physics, we set aside the broad implications outlined above, and simply note that a deconfined state (QGP) is needed to generate a true CME signal. Then we focus purely on the potential for our CME observable (the three-point correlator defined below) to flag the beam energy where QGP turns-off as we scan down in beam energy.

In a description of azimuthally anisotropic emission, $(dN_\pm/d\phi) \propto 1 + 2a_\pm \sin(\phi - \Psi_{\mathrm{RP}}) + \ldots$ the coefficient a represents the size of the CME signal, and the remaining terms (not shown explicitly) are the familiar cosines weighted by coefficients v_n for directed flow, elliptic flow, triangular flow, etc. The coefficient a averages to zero when integrated over many events. If chiral magnetism is present, a non-zero average signal can be obtained by forming a correlation between pairs of emitted particles relative to the reaction plane, in the form of the three-point correlator γ as defined in the expression along the vertical axis in Fig. 3, where α and β denote the particle type: $\alpha,\ \beta = +,\ -$. In that case, the CME signature is the non-zero difference between γ_{SS} (same charge, plotted as red stars) and γ_{OS} (opposite charge, plotted as blue circles) in Fig. 3.

As expected in the presence of CME, we indeed see a signal in the form of a significant positive $\gamma_{\mathrm{OS}} - \gamma_{\mathrm{SS}}$ which grows steadily as we move from central to peripheral collisions, except at the two lowest beam energies, where $\gamma_{\mathrm{OS}} - \gamma_{\mathrm{SS}}$ is consistent with zero at all centralities. One interpretation is that QGP is no longer produced at those lower energies.

Background effects, unrelated to CME, are argued to have a different centrality dependence and a different energy dependence to that shown by $\gamma_{\rm OS}$ and $\gamma_{\rm SS}$. However, specific theoretical calculations of the γ correlators are very challenging, and so far are lacking, and elliptic flow and conservation effects cannot be decisively ruled out as the origin of the signal $\gamma_{\rm OS} - \gamma_{\rm SS}$.

A future isobar measurement [53] in which collisions of ${}^{96}_{40}$Zr + ${}^{96}_{40}$Zr and ${}^{96}_{44}$Ru + ${}^{96}_{44}$Ru will be directly compared offers a robust and unambiguous way to settle this issue of possible elliptic-flow-related background effects in $\gamma_{\rm OS} - \gamma_{\rm SS}$. This measurement ought to be decisive, because the background scales with the total mass number of the initial system, which is the same for both pairs of incoming nuclei, while the true CME signal scales with the total electric charge of the initial system, which differs by 10% between Zr + Zr and Ru + Ru. This isobar measurement is scheduled to take data in 2018.

5. Search for the QCD critical point via net-proton fluctuations

Local fluctuations of conserved numbers like net electric charge or net baryon number increase near a critical point, and are considered to be among the most promising signatures. While such quantities cannot fluctuate globally (for an entire event), STAR measurements fall well short of being global, especially in regard to rapidity acceptance [54]. High moments like skewness, $S \propto \langle (N - \langle N \rangle)^3 \rangle$ and kurtosis, $\kappa \propto \langle (N - \langle N \rangle)^4 \rangle$ are of particular interest. Kurtosis is proportional to the seventh power of the correlation length, and notwithstanding the trivial increase of statistical errors with higher powers, it has been shown that high moments such as κ offer the best sensitivity to the fluctuations of interest [55, 56, 57, 16]. Another advantage of kurtosis and similar quantities is the fact that they can readily be calculated by lattice QCD [58].

High moments of net charge have been reported by both PHENIX [59] and STAR [23] at BES-I energies, but only monotonic behavior is observed. Experimental measurement of net baryons is very difficult, and net protons serve as a useful alternative observable. However, complicating factors such as resonance decays, hadronic scattering, etc., need to be taken into consideration when using net protons as a proxy for net baryons [60].

In Fig. 4 [61], it is evident that at centralities of 30-40% and 70-80%, $\kappa\sigma^2$ for net protons is

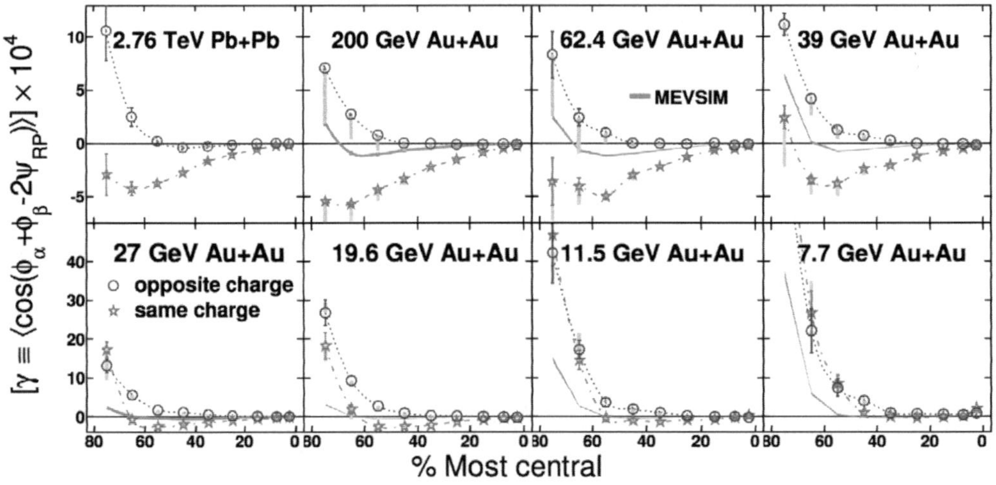

Figure 3. The three-point correlator γ as a function of centrality for Au+Au collisions from 200 GeV to 7.7 GeV [22]. ALICE results for Pb+Pb collisions at 2.76 TeV are also shown [52].

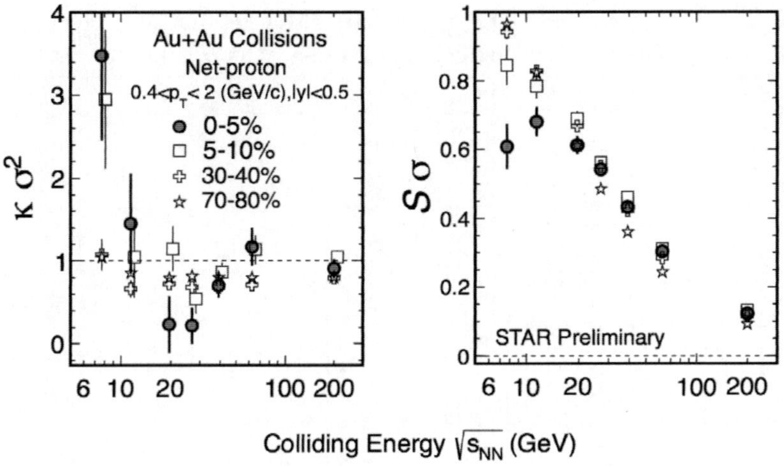

Figure 4. The dimensionless products $\kappa\sigma^2$ and $S\sigma$ for net protons as a function of beam energy in four centrality intervals [61]. The acceptance window for counting protons and antiprotons is $0.4 < p_T < 2.0$ GeV/c, $|y| < 0.5$. The kurtosis point for 0-5% centrality at 7.7 GeV is highly sensitive to efficiency corrections.

close to the baseline of 1.0 across all beam energies. However, for the red points (0-5% centrality), there is a minimum of moderate significance below the baseline centered near 20 GeV, with a steep rise as the beam energy is scanned further down, ending with a value about 2 standard errors above the baseline at 7.7 GeV. If this is a critical point signature of the kind predicted by Stephanov [16], $\kappa\sigma^2$ ought to return to the baseline value of unity at energies below 7.7 GeV. This observation strongly motivates exploration of beam energies below 7.7 GeV in the second phase of the beam energy scan (see Section 8 below).

6. Search for the QCD critical point via pion interferometry

Femtoscopy correlations [62] continue to evolve right up until kinetic freeze-out at the very end of the collision process. Nevertheless, this does not exclude the possibility that pion interferometry measurements provide sensitivity to EOS effects from a much earlier stage of the time evolution. Models suggest that such a sensitivity is present at a useful level [62].

The parameters $R_{\rm out}$, $R_{\rm side}$ and $R_{\rm long}$ are commonly used to orthogonally decompose the Gaussian radii for the spatial homogeneity regions from interferometry, and the difference $R^2_{\rm out} - R^2_{\rm side}$ is related to the time duration of emission [62]. Fig. 5 presents Lacey's compilation [63] of pion interferometry parameters $R^2_{\rm out} - R^2_{\rm side}$ in seven centrality intervals for Au + Au collisions at beam energies of 7.7, 11.5, 19.6, 27, 39, 62.4, and 200 GeV [24] and also for Pb + Pb collisions at 2.76 TeV [64]. From the observed peaks in $R^2_{\rm out} - R^2_{\rm side}$, one can infer a lengthening of the duration of pion emission, with the maxima occurring in the range of $\sqrt{s_{NN}} = 20$ to 40 GeV along the beam energy axis. It is also noteworthy that the beam energy of the peak drops smoothly as the collisions become less central. The peak also becomes smaller and its width increases as the collisions become less central.

Lacey [63] has interpreted these data under the assumption of Finite Size Scaling (FSS) behavior and critical exponents in the context of a 3D Ising model universality class [65, 66, 67]. In Ref. [63], it is argued that the data at all seven centralities vary according to the expected FSS behavior, as illustrated by all the points accurately mapping onto a common curve [63]. A chance occurrence of this scaling behavior would be highly improbable. This leads to the inference that the favored location of the critical point in this analysis is $(T, \mu_B)_{\rm crit} \sim (165, 95)$

MeV. This surprisingly low baryon chemical potential of 95 MeV at the inferred critical point does not appear to be compatible with lattice calculations, which rule out values below about 250 MeV [8, 9, 10, 11].

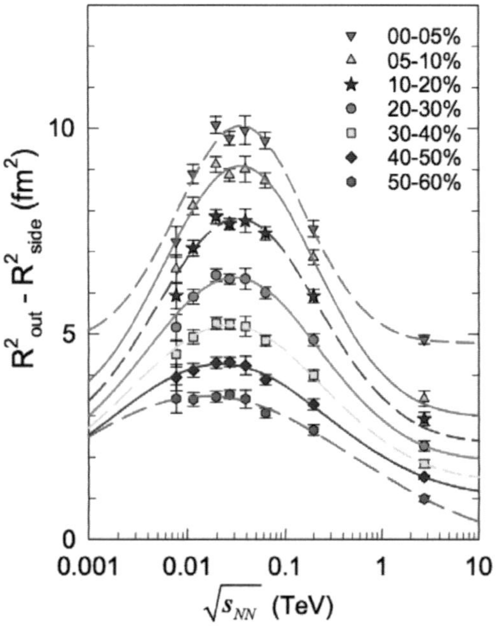

Figure 5. Pion femtoscopy parameters $R^2_{\rm out} - R^2_{\rm side}$ versus beam energy, covering Au + Au collisions at RHIC BES-I energies [24] through Pb + Pb at 2.76 TeV [64], in seven centrality intervals.

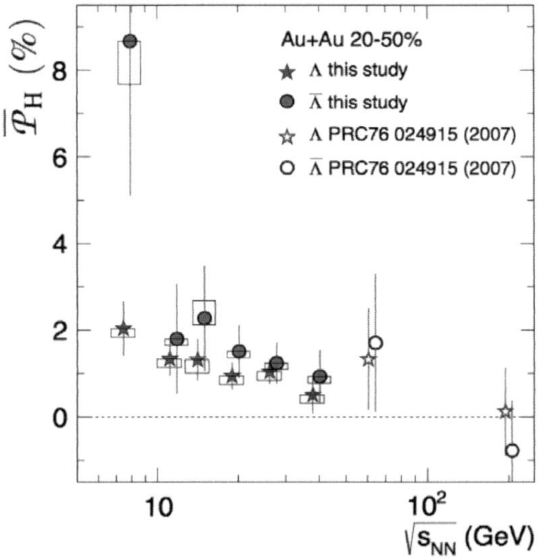

Figure 6. Average global polarization for Λ (stars) and anti-Λ (circles) produced in Au + Au collisions at 20-50% centrality as a function of beam energy. Results at 62.4 and 200 GeV, published by the STAR collaboration in 2007 [68], are presented along with recent measurements at BES-I energies [69]. Systematic errors are indicated by boxes.

7. Vorticity inferred from global polarization of Λ and anti-Λ

The Λ decay has the well-known 'self-analyzing' property whereby the daughter proton is preferentially emitted along the spin direction of the parent [70]. In *pp* collisions, this directly results in an observed Λ polarization at forward angles [71]. However, a 2007 search for the quite different phenomenon of global polarization in heavy-ion collisions at top RHIC energies yielded a null signal [68]. The term 'global' refers to an average over all (anti-)Λ momenta projected onto the angular momentum vector of each event before averaging over the available events. The reaction plane azimuth $\Psi_{\rm RP}$ is the basis for determining the direction of each event's angular momentum vector. As seen from Fig. 6, an increasingly significant non-zero global polarization of Λ and anti-Λ is observed at STAR as the beam energy is scanned down [69]. The significance of the signal considering all energies is just over 7σ for Λ and 4.6σ for $\bar{\Lambda}$. It has been theorized [72, 73, 74] that this dependence may result from the depolarizing effect of higher temperatures at higher beam energies, or from a different rapidity distribution of the fireball's angular momentum as the beam energy changes.

As a result of this observation of global polarization, the rotational properties of the medium created in nuclear collisions are experimentally accessible for the first time. Vorticity, which can be defined in terms of the curl of the medium's velocity vectors in the non-relativistic limit, is proportional in a hydrodynamic picture to the summed polarization of both Λ and $\bar{\Lambda}$, with appropriate corrections for Λ and $\bar{\Lambda}$ coming from the decay of heavier particles [69]. Given the fact that heavy-ion vortices are much smaller than those that arise in any other area of science, and of course are moving at relativistic speeds, it is not surprising that the the extracted energy-averaged vorticity, $\sim 9 \times 10^{21}$ s^{-1}, far exceeds any previous reported measurement (the closest is 10^7 s^{-1} for superfluid nanodroplets [75]). Models yield the right order of magnitude for the vorticity in nuclear collisions, but the calculated vorticity and its duration are subject to large uncertainties [76].

Particles with intrinsic magnetic moments interact with the intense magnetic field during a heavy-ion collision, and this is expected to result in a larger global polarization for $\bar{\Lambda}$ than for Λ [77]. In Fig. 6, a suggestion of such a difference is indeed observed, but averaged over all beam energies it is about a 2σ effect, and therefore STAR has deferred further probing of the heavy-ion magnetic field until the much larger statistics of BES-II become available.

8. Plans for Beam Energy Scan Phase-II

In order to revisit the several promising signals from BES-I analysis with greatly improved statistics, the follow-up BES-II program will take data in 2019 and 2020. The total beam time to be expended during BES-II will be comparable to that for BES-I, and the improved physics capability will come partly from three STAR detector upgrades, and partly from better RHIC luminosity.

The iTPC consists of new inner endcap sectors for the main STAR Time Projection Chamber. It replaces aged wires, and it changes the sparse pads of the old inner sectors to pads that cover the full area. This leads to better particle ID from dE/dx, and most important, it adds about 50% more pseudorapidity acceptance. A fourth iTPC benefit is extended acceptance at low transverse momentum, from $\sim$ 125 GeV/c to as low as 60 GeV/c. Another upgrade is the new Event Plane Detector (EPD). The EPD will have much improved event plane resolution compared with the Beam-Beam Counters currently used for that purpose, and will also be an excellent and flexible device for triggering and background rejection. The third upgrade is the endcap TOF detector (eTOF) on loan from the CBM collaboration at FAIR (Darmstadt). eTOF will improve particle ID in the new pseudorapidity acceptance region added by the iTPC.

The other main leg of the BES-II improvements comes from the RHIC machine. At $\sqrt{s_{NN}} \sim 12$ GeV and below, electron cooling will boost luminosity, while above that energy, there is a list of accelerator improvements involving bunch structure and β^*. The machine upgrades will deliver a nominal luminosity improvement over BES-I that ranges from a factor of $\sim$10 at 19.6 GeV to $\sim$25 at 7.7 GeV. Taking account of detector and RHIC upgrades, the STAR collaboration has documented many projected improvements in physics capabilities [78, 79].

RHIC luminosity in normal collider mode decreases like relativistic γ^3 at low energies, where the RHIC ring is used as a decelerator, and going below $\sqrt{s_{NN}} = 7.7$ GeV is not feasible in collider mode. However, allowing the halo of one of the RHIC beams to graze a fixed target inside the beam pipe at one end of the TPC leads to a reasonable acceptance over the forward half of rapidity space and allows enough events to be generated to easily saturate the DAQ bandwidth down to $\sqrt{s_{NN}} \sim 3$ GeV and below [80].

In a brief fixed-target test in 2015, it only required about a half hour of beam to acquire a million good Au + Au events with excellent trigger efficiency. This fixed target running therefore needs only a modest investment of beam time to extend μ_B coverage (at chemical freeze-out) from 420 MeV up to at least 720 MeV ($\sqrt{s_{NN}} = 3.0$ GeV or below), and in particular, to explore the crucial region where the net-proton kurtosis is predicted to drop back down to the baseline level (see Section 5).

9. Conclusions and outlook

BES-I at STAR has been very fruitful: final results are published in the case of 18 journal papers [21, 22, 23, 24, 25], with three further BES papers now in the refereeing process [19, 69, 81] and several more in earlier stages of the pipeline. These proceedings summarize two of the three BES-I goals, namely, the search for a 1st-order phase transition and critical point. Directed flow measurements point to a softest point in the QCD equation of state, but so far, overall model agreement with data is too poor for a clear conclusion about the nature of the phase transition. High-moment fluctuations and pion pair femtoscopy independently hint at a possible critical point; the former call for much better statistics and extended beam energy coverage on the low side, while conclusions from the latter require more study. There are excellent prospects for decisive progress in each of the above areas after the much enhanced data from BES-II in 2019 and 2020 become available.

These proceedings also touch on novel and exciting BES effects related to the heavy-ion magnetic field and vorticity, which were not even anticipated at the time when BES-II goals were formulated. With BES-II data, there are prospects for breakthrough-level progress in these topics, especially the first experimental determinations of the transient magnetic field in non-central heavy-ion collisions.

References

[1] Arsene I *et al.* (BRAHMS Collaboration) 2005 *Nucl. Phys.* A **757** 1
[2] Back B B *et al.* (PHOBOS Collaboration) 2005 *Nucl. Phys.* A **757** 28
[3] Adams J *et al.* (STAR Collaboration) 2005 *Nucl. Phys.* A **757** 102
[4] Adcox K *et al.* (PHENIX Collaboration) 2005 *Nucl. Phys.* A **757** 184
[5] Karsch F *et al.* 2004 *Nucl. Phys. Proc. Suppl.* **129** 614
[6] Aoki Y, Endrodi G, Fodor Z, Katz S D and Szabo K K 2006 *Nature* **443** 675
[7] Cheng M *et al.* 2009 *Phys. Rev.* D **79** 074505
[8] Fodor Z and Katz S D 2004 *J. High Energy Phys.* **04** 204
[9] Datta S, Gavai R V and Gupta S 2013 *PoS Lattice 2013* 202
[10] Kaczmarek O *et al.* 2011 *Phys. Rev.* D **83** 014504
[11] Bazavov A *et al.* 2012 *Phys. Rev.* D **85** 054503
[12] Ejiri S 2008 *Phys. Rev.* D **78** 074507
[13] Bowman E S and Kapusta J I 2009 *Phys. Rev.* C **79** 015202
[14] Rischke D H *et al.* 1995 *Heavy Ion Phys.* **1** 309
[15] Stöcker H 2005 *Nucl. Phys.* A **750** 121
[16] Stephanov M 2011 *J. Phys. G: Nucl. Part. Phys.* **38** 124147
[17] Ackermann K H *et al.* 2003 *Nucl. Instr. Meth.* A **499** 624
[18] Llope W for the STAR Collaboration 2012 *Nucl. Instr. Meth.* A **661** S110
[19] Adamczyk L *et al.* (STAR Collaboration) 2017 *Preprint* arXiv:1701.07065
[20] Cleymans J, Oeschler H, Redlich K and Wheaton S 2006 *Phys. Rev.* C **73** 034905
[21] Adamczyk L *et al.* (STAR collaboration) 2014 *Phys. Rev. Lett.* **112** 162301
[22] Adamczyk L *et al.* (STAR collaboration) 2014 *Phys. Rev. Lett.* **113** 052302
[23] Adamczyk L *et al.* (STAR collaboration) 2014 *Phys. Rev. Lett.* **113** 092301
[24] Adamczyk L *et al.* (STAR collaboration) 2015 *Phys. Rev.* C **92** 014904
[25] Adamczyk L *et al.* (STAR collaboration) 2016 *Phys. Rev.* C **94** 034908
Adamczyk L *et al.* 2016 *Phys. Rev. Lett.* **116** 112302
Adamczyk L *et al.* 2016 *Phys. Rev.* C **93** 014907
Adamczyk L *et al.* 2016 *Phys. Rev.* C **94** 024909
Adamczyk L *et al.* 2016 *Phys. Rev.* C **93** 021903
Adamczyk L *et al.* 2015 *Phys. Rev. Lett.* **114** 252302
Adamczyk L *et al.* 2015 *Phys. Lett.* B **750** 64
Adamczyk L *et al.* 2015 *Phys. Rev.* C **92** 021901
Adamczyk L *et al.* 2014 *Phys. Rev. Lett.* **112** 032302
Adamczyk L *et al.* 2013 *Phys. Rev.* C **88** 014902
Adamczyk L *et al.* 2013 *Phys. Rev. Lett.* **110** 142301
Adamczyk L *et al.* 2012 *Phys. Rev.* C **86** 054908
Agakishiev G *et al.* 2012 *Phys. Rev.* C **85** 014901

Abelev B I *et al.* 2010 *Phys. Rev.* C **81** 024911
[26] Konchakovski V P, Cassing W, Ivanov Y B and Toneev V D 2014 *Phys. Rev.* C **90** 014903
[27] Ivanov Yu B and Soldatov A A 2015 *Phys. Rev.* C **91** 024915
[28] Nara Y, Niemi H, Ohnishi A and Stöcker H 2016 *Phys. Rev.* C **94** 034906
[29] Singha S, Shanmuganathan P and Keane D 2016 *Adv. High Energy Phys.* **2016** 2836989
[30] Steinheimer J, Auvinen J, Petersen H, Bleicher M and Stöcker H 2014 *Phys. Rev.* C **89** 054913
[31] Nara Y, Niemi H, Steinheimer J and Stöcker H 2016 *Preprint* arXiv:1611.08023
[32] Voloshin S and Zhang Y 1996 *Z. Phys.* C **70** 665
[33] Poskanzer A M and Voloshin S A 1998 *Phys. Rev.* C **58** 1671
[34] Bilandzic A, Snellings R and Voloshin S 2011 *Phys. Rev.* C **83** 044913
[35] Bass S A *et al.* 1998 *Prog. Part. Nucl. Phys.* **41** 225
[36] Bleicher M *et al.* 1999 *J. Phys.* G **25** 1859
[37] Heinz U W 2010 *Relativistic Heavy Ion Physics* (Landolt-Boernstein New Series vol I/23) ed R Stock (New York: Springer Verlag)
[38] Sorge H 1997 *Phys. Rev. Lett.* **78** 2309
[39] Kolb P and Heinz U 2003 *Preprint* arXiv/nucl-th/0305084
[40] Huovinen P and Ruuskanen P V 2006 *Annu. Rev. Nucl. Part. Sci.* **56** 163
[41] Shanmuganathan P for the STAR Collaboration 2016 *Nucl. Phys.* A **956** 260
[42] Csernai L P and Stöcker H 2014 *J. Phys.* G **41** 124001
[43] Singha S for STAR Collaboration 2016 *Proc. Strangeness Quark Matter (Berkeley) Preprint* arXiv:1610.07423
[44] Kharzeev D 2006 *Phys. Lett.* B **633** 260
[45] Kharzeev D E, McLerran L D and Warringa H J 2008 *Nucl. Phys.* A **803** 227
[46] Buividovich P V, Chernodub M N, Luschevskaya E V and Polikarpov M I, 2009 *Phys. Rev.* D **80** 054503
[47] Bzdak A, Koch V and Liao J 2013 *Lect. Notes Phys.* **871** 503
[48] Kharzeev D E 2015 *Ann. Rev. Nucl. Part. Sci.* **65** 0000
[49] Miransky V A and Shovkovy I A 2015 *Phys. Rept.* **576** 1
[50] Huang X G 2016 *Rep. Prog. Phys.* **79** 076302
[51] D. Kharzeev D, Liao J, Voloshin S A and Wang G 2016 *Prog. Part. Nucl. Phys.* **88** 1 *Preprint* arXiv:1511.04050
[52] Abelev B I *et al.* (ALICE Collaboration) 2013 *Phys. Rev. Lett.* **110** 021301
[53] Deng W T, Huang X G, Ma G L and Wang G 2016 *Phys. Rev.* C **94** 041901
[54] Ling B and Stephanov M 2016 *Phys. Rev.* C **93** 034915
[55] Stephanov M A 2009 *Phys. Rev. Lett.* **102** 032301
[56] Asakawa M, Ejiri S and Kitazawa M 2009 *Phys. Rev. Lett.* **103** 262301
[57] Stephanov M A 2011 *Phys. Rev. Lett.* **107** 052301
[58] Gavai R V and Gupta S 2011 *Phys. Lett.* B **696** 459
[59] Adare A *et al.* (PHENIX Collaboration) 2016 *Phys. Rev.* C **93** 011901(R)
[60] Luo X, Mohanty B and Xu N 2014 *Nucl. Phys.* A **931** 808
[61] Luo X F and Xu N 2017 *Preprint* arXiv:1701.02105
[62] Lisa, M A, Pratt S, Soltz, R and Wiedemann U 2005 *Ann. Rev. Nucl. Part. Sci.* **55** 357
[63] Lacey R 2015 *Phys. Rev. Lett.* **114** 142301
[64] K. Aamodt *et al.* (ALICE Collaboration), *Phys. Lett.* B **696** 328
[65] Kleinert H 1999 *Phys. Rev.* D **60** 085001
[66] Andrea Pelissetto E V 2002 *Phys. Rept.* **368** 549 0012164.
[67] Ladrem M and Ait-El-Djoudi A 2005 *Eur. Phys. J.* C **44** 257
[68] Abelev B I *et al.* (STAR Collaboration) 2007 *Phys. Rev.* C **76** 024915
[69] Adamczyk L *et al.* (STAR Collaboration) 2017 *Preprint* arXiv:1701.06657
[70] Olive K A *et al.* 2014 *Chin. Phys.* C **38** 090001
[71] Bunce G *et al.* 1976 *Phys. Rev. Lett.* **36** 1113
[72] Betz B, Gyulassy M and Torrieri G 2007 *Phys. Rev.* C **76** 044901
[73] Jiang Y, Lin Z W and Liao J 2016 *Phys. Rev.* C **94** 044910
[74] Pang L G, Petersen H, Wang Q and Wang X N 2016 *Phys. Rev. Lett.* **117** 192301
[75] Gomez L F *et al.* 2014 *Science* **345** 906
[76] Becattini F *et al.* 2015 *Eur. Phys. J.* C **75** 406
[77] Becattini F, Karpenko I, Lisa M, Upsal I and Voloshin S 2016 *Preprint* arXiv:1610.02506
[78] STAR collaboration 2014 *STAR Note* 0598
[79] STAR Collaboration and CBM Collaboration eTOF Group 2016 *Preprint* arXiv:1609.05102
[80] Meehan K for the STAR Collaboration 2017 *Quark Matter presentation* https://indico.cern.ch/event/433345/contributions/2358275/
[81] Adamczyk L *et al.* (STAR Collaboration) 2017 *Preprint* arXiv:1701.06496

Applications of the Tsallis Statistics in High Energy Collisions

Trambak Bhattacharyya[1], Jean Cleymans[1], Prakhar Garg[2], Prateek Kumar[3], Sylvain Mogliacci[1], Raghunath Sahoo[3], Sushanta Tripathy[3]

[1] UCT-CERN Research Centre and Department of Physics, University of Cape Town, Rondebosch 7701, South Africa

[2] Department of Physics and Astronomy, Stony Brook University, SUNY, Stony Brook, New York 11794-3800, USA

[3] Discipline of Physics, School of Basic Sciences, Indian Institute of Technology Indore, M.P. 452020, India

E-mail: trambak.bhattacharyya@uct.ac.za

Abstract. We analytically investigate the thermodynamic variables of a hot and dense system in the framework of the Tsallis non-extensive classical statistics in the massless and in the massive cases. In addition to that, we study the effect of Tsallis Power Law distribution on the multi-particle production in the high-energy collisions. The effect of the Tsallis q parameter on the experimentally measured nuclear suppression factor has been investigated, and an attempt to describe the transverse momentum distribution of hadrons produced in high-energy collisions has been made.

1. Introduction

The transverse momentum distributions of hadrons at high energies is very often described by the Tsallis distribution [1]. The PHENIX and STAR collaborations [2, 3] at the Relativistic Heavy Ion Collider (RHIC) at BNL, and the ALICE, ATLAS and CMS collaborations [4, 5, 6, 7, 8] at the Large Hadron Collider (LHC) at CERN have made extensive use of the Tsallis distribution. Tsallis distribution has been very successful in explaining the experimental transverse momentum distribution, longitudinal momentum fraction distribution as well as the rapidity distribution of hadrons in e^+e^- as well as $p-p$ collisions [9, 10, 11, 12, 13, 14, 15].

Also, one can make use of the Tsallis distribution to parameterise the initial distribution of the high-energy particles which, due to interaction with the Quark Gluon Plasma (QGP) medium formed when two heavy ions collide with each other [16], lose energy. The modification of their distribution due to the energy loss sheds light on the characteristics of the QGP medium.

The Tsallis distribution is given by:

$$f = \left[1 + (q-1)\frac{E-\mu}{T}\right]^{-\frac{1}{q-1}}. \tag{1}$$

described by q, the Tsallis q parameter, and the Tsallis temperature T. In our work we use a Thermodynamically consistent form of the Tsallis distribution, f^q, described in detail in [17, 18]. The relevant thermodynamic quantities can be written as integrals over the Tsallis distribution.

2. Analytical calculations of the Tsallis Thermodynamic variables

2.1. Thermodynamic Relations

The entropy density, s, particle number density, n, energy density, ϵ, and the pressure, P in Tsallis thermodynamics are given by [18],

$$s = -g\int \frac{d^3p}{(2\pi)^3}\left[f^q \ln_q f - f\right], \tag{2}$$

$$n = g\int \frac{d^3p}{(2\pi)^3} f^q, \tag{3}$$

$$\epsilon = g\int \frac{d^3p}{(2\pi)^3}\, E\, f^q, \tag{4}$$

$$P = g\int \frac{d^3p}{(2\pi)^3}\frac{p^2}{3E}\, f^q. \tag{5}$$

where g is the degeneracy factor.

The function appearing in Eq. (2) is often referred to as q-logarithm, and is defined by [19]

$$\ln_q(x) \equiv \frac{x^{1-q}-1}{1-q}$$

The first and second laws of thermodynamics lead to the following two differential relations:

$$d\epsilon = T\, ds + \mu\, dn, \tag{6}$$

$$dP = s\, dT + n\, d\mu. \tag{7}$$

where, $s = S/V$ and $n = N/V$ are the entropy and particle number densities, respectively (V is the volume).

It is seen that if we use f^q instead of f to define the thermodynamic variables, the above equations satisfy the thermodynamic consistency conditions which require the following relations to be satisfied:

$$T = \left.\frac{\partial \epsilon}{\partial s}\right|_n, \tag{8}$$

$$\mu = \left.\frac{\partial \epsilon}{\partial n}\right|_s, \tag{9}$$

$$n = \left.\frac{\partial P}{\partial \mu}\right|_T, \tag{10}$$

$$s = \left.\frac{\partial P}{\partial T}\right|_\mu. \tag{11}$$

Eq. (8), in particular, shows that the variable T appearing in Eq. (1) can indeed be identified as a thermodynamic temperature. It is straightforward to show that these relations are indeed satisfied [18].

Analytical calculations of the Tsallis Thermodynamic variables can be approached in the following ways:

Method 1. Taylor's expansion method around $q-1$

Method 2. The massless limit

Method 3. Employing the Mellin-Barnes contour integration technique

Method 1, detailed in [20], results in very stringent conditions on the energy, temperature and q. In this section we will quote the results of the latter two methods, the details of which can be found in [21].

2.2. Tsallis Thermodynamic variables in the massless limit
Let us define a more general integral,

$$\mathcal{I}(\alpha,\beta) \equiv g \int \frac{\mathrm{d}^3\boldsymbol{p}}{(2\pi)^3} \frac{p^{\beta-2}}{\left[1+(q-1)\frac{p-\mu}{T}\right]^{\frac{\alpha}{(q-1)}}}, \tag{12}$$

where, α and β are nothing but handy variables. The former is set either to q or 1 at the end, in order to recover the thermodynamic variables. The above integral is built to converge, in three dimensions, upon some constraints on the various parameters. Those encompass the fact that the integrand shall only assume real values, and the usual infrared and ultraviolet convergence conditions apply. The conditions for the massless case turn out to be

$$1+\mathrm{Re}\,(\beta) > 0 \quad , \quad \mathrm{Re}\,(\alpha) > 0\,, \tag{13}$$

$$T > (q-1)\mu \quad , \quad (q-1) < \frac{\mathrm{Re}\,(\alpha)}{1+\mathrm{Re}\,(\beta)}, \tag{14}$$

for which we see that the first two are trivially satisfied, given the actual relevant set of integrals we wish to compute. The last two, on the other hand, are not trivial at all. Given the usual α and β values we are interested in, *e.g.*, some combinations of $\alpha = q,\ 1$ and $\beta = 2,\ 3$, we see that q must indeed be bounded at least by $q < 4/3$. This being said, the remaining constraint on the T and μ parameters is either not needed in the case $E_\mathrm{p} \geq \mu$, or fulfilled if $m < \mu$ with $q-1 < T/(\mu - E_\mathrm{p})$ and $E_\mathrm{p} < \mu$, only here for $m = 0$ in both cases. Notice that this last constraint brings an overall limit for the chemical potential, which is $\mu < 3T$. Thus, we see that this framework must be kept away from asymptotically dense systems. Having said so, we quote the results for different Tsallis Thermodynamic variables in the massless limit.

2.2.1. Number density In the massless limit, the number density can then be obtained from Eq. (12) by setting $\alpha = q$ and $\beta = 2$. Number density in the massless limit can, thus, be written as:

$$n \;=\; \frac{g\,T^3}{2\pi^2}\,\frac{\left(1-(q-1)\frac{\mu}{T}\right)^{\frac{2q-3}{q-1}}}{(2-q)\,\left(\frac{3}{2}-q\right)}. \tag{15}$$

From the above expression, wee see that the number density is divergent at $q = 3/2,\ 2$ which in the light of the previous discussion, implies that the condition of convergence for the corresponding integral is $0 \leq q < 3/2$.

2.2.2. Energy density Similarly, the massless energy density can be obtained from Eq. (12) by setting $\alpha = q$ and $\beta = 3$. Using the same tricks as above, we then arrive at

$$\epsilon = \frac{g\,T^4}{2\pi^2}\,\frac{\left(1-(q-1)\frac{\mu}{T}\right)^{\frac{3q-4}{q-1}}}{(2-q)\,(\frac{3}{2}-q)\,(\frac{4}{3}-q)}. \tag{16}$$

This time, from the above expression, wee see that the energy density diverges at $q = 4/3,\ 3/2,\ 2,$ which again shows that the condition of convergence for the corresponding integral is $0 \le q < 4/3$, precisely the physically relevant range that we just mentioned.

2.2.3. Pressure In the massless limit, the pressure can be obtained from Eq. (12) by setting $\alpha = q$ and $\beta = 3$, and dividing by 3. Using the same methods as above, this gives us

$$P = \frac{g\,T^4}{6\pi^2}\,\frac{\left(1-(q-1)\frac{\mu}{T}\right)^{\frac{3q-4}{q-1}}}{(2-q)\,(\frac{3}{2}-q)\,(\frac{4}{3}-q)}, \tag{17}$$

and we see that for a system of massless free classical particles following the Tsallis statistics, we indeed have $P = \epsilon/3$. In this case also, divergences arise at $q = 4/3,\ 3/2,\ 2,$ and the range of convergence is the physically relevant one $1 \le q < 4/3$.

2.3. Tsallis Thermodynamic variables for systems with massive particles

To compute the Tsallis thermodynamic pressure for a system of massive particles we take the help of the Mellin-Barnes contour integral representation ([22], [23] and [24]) which has been very helpful in Quantum Field Theory in describing the massive propagators in terms of those of the massless ones. Assuming $m \ge \mu$, the Tsallis pressure in the region $(q-1) > T/(m+\mu)$ is given by (for details, see [21]):

$$\begin{aligned} P_{\mathrm{U}} = \frac{g\,m^4}{16\pi^{\frac{3}{2}}}\left(\frac{T}{(q-1)\,m}\right)^{\frac{q}{q-1}} & \left[\frac{\Gamma\left(\frac{4-3q}{2(q-1)}\right)}{\Gamma\left(\frac{2q-1}{2(q-1)}\right)} \times {}_2F_1\left(\frac{q}{2(q-1)},\frac{4-3q}{2(q-1)},\frac{1}{2};\left(\frac{(q-1)\,\mu-T}{(q-1)\,m}\right)^2\right)\right. \\ & \left. +2\left(\frac{(q-1)\mu-T}{(q-1)\,m}\right)\times\frac{\Gamma\left(\frac{3-2q}{2(q-1)}\right)}{\Gamma\left(\frac{q}{2(q-1)}\right)} \times {}_2F_1\left(\frac{2q-1}{2(q-1)},\frac{3-2q}{2(q-1)},\frac{3}{2};\left(\frac{(q-1)\,\mu-T}{(q-1)\,m}\right)^2\right)\right] \end{aligned} \tag{18}$$

Pressure in the region $(q-1) < T/(m+\mu)$ can be obtained (see [21]) by analytically continuing the hypergeometric functions using [25]. Other thermodynamic variables for the system of massive particles can be obtained using the Thermodynamic relations and the chain rules for the partial derivative of the hypergeometric functions.

3. Description of multi-particle production using the Tsallis distribution

In this part of the work, we represent the initial distribution of the energetic particles with the help of the Tsallis power law distribution parameterized by the Tsallis q parameter and the Tsallis temperature T , remembering the fact that their genesis is due to very hard scatterings. We plug the initial distribution (f_{in}) in the Boltzmann Transport Equation (BTE) and solve it with the help of the Relaxation Time Approximation (RTA) of the collision term to find out the final distribution (f_{fin}). The ratio expressible in terms of q and T can be computed and compared with the experimentally observed values of R_{AA} which is theoretically represented as:

$$R_{\mathrm{AA}} = \frac{f_{fin}}{f_{in}} \tag{19}$$

and can be compared with experimentally measured R_{AA} given by:

$$R_{AA}(p_T) = \frac{(1/N_{AA}^{evt})d^2N_{AA}/dydp_T}{(\langle N_{coll}\rangle/\sigma_{NN}^{inel}) \times d^2\sigma_{pp}/dydp_T}, \tag{20}$$

where $d^2N_{AA}/dydp_T$ is the yield in A+A collisions, $\langle N_{coll}\rangle$ is the number of binary nucleon-nucleon collisions averaged over the impact parameter range of the corresponding centrality bin calculated by Glauber Monte-Carlo simulation [26]. σ_{NN}^{inel} is the inelastic cross section and $d^2\sigma_{pp}/dydp_T$ is the differential cross section for inelastic $p+p$ collisions. N_{AA}^{evt} is the number of events in A+A collisions.

The evolution of the particle distribution owing to its interaction with the medium particles can be studied through BTE,

$$\frac{df(\vec{x},\vec{p},t)}{dt} = \frac{\partial f}{\partial t} + \vec{v}.\vec{\nabla}_x f + \vec{F}.\vec{\nabla}_p f = C[f], \tag{21}$$

where $f(\vec{x},\vec{p},t)$ is the distribution of particles which depends on position, momentum and time. $\vec{v}$ is the velocity and $\vec{F}$ is the external force. $\vec{\nabla}_x$ and $\vec{\nabla}_p$ are the partial derivatives with respect to position and momentum, respectively. $C[f]$ is the collision term which encodes the interaction of the probe particles with the medium. Assuming homogeneity of the system ($\vec{\nabla}_x f = 0$) and absence of external force ($\vec{F} = 0$), the second and third terms of the above equation become zero and Eq. (21) becomes,

$$\frac{df(\vec{x},\vec{p},t)}{dt} = \frac{\partial f}{\partial t} = C[f] \tag{22}$$

In the relaxation time approximation [27], the collision term can be expressed as,

$$C[f] = -\frac{f - f_{eq}}{\tau} \tag{23}$$

where f_{eq} is Boltzmann local equilibrium distribution characterized by a temperature T_{eq}. τ is the relaxation time, the time taken by a non-equilibrium system to reach equilibrium. With the ansatz in Eq. (23), Eq. (22) becomes,

$$\frac{\partial f}{\partial t} = -\frac{f - f_{eq}}{\tau} \tag{24}$$

Solving the above equation in view of the initial conditions i.e. at $t = 0, f = f_{in}$ and at $t = t_f, f = f_{fin}$ leads to,

$$f_{fin} = f_{eq} + (f_{in} - f_{eq})e^{-\frac{t_f}{\tau}}, \tag{25}$$

where t_f is the freeze-out time. Using Eq. (25), the nuclear modification factor can be expressed as,

$$R_{AA} = \frac{f_{fin}}{f_{in}} = \frac{f_{eq}}{f_{in}} + \left(1 - \frac{f_{eq}}{f_{in}}\right) e^{\frac{-t_f}{\tau}} \tag{26}$$

IOP Conf. Series: Journal of Physics: Conf. Series **878** (2017) 012016 doi:10.1088/1742-6596/878/1/012016

Eq.(26) is the derived nuclear modification factor after incorporating relaxation time approximation [27], which is the basis of our analysis. It involves the (power law-like) initial distribution and the equilibrium distribution. In this analysis, the initial distribution is parameterized using the thermodynamically consistent Tsallis distribution [18] at $\mu = 0$.

$$f_{in} = \frac{gV}{(2\pi)^2}\, p_T\, m_T \left[1 + (q-1)\frac{m_T}{T}\right]^{-\frac{q}{q-1}}, \tag{27}$$

where $m_T = \sqrt{p_T^2 + m^2}$, for a particle with mass m, g is the degeneracy and V is the volume of the system. The nuclear modification factor can be expressed as,

$$R_{AA} = \frac{e^{-\frac{m_T}{T_{eq}}}}{(1+(q-1)\frac{m_T}{T})^{-\frac{q}{q-1}}} + \left[1 - \frac{e^{-\frac{m_T}{T_{eq}}}}{(1+(q-1)\frac{m_T}{T})^{-\frac{q}{q-1}}}\right] e^{-\frac{t_f}{\tau}} \tag{28}$$

This will be used as the fitting function to the experimental results in the next section.

3.1. Results and Discussion

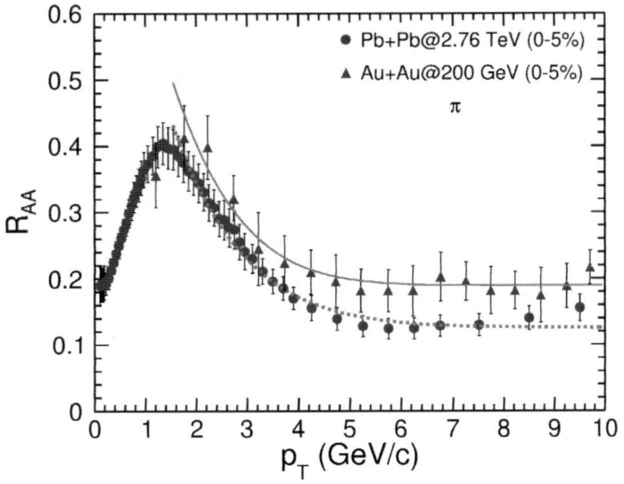

Figure 1. Fitting of experimental data for nuclear modification factor with our proposed model for π^0[29] (blue triangles) in Au+Au collisions at $\sqrt{s_{NN}}$= 200 GeV, and $\pi^+ + \pi^-$[30] in Pb+Pb collisions at $\sqrt{s_{NN}}$= 2.76 TeV (blue dots). The solid red line shows the fitting for blue triangles and the dotted red line shows the fitting for blue dots.

We now proceed to the more detailed analysis of the experimental data with the model proposed above. Keeping all the parameters free, we fit the spectra for different particles in different centralities for Pb+Pb and Au+Au collisions using the TMinuit class available in the ROOT library [28] to get a convergent solution. The convergent solution is obtained by χ^2 minimization technique. Here T, q and t_f/τ are the fitting parameters for the experimental data. The equilibrium temperature T_{eq} is fixed to 160 MeV throughout the analysis.

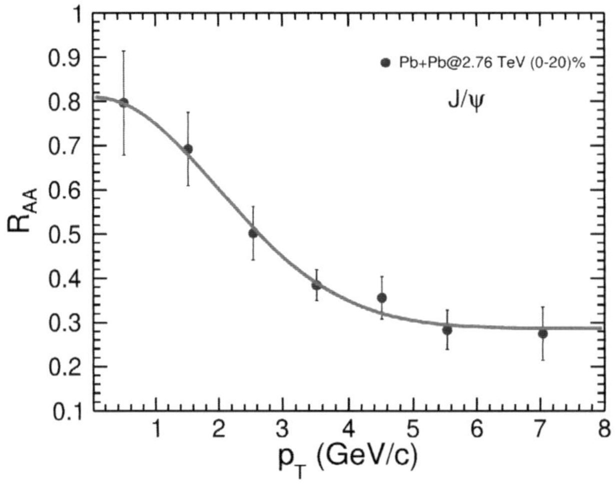

Figure 2. Fitting of experimental data for nuclear modification factor with our proposed model for J/ψ [31] in Pb+Pb collisions at $\sqrt{s_{NN}}$= 2.76 TeV (blue dots) with centrality (0-90)%. The solid red line shows the fitting to the experimental data.

Fig. (1) shows the fitting of experimental data using Eq. (28) for the π^0 meson in the most central Au+Au collisions at $\sqrt{s_{NN}}$= 200 GeV. The derived expression for R_{AA} fits the data in intermediate to high-p_T range. Also, in Fig. (1) we show the fitting of experimental data for $\pi^+ + \pi^-$ in most central Pb+Pb collisions at $\sqrt{s_{NN}}$= 2.76 TeV. The model considered here, fits the data from intermediate to high-p_T range with the parameter values: $q = 1.00151 \pm 0.00149$, $T = 0.17854 \pm 0.00143$, $t_f/\tau = 2.07313 \pm 0.061906$.

In our analysis, it is observed that the fitting of low p_T for light flavour particles fails due to the reason that it involves different physical processes such as regeneration, coalescence, shadowing etc., which are out of the scope of the present formalism. But in heavy flavour particles, no such processes are involved at the discussed energies. Thus, our proposed model explains successfully the heavy flavour R_{AA} data (J/ψ for example) with the fitting parameters $q = 1.01252 \pm 0.00998$, $T = 0.14676 \pm 0.01408$, $t_f/\tau = 1.06248 \pm 0.157049$. Analysis for the other particle species can be found in [16].

4. Summary and Conclusion

In the present paper, we have discussed about the two different applications of the Tsallis distribution in the field of high energy collisions. In the first part, we have discussed about the analytical calculation of the Tsallis Thermodynamic variables in the massless limit, and in the massive limit. In the latter case, we have used the Mellin-Barnes contour integral representation which is used in Quantum Field Theory to represent the massive propagators in terms of the massless ones. The calculations may be particularly important in reducing the cumbersome momentum integrals, so far performed numerically, into simpler analytic expressions.

In the second part of the paper, we have discussed about the application of the Tsallis Statistics in the multi-particle production and their nuclear modification factor. We have seen that the J/ψ data is well explained once we parameterize the initial distributions using the Tsallis distribution. The same inference can be drawn for the other heavy flavours as well.

IOP Conf. Series: Journal of Physics: Conf. Series **878** (2017) 012016 doi:10.1088/1742-6596/878/1/012016

Acknowledgements

T. B. acknowledges the University of Cape Town Research Committee (South Africa) for support. S. M. would like to acknowledge the financial support from the Claude Leon Foundation (South Africa).

References

[1] Tsallis C 1988 *J. Statist. Phys.* **52** 479
[2] Abelev B I *et al.* (STAR collaboration) 2007 *Phys. Rev. C* **75** 064901
[3] Adare A *et al.* (PHENIX collaboration) 2011 *Phys. Rev. D* **83** 052004
[4] Aamodt K *et al.* (ALICE collaboration) 2010 *Phys. Lett. B* **693** 53
[5] Aamodt K *et al.* (ALICE collaboration) 2011 *Eur. Phys. J C* **71** 1655
[6] Abelev B *et al.* (ALICE collaboration) 2012, Phys. Rev. Letts. **109** 252301
[7] Khachatryan V *et al.* (CMS collaboration) 2010 *J. of High Eng. Phys.* **02** 041
[8] Aad G *et al.* (ATLAS collaboration) 2011 *New J. Phys.* **13** 053033
[9] Bediaga I, Curado E M F and Miranda J M de 2000 *Physica A* **286** 156.
[10] Wilk G and Włodarczyk Z 2015 *Acta Phys. Polon. B* **46** 1103
[11] Ürmössy K, Barnaföldi G G and Biró T S 2011 *Phys. Lett. B* **701** 111
[12] Ürmössy K, Barnaföldi G G, and Biró T S 2012 *Phys. Lett. B* **718** 125.
[13] Khandai P K, Sett P, Shukla P and Singh V 2013 *Int. Jour. Mod. Phys. A* **28** 1350066
[14] Li B -C, Wang Y -Z and Liu F -H 2013 *Phys. Lett. B* **725** 352
[15] Marques L, Cleymans J and Deppman A 2015 *Phys. Rev. D* **91** 054025
[16] Tripathy S, Bhattacharyya T, Garg P, Kumar P, Sahoo R, and Cleymans J 2016 *Eur. Phys. J A* **52** 289
[17] Cleymans J and Worku D 2012 2012 *J. Phys. G* **39** 025006
[18] Cleymans J and Worku D 2012 *Eur. Phys. Jour. A* **48** 160
[19] Tsallis C 2009 *Introduction to Nonextensive Statistical Mechanics* (New York: Springer Science+Business Media)
[20] Bhattacharyya T, Cleymans J, Khuntia A, Pareek P and Sahoo R 2016 *Eur. Phys. J* A **52** 30
[21] Bhattacharyya T, Cleymans J and Mogliacci S 2016 *Phys. Rev. D* **94** 094026
[22] Smirnov V A 2004 *Evaluating Feynman Integrals* (Berlin: Springer-Verlag) .
[23] Boos E E and Davydychev A I 1991 Theor. Math. Phys. **89** 1052
[24] Davydychev A I and Tausk J B 1993 Nucl. Phys. B **397** 123
[25] Bateman H 1953 *Higher Transcendental Functions* (New York: McGraw-Hill Book Company) (See Section 2.10 formula (4), and the conditions therein)
[26] Glauber R J and Matthiae G 1970 Nucl. Phys. B **21** 135.
[27] Balescu R 1975 *Equilibrium and Non-Equilibrium Statistical Mechanics* (USA: John Wiley and Sons).
[28] CERN ROOT V.5.34/32 (June 23, 2015) Package: http://root.cern.ch.
[29] Adare A *et al.* (PHENIX Collaboration) 2008 Phys. Rev. Lett. **101** 232301
[30] Abelev B B *et al.* (ALICE Collaboration) 2014 Phys. Lett. B **736** 196
[31] Abelev B B *et al.* (ALICE Collaboration) 2014 Phys. Lett. B **734** 314

The Elastic $q\bar{q}$ Cross Section in the Nambu–Jona-Lasinio Model

Rafael Chapelle[1], **Joerg Aichelin**[1,2], **Juan M. Torres-Rincon**[2]

[1] Subatech, UMR 6457, IN2P3/CNRS, Université de Nantes, École de Mines de Nantes, 4 rue Alfred Kastler 44307, Nantes, France

[2] Frankfurt Institute for Advances Studies. Johann Wolfgang Goethe University, Ruth-Moufang-Str. 1, 60438, Frankfurt am Main, Germany

Résumé. We discuss the quark masses and the elastic $q\bar{q}$ cross sections at finite chemical potential in the Nambu–Jona-Lasinio model. We comment the generic features of the cross sections as functions of the chemical potential, temperature and collision energy. Finally, we discuss their relevance in the construction of a relativistic transport model for heavy-ion collisions based on this effective Lagrangian.

1. Introduction

The Nambu–Jona-Lasinio (NJL) model has been extensively used in the context of strong interactions due to its ability to account for several key phenomena of the Quantum Chromodynamics (QCD), like the spontaneous symmetry breaking (together with the generation of Goldstone bosons) and its restoration at high temperatures and densities. This model works as an effective realization of QCD at low energies, and allows us performing studies in a much simpler way in the regime where QCD is too difficult to solve, or computationally expensive like in lattice-QCD calculations. This fact we use here to derive elastic $q\bar{q}$ cross sections in the non-perturbative region. These cross sections are needed if one wants to simulate the expansion of a plasma created in ultrarelativistic heavy-ion collisions.

We consider the Lagrangian of the NJL model with (color neutral) pseudoscalar and scalar interactions (neglecting the vector and axial-vector vertices for simplicity) [1],

$$\begin{aligned}\mathscr{L}_{NJL} &= \sum_i \bar{\psi}_i (i\partial\!\!\!/ - m_{0i} + \mu_i \gamma_0)\psi_i \\ &+ G\sum_a \sum_{ijkl} \left[(\bar{\psi}_i\, i\gamma_5 \tau^a_{ij} \psi_j)\,(\bar{\psi}_k\, i\gamma_5 \tau^a_{kl}\psi_l) + (\bar{\psi}_i \tau^a_{ij}\psi_j)\,(\bar{\psi}_k \tau^a_{kl}\psi_l)\right] \\ &- K \det_{ij}\left[\bar{\psi}_i\,(\mathbb{1}-\gamma_5)\psi_j\right] - K\det_{ij}\left[\bar{\psi}_i\,(\mathbb{1}+\gamma_5)\psi_j\right] \end{aligned} \tag{1}$$

where the flavor indices $i, j, k, l = 1,2,3$ and τ^a $(a = 1,...,8)$ being the $N_f = 3$ flavor generators with normalization

$$\mathrm{tr}_f\,(\tau^a \tau^b) = 2\delta^{ab}\,, \tag{2}$$

with tr_f denoting the trace in flavor space. In the Lagrangian (1) the bare quark masses are represented by m_{0i} and their chemical potential by μ_i. The coupling constant for the scalar and pseudoscalar interaction G is taken as a free parameter (fixed e.g. by the pion mass in vacuum). The third term

IOP Conf. Series: Journal of Physics: Conf. Series **878** (2017) 012017 doi:10.1088/1742-6596/878/1/012017

of Eq. (1) is the so-called 't Hooft Lagrangian, which mimics the effect of the axial $U(1)$ anomaly, accounting for the physical splitting between the η and the η' meson masses. K is an unknown coupling constant (fixed by the value of $m_{\eta'} - m_\eta$) and $\mathbb{I}$ is the identity matrix in Dirac space. As the NJL model is non-renormalizable, it also requires an ultraviolet regulator, which we introduce in the form of a cutoff Λ.

This Lagrangian has been widely used to study strongly interacting systems in the vacuum and at finite temperature. It contains 5 parameters. For all details we refer to the reviews [2–6].

2. Masses of u, d and s quarks as functions of temperature and chemical potential

In the SU(3) version of the NJL Lagrangian the mass of a quark of flavor i is given by

$$m_i = m_{0i} - 4G\langle\overline{q}_i q_i\rangle + 2K\langle\overline{q}_j q_j\rangle\langle\overline{q}_k q_k\rangle\,, \tag{3}$$

where m_{0i} is the bare mass, $i \neq j \neq k$ are the u, d, s quarks, $\langle\overline{q}_i q_i\rangle$ the scalar condensate which, with the thermal distribution function

$$f_i^{\pm}(E_i, T, \mu_i) = \frac{1}{1+\exp\left(\frac{E_i \mp \mu_i}{T}\right)}\,, \tag{4}$$

is given by

$$\langle\bar{q}_i q_i\rangle = -2N_c \int_0^{\Lambda} \frac{m_i}{\sqrt{p^2+m_i^2}}(1 - f_i^+ - f_i^-)\frac{d^3p}{(2\pi)^3}\,. \tag{5}$$

In the present calculation we employ the parameters of Table 1, determined by vacuum meson masses

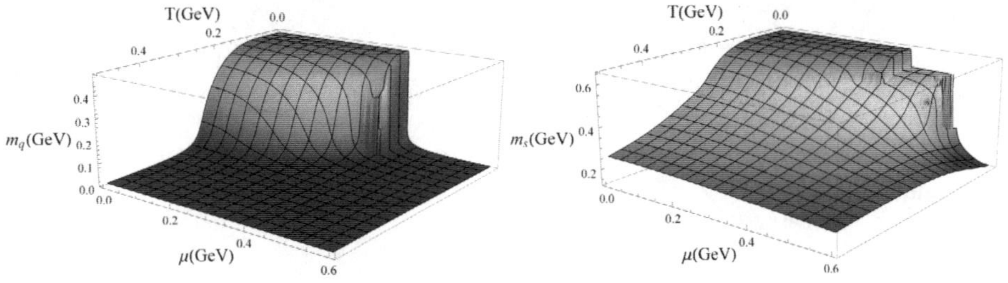

FIGURE 1: The mass of u, d (left) and the s (right) quark as functions of the temperature and the chemical potential.

and decay constants.

TABLE 1: Parameters used in the calculation and the critical chemical potential μ_{crit}, where the chiral restoration occurs as a first order phase transition.

Λ	G	K	m_{0u}	m_{0s}	μ_{crit}
569 MeV	$2.3/\Lambda^2$	$11/\Lambda^5$	5.5 MeV	134 MeV	338 MeV

For $\mu = 0$ (μ denotes the light chemical potential, $\mu_u = \mu_d$) we observe for all quarks a smooth transition of the mass as a function of the temperature. This behavior continues for u and d quarks along the transition line between the quark plasma and hadrons until μ_{crit} is reached. Then the crossover becomes a first order phase transition and we see a sudden change of the masses. For the s quark the situation is more complicated. We see a first sudden but continuous change at the phase transition for the u and d quarks because both are related (Eq. 3) and then a second first order transition when the s quark mass become discontinuous.

3. The $q\bar{q}$ cross sections

The $q\bar{q}$ cross section can be obtained in the standard way from the Lagrangian [7]. The Feynman diagrams for the two channels lead to the following matrix elements,

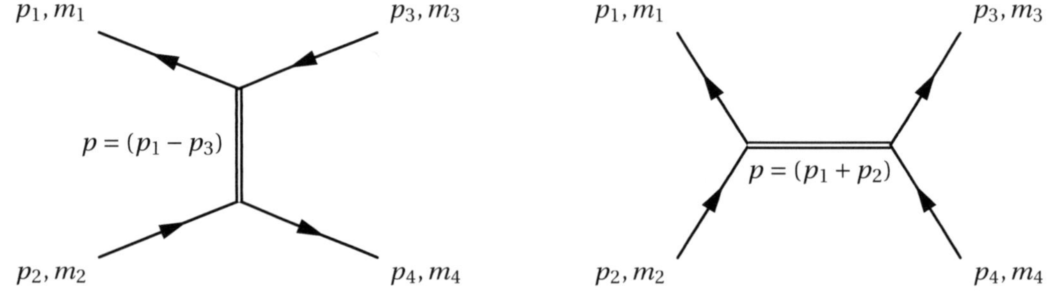

FIGURE 2: $t-$ and $s-$channel Feynman diagrams for elastic $q\bar{q}$ collisions.

$$\begin{aligned} -i\mathcal{M}_t = &\delta_{c1,c3}\delta_{c2,c4}\bar{u}(p_3)Tv(p_1)[i\mathcal{D}_t^S(p_1-p_3)]\bar{v}(p_2)Tu(p_4) \\ &+\delta_{c1,c3}\delta_{c2,c4}\bar{u}(p_3)(i\gamma_5 T)v(p_1)[i\mathcal{D}_t^P(p_1-p_3)]\bar{v}(p_2)(i\gamma_5 T)u(p_4)\,, \end{aligned} \tag{6}$$

$$\begin{aligned} -i\mathcal{M}_s = &\delta_{c1,c2}\delta_{c3,c4}\bar{v}(p_2)Tu(p_1)[i\mathcal{D}_s^S(p_1+p_2)]\bar{u}(p_3)Tv(p_4) \\ &+\delta_{c1,c2}\delta_{c3,c4}\bar{v}(p_2)(i\gamma_5 T)u(p_1)[i\mathcal{D}_s^P(p_1-p_4)]\bar{u}(p_3)(i\gamma_5 T)v(p_4)\,. \end{aligned} \tag{7}$$

u, $\bar{u}$ and v, $\bar{v}$ are the Dirac spinors. $\delta_{c3,c4}$ impose the color conservation. The symbolic expression $T\mathcal{D}_i^{S/P}T$ stands for the exchange of all possible scalar and pseudoscalar mesons, respectively. We limit us in this calculation to the exchange of color neutral mesons and the s-channel is taken in first order in N_c. The square of the matrix element, averaged over spin and color in the entrance and summed over in the exit channel,

$$\frac{1}{4N_c^2}\sum_{s,c}|\mathcal{M}_{total}|^2 \tag{8}$$

gives the cross section

$$\frac{d\sigma}{dt} = \frac{1}{16\pi[s-(m_1+m_2)^2][s-(m_1-m_2)^2]}\frac{1}{4N_c^2}\sum_{s,c}|\mathcal{M}_{total}|^2. \tag{9}$$

In a thermal heat bath some of the exit states are partially blocked. Introducing the Pauli blocking factors we obtain

$$\sigma = \int \frac{d\sigma}{dt}(1-f^{\pm}(E_3,T,\mu_3))(1-f^{\pm}(E_4,T,\mu_4))dt. \tag{10}$$

The integration limits for t are $-(s-\sum_i m_i^2)$ and 0. In order to calculate the transition amplitudes one has to know the mesons which can be exchanged and the corresponding flavor factors. The exchanged mesons are either scalars σ_π, σ_K, σ and σ' or pseudoscalars π ,K, η and η'. They are displayed in Table 2. The Gell-Mann matrices for the different flavors are

$$\begin{cases} \pi^0, \lambda_3 \\ \pi^{\pm}, \frac{1}{\sqrt{2}}(\lambda_1 \pm i\lambda_2) \\ K^0, \bar{K}^0, \frac{1}{\sqrt{2}}(\lambda_6 \pm i\lambda_7) \\ K^{\pm}, \frac{1}{\sqrt{2}}(\lambda_4 \pm i\lambda_5). \end{cases} \tag{11}$$

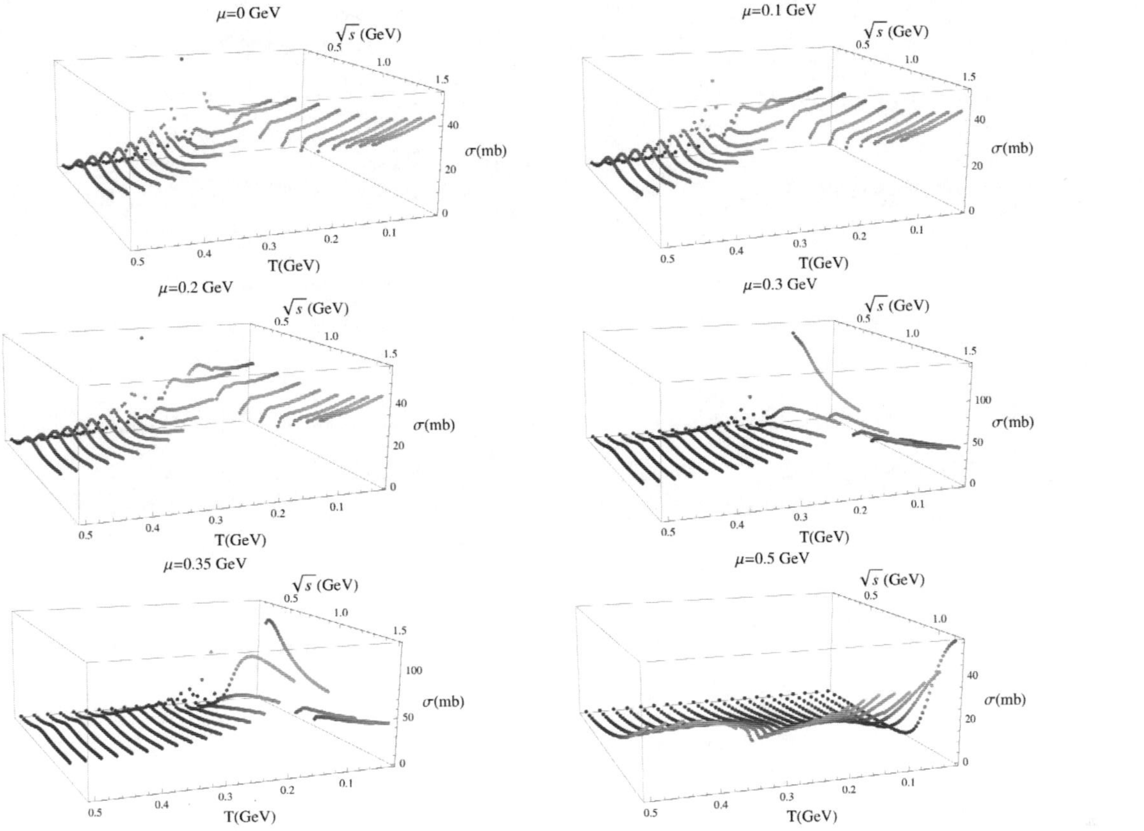

FIGURE 3: $u\bar{u} \to u\bar{u}$ cross section as a function of the temperature and of the center of mass energy for different chemical potentials μ.

TABLE 2: The mesons which can be exchanged in the different channels for $q\bar{q} \to q\bar{q}$

Process	Mesons exchanged in the s-channel	Mesons exchanged in the t-channel
$u\bar{u} \to u\bar{u}$	$\pi,\eta,\eta',\sigma_\pi,\sigma,\sigma'$	$\pi,\eta,\eta',\sigma_\pi,\sigma,\sigma'$
$u\bar{u} \to u\bar{d}$	$\pi,\eta,\eta',\sigma_\pi,\sigma,\sigma'$	π,σ_π
$u\bar{d} \to u\bar{d}$	π,σ_π	$\pi,\eta,\eta',\sigma_\pi,\sigma,\sigma'$
$u\bar{s} \to u\bar{s}$	K,σ_K	$\eta,\eta',\sigma,\sigma'$
$u\bar{u} \to s\bar{s}$	$\eta,\eta',\sigma,\sigma'$	K,σ_K
$s\bar{s} \to u\bar{u}$	$\eta,\eta',\sigma,\sigma'$	K,σ_K
$s\bar{s} \to s\bar{s}$	$\eta,\eta',\sigma,\sigma'$	$\eta,\eta',\sigma,\sigma'$

To calculate which mesons can be exchanged one has to calculate the corresponding flavor matrices. As an example we calculate $ud \to ud$ with the exchange of a π^0 in the t-channel. Here one finds

$$\bar{u}\lambda_3 u \times \bar{d}\lambda_3 d = \begin{pmatrix}1 & 0 & 0\end{pmatrix}\begin{pmatrix}1 & 0 & 0\\ 0 & -1 & 0\\ 0 & 0 & 0\end{pmatrix}\begin{pmatrix}1\\0\\0\end{pmatrix} \times \begin{pmatrix}0 & 1 & 0\end{pmatrix}\begin{pmatrix}1 & 0 & 0\\ 0 & -1 & 0\\ 0 & 0 & 0\end{pmatrix}\begin{pmatrix}0\\1\\0\end{pmatrix} = -1. \qquad (12)$$

Hence a π_0 can be exchanged, whereas for a π^+ we find zero. If we apply this for all possible combinations we find the mesons of Table 2.

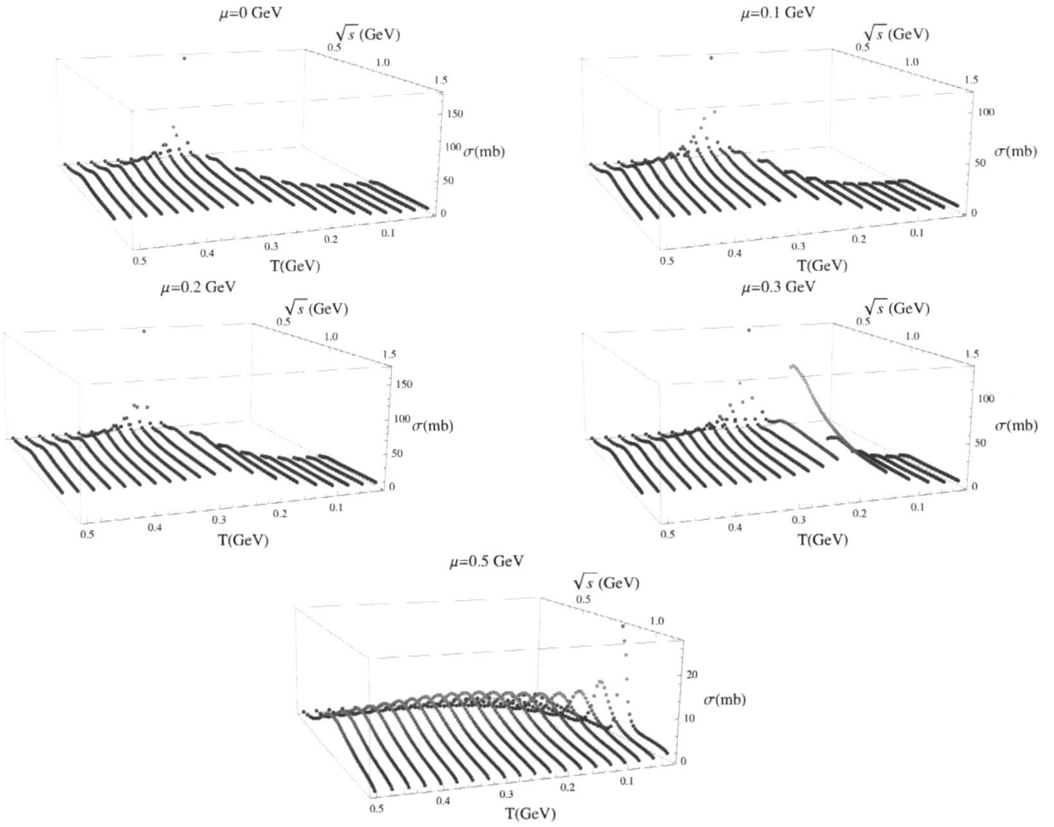

FIGURE 4: Cross section for $u\bar{d} \to u\bar{d}$ as a function of the temperature T and the center of mass energy $\sqrt{s}$ for different chemical potentials μ.

4. Results

The $q\bar{q}$ cross section has a quite different behavior as compared to the qq cross section due to the s-channel contribution. The qq cross section is always ≤20 mb. In the s channel it happens that the incoming $q\bar{q}$ pair is in resonance with the meson which it produces. In this case the denominator of the meson propagator becomes small and hence the matrix element large. This resonance between the incoming $q\bar{q}$ pair appears close to the temperature where the mass of the meson is close to the sum of the masses of the two valence quarks. Therefore the peak of the cross section moves to lower temperatures when the chemical potential increases.

4.1. $u\bar{d} \to u\bar{d}$ and $u\bar{u} \to u\bar{u}$

Close to the critical chemical potential, μ_{crit}, (see Table 1) the cross section becomes maximal being around 100 mb for the $u\bar{u}$ channel at the grid points which we calculated. At larger chemical potentials the cross sections are smaller again and arrive at a maximal value of 25 mb at $\mu = 0.5$ MeV (see Fig.3). For even higher chemical potentials the cross section is reduced to a couple of millibarn because then a transition between the plasma and the hadronic world does not exist anymore in the NJL approach. There we are for all temperatures in the deconfined phase.

If one compares the $u\bar{u}$ and $u\bar{d}$ elastic cross sections, Figs. 3 and 4, one sees that the latter is about two to four times larger. This is a consequence of the flavor factor which doubles the s-channel contribution and gives a relative minus sign to the t-channel contribution. The form of the cross sections are rather similar, only at large values of $\sqrt{s}$ the fact that the η and η' mesons are not allowed in the s channel of $u\bar{d}$ makes the $u\bar{u}$ cross section relatively larger.

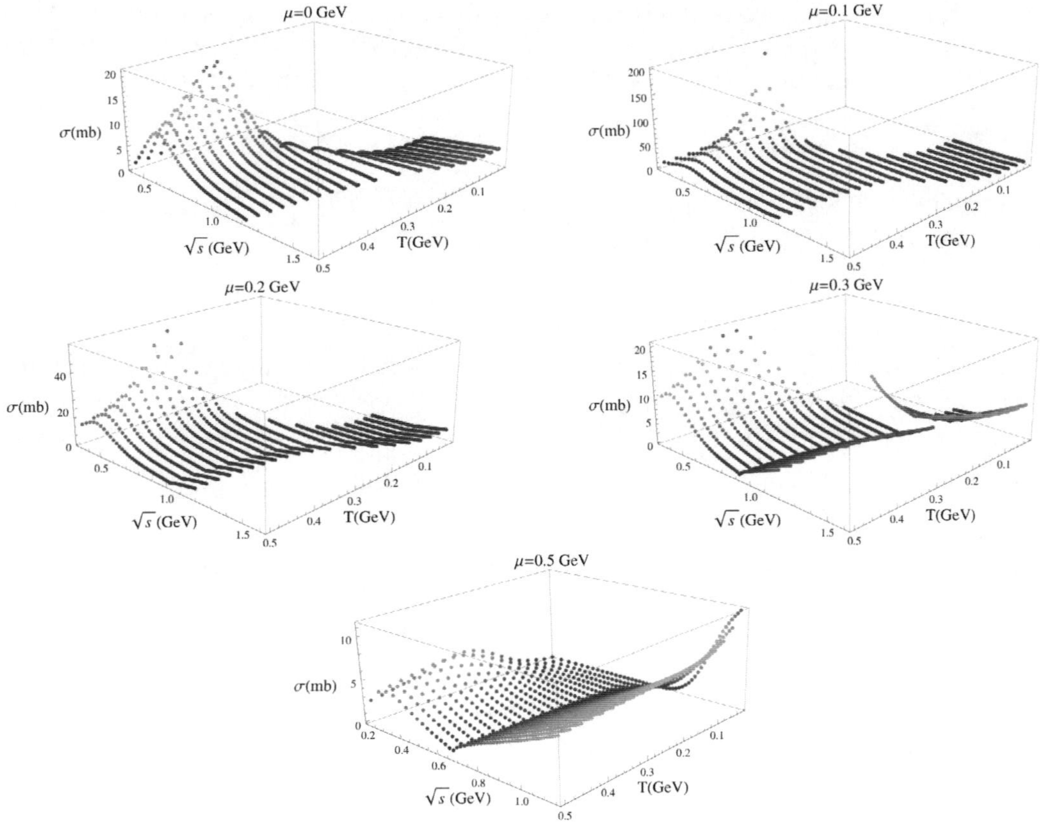

FIGURE 5: Cross section for $u\bar{s} \to u\bar{s}$ as a function of the temperature T and the center of mass energy $\sqrt{s}$ for different chemical potentials μ.

4.2. $u\bar{s} \to u\bar{s}$

The $u\bar{s}$ cross section is displayed in Fig. 5. As compared to the other cross sections the maximum of the cross section is obtained at a higher temperature. This is a consequence that now in the s channel strange mesons are exchanged which have a mass well above the mass of the quarks in the entrance channel. Therefore more energetic particles are needed to be resonant with the exchanged meson.

5. Conclusions

This calculation of the cross sections at finite μ is the first step towards a transport theory based on the NJL Lagrangian, for finite chemical potentials, extending the work of Ref. [8, 9]. We see that also at a finite chemical potential close to the transition between the partonic and hadronic world the elastic cross sections become very large and therefore are very effective to equilibrate the system when it has expanded to the density where the phase transition takes place. Before the cross sections are too low to keep a local equilibrium. Consequently, for finite chemical potentials we expect the same generic behavior which we have observed at zero chemical potential, even if the numerical values of the cross sections differ in details. Therefore the NJL transport approach may be extended to a finite chemical potential to study how the change of the structure of the phase transition is seen in the observables.

IOP Conf. Series: Journal of Physics: Conf. Series **878** (2017) 012017 doi:10.1088/1742-6596/878/1/012017

Acknowledgement

This work has been funded by the program TOGETHER from Région Pays de la Loire and EU Integrated Infrastructure Initiative HadronPhysics3 Project under Grant Agreement n. 283286. JMTR thanks funding from a Helmholtz Young Investigator Group VH-NG-822 from the Helmholtz Association and GSI, and the Project FPA2013-43425-P from Ministerio de Ciencia e Innovación (Spain).

References

[1] Torres-Rincon J M, Sintes B and Aichelin J 2015 *Phys. Rev.* **C91** 065206

[2] Vogl U and Weise W 1991 *Prog. Part. Nucl. Phys.* **27** 195

[3] Klevansky S P 1992 *Rev. Mod. Phys.* **64** 649

[4] Hatsuda T and Kunihiro T 1994 *Phys. Rept.* **247** 221

[5] Alkofer R and Reinhardt H 1995 Chiral quark dynamics Berlin, Germany : Springer (Lecture notes in physics)

[6] Buballa M 2005 *Phys. Rept.* **407** 205

[7] Rehberg P, Klevansky S P and Hufner J 1996 *Nucl. Phys.* A **608**, 356

[8] Marty R and Aichelin J 2013 *Phys. Rev.* C **87** 034912

[9] Marty R, Bratkovskaya E, Cassing W and Aichelin J 2015 *Phys. Rev.* **C92** 015201

IOP Conf. Series: Journal of Physics: Conf. Series **878** (2017) 012018 doi:10.1088/1742-6596/878/1/012018

Chiral symmetry restoration versus deconfinement in heavy-ion collisions at high baryon density

E. L. Bratkovskaya[1,2], A. Palmese[3], W. Cassing[3], E. Seifert[3], T. Steinert[3], and P. Moreau[1]

[1] Institute for Theoretical Physics, University of Frankfurt, Frankfurt, Germany
[2] GSI Helmholtzzentrum für Schwerionenforschung GmbH, Darmstadt, Germany
[3] Institute for Theoretical Physics, University of Giessen, Giessen, Germany

Abstract. The effect of the chiral symmetry restoration (CSR) on observables from heavy-ion collisions is studied in the energy range $\sqrt{s_{NN}}$=3–20 GeV within the Parton-Hadron-String Dynamics (PHSD) transport approach. The PHSD includes the deconfinement phase transition as well as essential aspects of CSR in the dense and hot hadronic medium, which are incorporated in the Schwinger mechanism for the hadronic particle production. We adopt different parametrizations of the nuclear equation of state from the non-linear $\sigma - \omega$ model, which enter in the computation of the quark scalar density for the CSR mechanism, in order to estimate the uncertainty in our calculations. For the pion-nucleon Σ-term we adopt $\Sigma_\pi \approx$ 45 MeV which corresponds to some 'world average'. Our systematic studies show that chiral symmetry restoration plays a crucial role in the description of heavy-ion collisions at $\sqrt{s_{NN}}$=3–20 GeV, realizing an increase of the hadronic particle production in the strangeness sector with respect to the non-strange one. We identify particle abundances and rapidity spectra to be suitable probes in order to extract information about CSR, while transverse mass spectra are less sensitive. Our results provide a microscopic explanation for the "horn" structure in the excitation function of the K^+/π^+ ratio: the CSR in the hadronic phase produces the steep increase of this particle ratio up to $\sqrt{s_{NN}} \approx 7$ GeV, while the drop at higher energies is associated to the appearance of a deconfined partonic medium.

1. Introduction

The main goal of heavy-ion collision (HIC) physics is the study of nuclear matter at high temperatures and/or high densities. According to Quantum-Chromo-Dynamics (QCD), at large temperatures and densities the hadrons cannot survive anymore as bound states and they dissolve forming the Quark-Gluon-Plasma (QGP). At Relativistic Heavy-Ion Collider (RHIC) energies for the first time the creation of a QGP, identified as an almost perfect fluid, has been proven experimentally. Actually, the properties of this deconfined state of matter are still debated as well as the phase boundaries to the hadronic phase. In order to shed some light on this issue, many heavy-ion experiments are performed at the Super-Proton Synchrotron (SPS), RHIC, the Large Hadron Collider (LHC) and will be performed at the future Facility for Antiproton and Ion Research (FAIR) as well as the Nuclotron-based Ion Collider fAcility (NICA). The crucial challenge is to identify in the final particle distributions those signatures, which allow to disentangle the QGP contribution that is impossible to observe directly or independently.

The strange particle production has always been suggested as one of the most sensitive observables that could spot out the creation of a QGP during the early stages of a HIC. The earliest suggested signature is the strangeness enhancement in A+A collisions with respect to elementary p+p collisions [1, 2]. Later on, Gazdzicki and Gorenstein [3] proposed that a sharp rise and drop in the excitation function of the K^+/π^+ ratio (so called "horn") should show up due to the appearance of a QGP phase at a center-of-mass energy $\sqrt{s_{NN}} \sim 7\,\mathrm{GeV}$ [2]. Several statistical models [4, 5, 6] have succeeded in reproducing the trend of the experimental observation of the K^+/π^+ ratio and other strange to non-strange particle ratios, but they can provide only a statistical description of the heavy-ion collision process. On the other hand there was no conclusive interpretation of the "horn" from dynamical approaches for HIC, like microscopic transport models [7, 8, 9]. Only recently, the Parton-Hadron-String Dynamics (PHSD), a transport approach describing HIC on the basis of partonic, hadronic and string degrees-of-freedom, obtained a striking improvement on this issue when including chiral symmetry restoration (CSR) in the string decay for hadronic particle production [10, 11].

Apart from deconfinement the chiral symmetry restoration addresses another aspect of the QCD phase diagram in the (T, μ_B)-plane as an additional transition between a phase with broken and a phase with restored chiral symmetry. As in case of the QCD deconfinement phase transition, the boundaries of the CSR phase transition line are not well known. Lattice QCD (lQCD) calculations show that at vanishing baryon chemical potential μ_B=0 the CSR takes place at roughly the same critical temperature and energy density as the deconfinement phase transition which is a crossover. At finite baryon chemical potential lQCD calculations cannot be performed due to the sign problem and one must rely on effective models (or extrapolations) in order to study the QCD phase transitions [12, 13, 14, 15, 16]. Different models support the idea that at finite chemical potential a partially restored phase is achieved before the deconfinement occurs [17, 18, 19]. In order to distinguish the two phases of such a transition, effective models use the scalar quark condensate $\langle \bar{q}q \rangle$ as an order parameter. As the baryon density and temperature increase, the scalar quark condensate $\langle \bar{q}q \rangle$ is expected to decrease from a non-vanishing value in the vacuum to $\langle \bar{q}q \rangle \approx 0$ which corresponds to CSR. Since $\langle \bar{q}q \rangle$ is not a measurable quantity, it is crucial to determine experimental observables which are sensitive to this feature. Since long the dilepton spectroscopy has been in the focus in this respect since in a chirally restored phase the spectral functions of the the ρ- and the a_1-meson should become identical. However, no clear evidence has been achieved so far [20]. On the other hand, the strangeness production at Alternating-Gradient Synchrotron (AGS) and lower SPS energies was suggested to be a signature of CSR [10].

In this contribution we will report on the main results from Refs. [10, 11] where a systematic study has been performed on effects of the CSR on final particle distributions within the PHSD approach.

2. Reminder of the PHSD transport approach

The Parton-Hadron-String Dynamics (PHSD) is a microscopic covariant dynamical approach for strongly interacting systems in and out-of equilibrium [21, 22, 23]. It is a transport approach which goes beyond the quasi-particle approximation, since it is based on Kadanoff-Baym equations for the Green's functions in phase-space representation in first-order gradient expansion [24, 25]. Including both a hadronic and a partonic phase as well as a transition between the effective degrees-of-freedom, PHSD is capable to describe the full time evolution of a relativistic heavy-ion collision. The theoretical description of the partonic degrees-of-freedom (quarks and gluons) is realized in line with the Dynamical-Quasi-Particle Model (DQPM) [25] which reproduces lQCD results in thermodynamical equilibrium and provides the properties of the partons, i.e. masses and widths in their spectral functions. In equilibrium the PHSD reproduces the partonic transport coefficients such as shear and bulk viscosities or the electric

[2] In this work we adopt natural units, hence $\hbar = c = 1$.

conductivity from lQCD calculations as well. For a review on PHSD we refer the reader to Ref. [26]. We further note that the pure hadronic phase in PHSD is equivalent to the Hadron-Strings Dynamics (HSD) model [27]. Accordingly, the comparison between PHSD and HSD calculations allows us to disentangle the role of the QGP phase in heavy-ion collisions.

The PHSD approach has been tested for different colliding systems (p+p, p+A, A+A) in a wide range of bombarding energy, from AGS to LHC energies, and has been able to describe a large number of experimental observables, such as charged particle spectra, collective flow coefficients v_n as well as electromagnetic probes such as photons and dileptons [26]. More recently, it has been also shown to provide a microscopic description of the maximum in the K^+/π^+ ratio in central nucleus-nucleus collisions [10, 11].

2.1. String fragmentation in PHSD

The string formation and decay represents the dominant particle production mechanism in nucleus-nucleus collisions for bombarding energies from 2 AGeV to 160 AGeV. In PHSD, the primary hard scatterings between nucleons are described by string formation and decay in the FRITIOF Lund model [28]. A string is an excited color-singlet state, which is composed of two string ends corresponding to the leading constituent quarks of the colliding hadrons and a color flux tube in between. As the string ends recede, virtual $q\bar{q}$ or $qq\bar{q}\bar{q}$ pairs are produced in the uniform color field, causing the breaking of the string. Finally, the string decays into mesons or baryon-antibaryon pairs with formation time $\tau_f \sim 0.8\,\mathrm{fm/c}$ (in the rest-frame of the string).
In the string decay, the flavor of the produced quarks is determined via the Schwinger formula [29], which defines the production probability of massive $s\bar{s}$ pairs with respect to light flavor production ($u\bar{u}, d\bar{d}$) pairs:

$$\frac{P(s\bar{s})}{P(u\bar{u})} = \frac{P(s\bar{s})}{P(d\bar{d})} = \gamma_s = \exp\Big(-\pi\frac{m_s^2 - m_{u,d}^2}{2\kappa}\Big), \tag{1}$$

with $\kappa \approx 0.176\,\mathrm{GeV}^2$ representing the string tension and $m_{u,d,s}$ denoting the constituent quark masses for strange and light quarks. For the constituent quark masses $m_u \approx 0.35\,\mathrm{GeV}$ and $m_s \approx 0.5\,\mathrm{GeV}$ in the vacuum, the production of strange quarks is suppressed by a factor of $\gamma_s \approx 0.3$ with respect to the light quarks, which is the default setting in the FRITIOF routines.

2.2. Modeling of the chiral symmetry restoration

In Refs. [10, 11] the PHSD has been extended to include CSR in the string decay in a hadronic environment of finite baryon and meson density. Here we recall the main aspect of this extension which is based on the Hellman-Feynman theorem for the scalar quark condensate [30]. Accordingly, a linear decrease of the scalar quark condensate $\langle\bar{q}q\rangle$ – which is nonvanishing in the vacuum due to a spontaneous breaking of chiral symmetry – is expected with baryon density ρ_B towards a chiral symmetric phase characterized by $\langle\bar{q}q\rangle \approx 0$ [31, 32]. This decrease of the scalar quark condensate is expected also to lead to a change of the hadron properties with density and temperature, i.e. in a chirally restored phase the vector and axial vector currents should become equal [33]; the latter implies that e.g. the ρ and a_1 spectral functions should become identical (as addressed above in the context of dilepton production). Since the scalar quark condensate $\langle\bar{q}q\rangle$ is not a direct observable, its manifestations should also be found indirectly in different hadronic abundances and spectra or particle ratios like K^+/π^+, $(\Lambda + \Sigma^0)/\pi^-$ etc. as advocated in Ref. [10].

In leading order the scalar quark condensate $\langle\bar{q}q\rangle$ can be evaluated in a dynamical calculation as follows [34]:

$$\frac{\langle\bar{q}q\rangle}{\langle\bar{q}q\rangle_V} = 1 - \frac{\Sigma_\pi}{f_\pi^2 m_\pi^2}\rho_S - \sum_h \frac{\sigma_h \rho_S^h}{f_\pi^2 m_\pi^2}, \tag{2}$$

where σ_h stands for the σ-commutator of the relevant mesons h, $\langle \bar{q}q \rangle_V$ represents the vacuum condensate, $\Sigma_\pi \approx 45\,\mathrm{MeV}$ is the pion-nucleon Σ-term and f_π and m_π are the pion decay constant and pion mass, respectively.

In Eq. (2), the quantities ρ_S and ρ_S^h denote the nucleon scalar density and the scalar density for a meson of type h, respectively. The scalar density of mesons h is evaluated in the independent-particle approximation as:

$$\rho_S^h(x) = \frac{(2s+1)(2\tau+1)}{(2\pi)^3} \int \mathrm{d}^3p \frac{m_h}{\sqrt{\mathbf{p}^2 + m_h^2}} f_h(x, \mathbf{p}) , \tag{3}$$

where $f_h(x, \mathbf{p})$ denotes the meson phase-space distribution ($x = (\mathbf{r}, t)$) and s, τ refer to the discrete spin and isospin quantum numbers, respectively. Moreover, the vacuum scalar condensate $\langle \bar{q}q \rangle_V = \langle \bar{u}u \rangle_V + \langle \bar{d}d \rangle_V \approx 2\langle \bar{u}u \rangle_V$ can be computed according to the Gell-Mann-Oakes-Renner (GOR) relation [35, 36],

$$f_\pi^2 m_\pi^2 = -\frac{1}{2}(m_u^0 + m_d^0)\langle \bar{q}q \rangle_V , \tag{4}$$

and gives $\langle \bar{q}q \rangle_V \approx -3.2\,\mathrm{fm}^{-3}$ for the bare quark masses $m_u^0 = m_d^0 \approx 7\,\mathrm{MeV}$. Finally, in Eq. (2) the nucleon scalar density ρ_S has to be determined in a suitable model with interacting degrees-of-freedom in order to match our knowledge on the nuclear EoS at low temperature and finite density. A proper (and widely used) approach is the non-linear $\sigma - \omega$ model for nuclear matter where ρ_S is defined as:

$$\rho_S(x) = \frac{d}{(2\pi)^3} \int \mathrm{d}^3p \frac{m_N^*}{\sqrt{\mathbf{p}_N^{*2} + m_N^{*2}}} f_N(x, \mathbf{p}) , \tag{5}$$

where m_N^* and p_N^* denote the effective mass and momentum, respectively, and $f_N(x, \mathbf{p})$ the phase-space occupation of a nucleon while the degeneracy factor is d=4. In fact, in the non-linear $\sigma - \omega$ model the nucleon mass is modified due to the scalar interaction with the medium:

$$m_N^*(x) = m_N^v - g_s\sigma(x) , \tag{6}$$

where m_N^v denotes the nucleon mass in vacuum and $\sigma(x)$ is the scalar field which mediates the interaction between the nucleons and the medium with the coupling g_s. In order to calculate ρ_S, we need to determine the value of the scalar field $\sigma(x)$ at each space-time point x. This is done via the non-linear gap equation [37, 38]:

$$m_\sigma^2\sigma(x) + B\sigma^2(x) + C\sigma^3(x) = g_s\rho_S(x) = g_s d \int \frac{\mathrm{d}^3p}{(2\pi)^3} \frac{m_N^*(x)}{\sqrt{\mathbf{p}^2 + m_N^{*2}}} f_N(x, \mathbf{p}) , \tag{7}$$

since for matter at rest we have $\mathbf{p}^* = \mathbf{p}$. In Eq. (7) the self-interaction of the σ-field is included up to the forth order. The parameters g_s, m_σ, B, C are fixed in order to reproduce the values of the nuclear matter quantities at saturation, i.e. the saturation density, the binding energy per nucleon, the compression modulus, and the effective nucleon mass. Actually, there are different sets for these quantities (NL1,NL2,NL3) that lead to slightly different saturation properties. For details we refer the reader to Ref. [11].

The main idea in Ref. [10] is to consider effective masses for the dressed quarks in the Schwinger formula (1) for the string decay in a hot and dense medium. The effective quark masses can be expressed in terms of a scalar coupling to the quark condensate $\langle\bar{q}q\rangle$ in first order as follows:

$$m_s^* = m_s^0 + (m_s^v - m_s^0)\frac{\langle\bar{q}q\rangle}{\langle\bar{q}q\rangle_V}, \qquad m_q^* = m_q^0 + (m_q^v - m_q^0)\frac{\langle\bar{q}q\rangle}{\langle\bar{q}q\rangle_V}, \tag{8}$$

with $m_s^0 \approx 100\,\text{MeV}$ and $m_q^0 \approx 7\,\text{MeV}$ for the bare quark masses. In Eq. (8) the effective masses decrease from the vacuum values with decreasing scalar condensate $\langle\bar{q}q\rangle$ to the constituent masses. This adaptation of the Schwinger formula in case of a hot and dense medium implies a modification of the flavor production factors in Eq. (1). In an actual nucleus-nucleus collision, PHSD incorporates a dynamical calculation of all these features for each cell in space-time:

- the scalar density ρ_S is determined by solving the gap equation (7) for the σ-field;
- the scalar condensate $\langle\bar{q}q\rangle$ is then computed via Eq. (2);
- the effective masses m_q^*, m_s^* are calculated according to Eqs. (8) and plugged in the Schwinger formula (1) in order to compute the flavor production ratios for the string decay.

We stress that, once the nucleon scalar density ρ_S and $\Sigma_\pi(\approx 45$ MeV) are fixed, there is no additional 'parameter' in the PHSD3.3 compared to the previous version PHSD3.2 that has been employed for a couple years for the analysis of relativistic heavy-ion reactions [26]. We note that the sets NL1 and NL3 have the same compression modulus K but differ in the effective mass m^*/m at saturation density whereas NL1 and NL2 have the same effective mass but differ in the compression modulus K. By comparing the results from NL1, NL2 and NL3 we will be able to explore separately the effects from the effective mass and compression modulus. For a detailed discussion of the flavor ratio γ_s versus energy density for the different parameter sets NL1, NL2 and NL3 we refer the reader to Ref. [11]. We only recall that in the partonic phase the s/u ratio remains constant as a function of the energy density since the strings are no longer formed in the QGP.

3. Application to nucleus-nucleus collisions

In this section we present results for the rapidity distribution of the most abundant particles at AGS and SPS energies for different nuclear equations of state in order to estimate the uncertainties of our approach. In addition we also explore the impact of CSR on the transverse dynamics by calculating the transverse mass spectra for protons, pions, kaons and antikaons in comparison to available data. We recall that in particular the transverse slopes of the kaon spectra had been clearly underestimated in the earlier HSD studies (without a partonic phase) [8].

3.1. Rapidity spectra at AGS and SPS energies

We present in Fig. 1 the PHSD results for the rapidity distribution of protons, $(\Lambda + \Sigma^0)$'s, pions and kaons for central nucleus-nucleus collisions at different energies (from AGS to top SPS energies) in comparison to the experimental data from Refs. [39, 40, 41, 42]. The following scenarios will be explored at different energies:

- default PHSD calculations without CSR, represented by the dotted blue lines;
- PHSD calculations including CSR with NL3 as parameter set for the nuclear EoS, represented by the solid red lines;
- PHSD calculations including CSR with NL1 as parameter set for the nuclear EoS, represented by the dashed green lines.

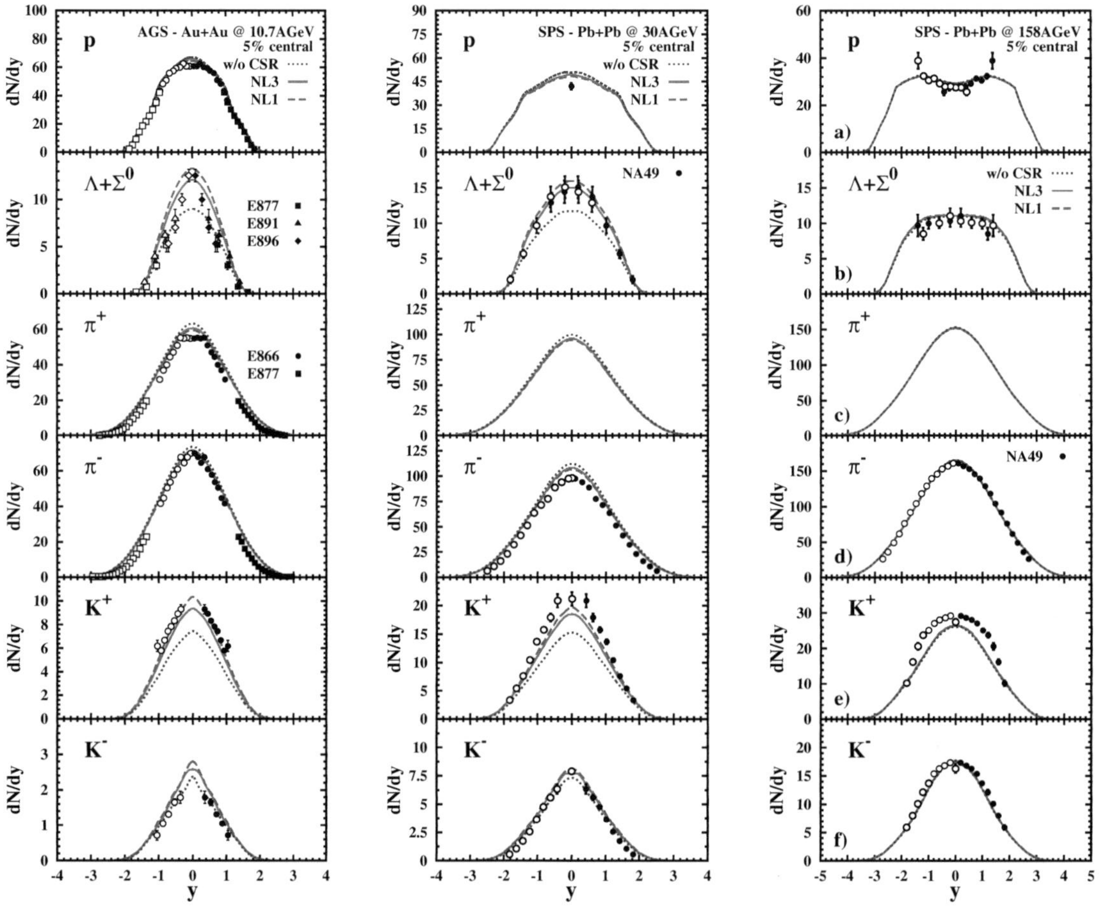

Figure 1. (l.h.s.) The rapidity distribution of protons, $(\Lambda + \Sigma^0)$'s, pions and kaons for 5% central Au+Au collisions at 10.7 AGeV in comparison to the experimental data from Refs. [39, 40]. The solid (red) lines show the results from PHSD including CSR with NL3 parameters, the dashed green lines show the results from PHSD including CSR with NL1 parameters and the blue dotted lines show the result from PHSD without CSR. (middle panel) The rapidity distribution of protons, $(\Lambda+\Sigma^0)$'s, pions and kaons for 5% central Au+Au collisions at 30 AGeV in comparison to the experimental data from Ref. [41]. (r.h.s.) The rapidity distribution of protons, $(\Lambda+\Sigma^0)$'s, pions and kaons for 5% central Au+Au collisions at 158A GeV in comparison to the experimental data from Ref. [42].

We note that PHSD calculations for the parameter set NL2 are not shown explicitly since the results are in between those for NL1 and NL3 [11].

First, we compare the results with and without CSR at $E_{Lab} = 10.7$ AGeV (Fig. 1, l.h.s.), to point out the general effect of this mechanism on the final particle rapidity distributions and in particular show the dependence on the parameter sets NL1 and NL3 for the non-linear $\sigma - \omega$ model for the nuclear EoS. The restoration of chiral symmetry gives an enhancement of the strange particle yields both for mesons and baryons. On the other hand, it produces a slight decrease in the number of pions at midrapidity due to the suppression of pions in the string decays in favor of strange hadrons. The proton rapidity spectra do not present any sensible variation, in fact the CSR as implemented in PHSD modifies essentially the chemistry

IOP Conf. Series: Journal of Physics: Conf. Series **878** (2017) 012018 doi:10.1088/1742-6596/878/1/012018

of the newly produced particles in the string decay and has a minor impact the dynamics of the nucleons, which in the string picture are associated to the string ends of the primary interactions in the system. The inclusion of the CSR is essential in order to correctly reproduce the strange particle rapidity spectra, as we can see especially for $(\Lambda + \Sigma^0)$ hyperons and K^+ mesons. Furthermore, our calculations for the proton rapidity spectra are in good agreement with experimental observation.

Next, we discuss the results from PHSD with CSR using two different parametrization for the nuclear equation of state, i.e. NL3 and NL1. The general features of the strangeness enhancement hold for both parametrizations; in particular the NL1 set provides larger values for all strange particle rapidity spectra at midrapidity since the scalar nucleon density is larger in this case. The difference between the two parametrizations represents the uncertainty of our results related to CSR as implemented in PHSD.

In Fig. 1 (middle panel) the rapidity spectra of various hadrons at $E_{Lab} = 30$ AGeV are shown. We find the same features as for $E_{Lab} = 10.7$ AGeV concerning the strangeness enhancement and the comparison between the two parameter sets for the equation of state; the differences are slightly smaller at this energy. At midrapidity, both protons and pions are very slightly over-estimated in all explored scenarios which suggests that the nuclear stopping is still a bit overestimated. Finally, at the top SPS energy $E_{Lab} = 158$ AGeV (Fig. 1, r.h.s.) the CSR does not play a significant role, since the dynamics is dominated by the QGP phase. Thus, there is no appreciable difference between the results with and without CSR for the two different EoS. Our results for K^+ are lower with respect to the experimental data, however, the $(\Lambda + \Sigma^0)$, π^- and K^- as well as the protons are correctly reproduced. It is presently unclear where these final differences stem from, since strangeness conservation is exactly fulfilled in the PHSD calculations.

3.2. Transverse mass spectra at AGS and SPS energies

We recall that in earlier HSD calculations (without a partonic phase) the slopes of the transverse mass distributions have been severely underestimated [8]. In Fig. 2 we display the transverse mass spectra for pions and kaons in central Au+Au collisions at AGS energies (l.h.s.) and in Pb+Pb collisions at SPS energies (r.h.s.), respectively. We focus on the role played by the CSR on the mesons transverse mass spectra. At the lower energies, $E_{Lab} = 2$ AGeV there is no appreciable difference between the calculation with and without CSR, since the energy density reached by the system is not high enough to produce a vanishing scalar quark condensate. Instead, in the energy range $E_{Lab} = 4 - 40$ AGeV, we notice a small difference between the scenarios with and without CSR. As already mentioned, the CSR acts directly on the chemistry and not so much on the dynamics of the Schwinger mechanism, thus the effect of the partial restoration of chiral symmetry is rather small on the transverse mass spectra. The kaon spectra are harder when CSR is included, while the pion spectra remain essentially unchanged. At the higher SPS energies $E_{Lab} = 80, 158$ AGeV the dynamics of the system is ruled dominantly by the QGP phase and our calculations do not show any sensitivity on the inclusion of CSR. The agreement of our PHSD calculations with the data in Fig. 2 is good in all cases studied.

3.3. Strange particle abundances and ratios

In this subsection we study the excitation function of the particle ratios K^+/π^+, K^-/π^- and $(\Lambda + \Sigma^0)/\pi$ at midrapidity from 5% central Au+Au collisions. In Fig. 3 (l.h.s.) we show the calculations for the following three scenarios: the default PHSD without CSR (blue dotted line), PHSD including CSR with NL3 and NL1 as parameter sets for the nuclear EoS from the non-linear $\sigma - \omega$ model (red solid and green dashed lines, respectively). The shaded area displays the uncertainties of our calculations from the two scenarios for the nuclear EoS since the results from the parameter set NL2 are always in between those from NL1 and NL3 [11].

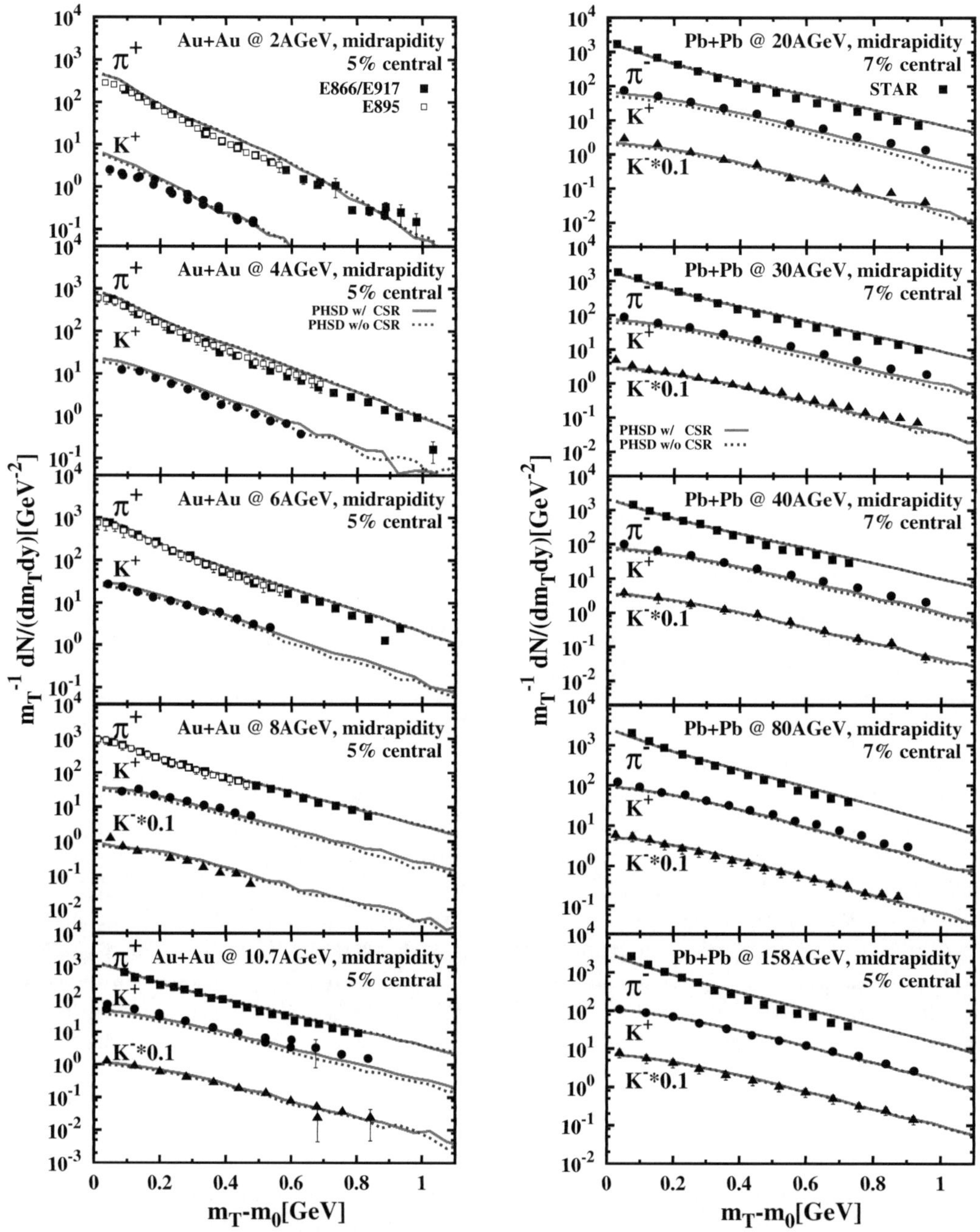

Figure 2. (l.h.s.) The transverse mass spectra of pions and kaons for 5% central Au+Au collisions at 2, 4, 6, 8, 10.7 AGeV in comparison to the experimental data from Refs. [43, 44]. We show the results from PHSD including CSR with NL3 parameters by solid (red) lines and those from PHSD without CSR by dotted (blue) lines. (r.h.s.) The transverse mass spectra of pions and kaons for 5% and 7% central Pb+Pb collisions at 20, 30, 40, 80, 158 AGeV in comparison to the experimental data from Ref. [45].

IOP Conf. Series: Journal of Physics: Conf. Series **878** (2017) 012018 doi:10.1088/1742-6596/878/1/012018

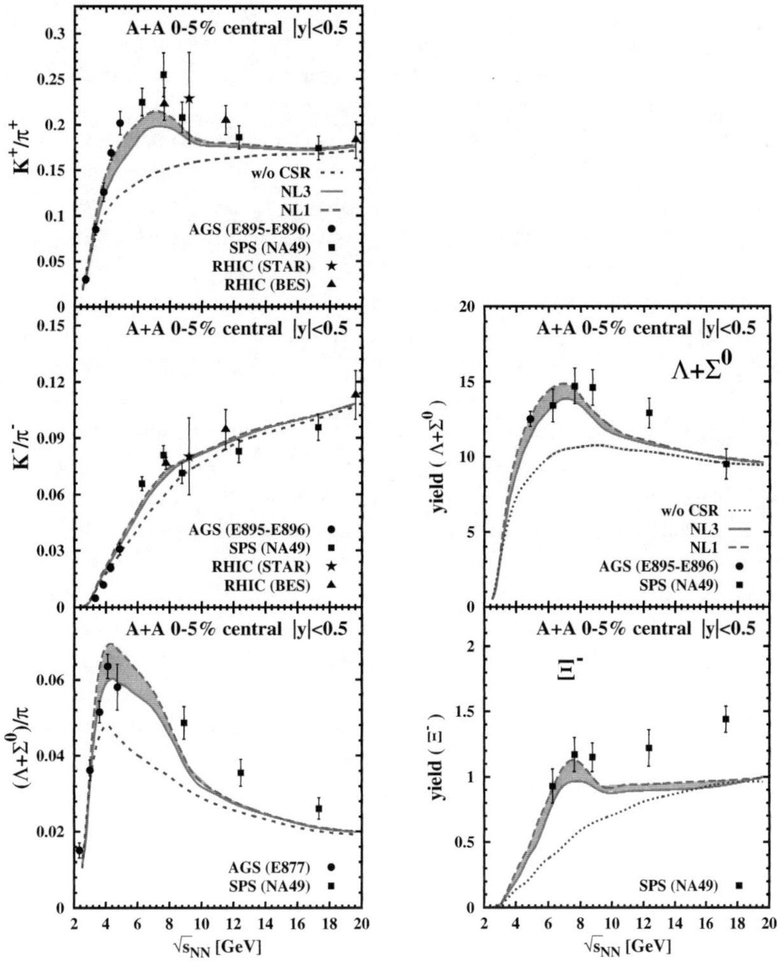

Figure 3. (l.h.s.) The ratios K^+/π^+, K^-/π^- and $(\Lambda + \Sigma^0)/\pi$ at midrapidity from 5% central Au+Au collisions as a function of the invariant energy $\sqrt{s_{NN}}$ up to the top SPS energy in comparison to the experimental data from [44, 46, 47]. The coding of the lines is the same as in Fig. 1. The grey shaded area represents the results from PHSD including CSR taking into account the uncertainty from the parameters of the $\sigma - \omega$-model for the EoS. (r.h.s.) The yields of $(\Lambda + \Sigma^0)$ and Ξ^- at midrapidity from 5% central Au+Au collisions as a function of the invariant energy $\sqrt{s_{NN}}$ up to the top SPS energy in comparison to the experimental data from Refs. [40, 47].

As described in Ref. [10], the inclusion of CSR in PHSD is responsible for the strong strangeness enhancement at AGS and low SPS energies. The experimental observations of the ratios K^+/π^+ and $(\Lambda + \Sigma^0)/\pi$ show the well-known "horn" structure, which is reproduced by the PHSD calculations with CSR. In fact, CSR gives rise to a steep increase of these ratios at energies lower than $\sqrt{s_{NN}} \approx 7\,\mathrm{GeV}$, while the drop at larger energies is associated to the appearance of a deconfined partonic medium. The NL1 parameter set produces a sharper peak both in the K^+/π^+ and in the $(\Lambda + \Sigma^0)/\pi$ excitation functions with a $\approx 10\%$ maximum increase with respect to the NL3 result. We point out that even adopting different parametrizations for the $\sigma - \omega$ model, we recover the same "horn" feature. This supports the reliability of the CSR mechanism as implemented in the PHSD model.

At AGS energies, the energy dependencies of the ratios K^+/π^+ and $(\Lambda + \Sigma^0)/\pi$ are closely connected, since K^+ and Λ (or Σ^0) are mostly produced in pairs due to strangeness conservation. On the other hand, the excitation function of the K^-/π^- ratio does not show any peak, but smoothly increases as a function of $\sqrt{s_{NN}}$. In fact, especially at AGS energies, the antikaon production differs substantially from the production of K^+ and Λ, which occurs predominantly via string formation. In fact, the antikaons are produced dominantly via secondary meson-baryon interactions by flavor exchange and their production is suppressed with respect to the Λ hyperons that carry most of the strange quarks. This is the reason why the inclusion of chiral symmetry restoration provides a substantial enhancement of the K^+/π^+ and $(\Lambda + \Sigma^0)/\pi$ excitation functions and a smaller change on the K^-/π^- ratio. We also notice that there is no sizeable difference between the NL1 and NL3 results for the K^-/π^- ratio. At top SPS energies the strangeness is produced predominantly by the hadronization of partonic degrees-of-freedom, thus our results for all the ratios do not show an appreciable sensitivity to the nuclear EoS and the calculations with and without CSR tend to merge at $\sqrt{s_{NN}} \approx 20\,\text{GeV}$.

Finally, in Fig. 3 (r.h.s.) we present the yields of $(\Lambda + \Sigma^0)$ and Ξ^- at midrapidity from 5% central Au+Au collisions as a function of the invariant energy $\sqrt{s_{NN}}$ in comparison to the available data from Refs. [40, 47]. We recover a "horn" structure, similar to that shown in Fig. 3 (l.h.s.) for the energy dependence of the strange to non-strange particle ratios. A sensitivity on the nuclear model parametrizations persists at low energy, while in the top SPS energy regime the results corresponding to the different scenarios merge. The comparison with the available data at $\sqrt{s_{NN}} < 8\,\text{GeV}$ supports the validity of the CSR picture, while at larger energies we under-estimate the experimental observations. We mention that this discrepancy is not due to the CSR mechanism, since it does not play an essential role in the high-energy regime as pointed out above.

4. Summary

We point out that our PHSD calculations provide a microscopic interpretation of the "horn" structure in the excitation function of the K^+/π^+ ratio in central Au+Au (or Pb+Pb) collisions. The steep rise of this ratio at AGS energies is associated to chiral symmetry restoration (CSR), while the drop at higher SPS energies is due to the appearance of the QGP phase in an increasing volume of the interaction region. We have found an analogous energy dependence for the $(\Lambda + \Sigma^0)/\pi$ ratio, while the excitation function of the K^-/π^- ratio does not show any explicit peak. In general, the PHSD results obtained with the inclusion of CSR are in a good agreement with the available data for all observables analyzed, while calculations without CSR fail substantially.

The microscopic PHSD studies support the idea that CSR occurs in hadronic systems with high temperatures and densities before the deconfinement phase transition takes over. We suggest that the strange particle spectra and yields are suitable signatures to study the properties of CSR in HICs in future also as a function of system size and centrality as advocated in Ref. [11].

Acknowledgments

The authors acknowledge inspiring discussions with J. Cleymans, M. Gazdzicki, M. Gorenstein and O. Linnyk and thank the Helmholtz International Center for FAIR (HIC for FAIR), the Helmholtz Graduate School for Hadron and Ion Research (HGS-HIRe), the Helmholtz Research School for Quark Matter Studies in Heavy-Ion Collisions (H-QM), and the Bundesministerium für Bildung und Forschung (BMBF) for support. The computational resources have been provided by the LOEWE-CSC.

IOP Conf. Series: Journal of Physics: Conf. Series **878** (2017) 012018 doi:10.1088/1742-6596/878/1/012018

References

[1] Rafelski J and Müller B 1982 *Phys. Rev. Lett.* **48** 106
[2] Stock R 2002 *J. Phys.* **G 28** 1517
[3] Gazdzicki M and Gorenstein M I 1999 *Acta Phys. Polon.* **B 30** 2705
[4] Cleymans J *et al.* 2005 *Phys. Lett.* **B 615** 50
[5] Andronic A Braun-Munzinger P and Stachel J 2009 *Phys. Lett.* **B 673** 142
[6] Bugaev K A *et al.* 2013 *Europhys. Lett.* **104** 22002
[7] Geiss J Cassing W and Greiner C 1998 *Nucl. Phys.* **A 644** 107
[8] Bratkovskaya E L *et al.* 2004 *Phys. Rev.* **C 69** 054907
[9] Bratkovskaya E L Soff S Stöcker H van Leeuwen M and Cassing W 2004 *Phys. Rev. Lett.* **92** 032302
[10] Cassing W Palmese A Moreau P and Bratkovskaya E L 2016 *Phys. Rev.* **C 93** 014902
[11] Palmese A *et al.* 2016 *Phys. Rev.* **C 94** 044912
[12] Senger P *et al.* 2011 *Lect. Notes Phys.* **814** 681
[13] Fischer C S Luecker J and Welzbacher C 2014 *Phys. Rev.* **D 90** 034022
[14] Fischer C S Fister L Luecker J and Pawlowski J M 2014 *Phys. Lett.* **B 732** 273
[15] Eichmann G Fischer C S and Welzbacher C 2016 *Phys. Rev.* **D 93** 034013
[16] Herbst T K Pawlowski J M and Schaefer B J 2013 *Phys. Rev.* **D 88** 014007
[17] McLerran L and Pisarski R D 2007 *Nucl. Phys.* **A 796** 83
[18] Nambu Y and Jona-Lasinio G 1961 *Phys. Rev.* **122** 345
[19] Klevansky S P 1992 *Rev. Mod. Phys.* **64** 649
[20] Rapp R 2013 *Adv. High Energy Phys.* **2013** 148253
[21] Cassing W and Bratkovskaya E L 2009 *Nucl. Phys.* **A 831** 215
[22] Bratkovskaya E L Cassing W Konchakovski V P and Linnyk O 2011 *Nucl. Phys.* **A 856** 162
[23] Konchakovski V P *et al.* 2012 *Phys. Rev. C* **85** 044922
Konchakovski V P *et al.* 2012 *Phys. Rev.* **C 85** 011902
[24] Kadanoff L P and Baym G 1962 *Quantum Statistical Mechanics*, Benjamin, New York
[25] Cassing W 2009 *Eur. Phys. J. ST* **168** 3
Cassing W 2007 *Nucl. Phys.* **A 795** 70
[26] Linnyk O Bratkovskaya E and Cassing W 2016 *Prog. Part. Nucl. Phys* **87** 50
[27] Cassing W and Bratkovskaya E L 1999 *Phys. Rep.* **308** 65
[28] Andersson B Gustafson G and Pi H 1993 *Z. Phys.* **C 57** 485
[29] Schwinger J 1951 *Phys. Rev.* **83** 664
[30] Cohen T D Furnstahl R J and Griegel D K 1992 *Phys. Rev.* **C 45** 1881
[31] Vogl U and Weise W 1991 *Prog. Part. Nucl. Phys.* **27** 195
[32] Birse M C 1994 *J. Phys.* **G 20** 1537
[33] Koch V 1997 *Int. J. Mod. Phys.* **E 6** 203
[34] Friman B Nörenberg W and Toneev V D 1998 *Eur. Phys. J.* **A 3** 165
[35] Bando M Kugo T and Yamawaki K 1988 *Phys. Rep.* **164** 217
[36] Cohen T D Furnstahl R J Griegel D K and Jin X 1995 *Prog. Part. Nucl. Phys.* **35** 221
[37] Boguta J and Bodmer A R 1977 *Nucl. Phys.* **A 292** 413
[38] Lang A *et al.* 1991 *Z. Phys.* **A 340** 287
[39] Stachel J 1996 *Nucl. Phys.* **A 610** 509C
[40] Albergo S *et al.* 2002 *Phys. Rev. Lett.* **88** 062301
[41] Alt C *et al.* 2008 *Phys. Rev.* **C 77** 024903 2006 *Phys. Rev.* **C 73** 044910 2008 *Phys. Rev.* **C 78** 034918
[42] Afanasiev S V *et al.* 2002 *Phys. Rev.* **C 66** 054902
Anticic T *et al.* 2011 *Phys. Rev.* **C 83** 014901 2012 *Phys. Rev.* **C 86** 054903
[43] Ahle L *et al.* 2000 *Phys. Lett.* **B 476** 1 (2000) 1998 *Phys. Rev.* **C 58** 3523
[44] Klay J L *et al.* 2003 *Phys. Rev.* **c 68** 054905
[45] Alt V *et al.* 2004 *J. Phys.* **G 30** 119 2008 *Phys. Rev.* **C 77** 024903
Afanasiev S V *et al.* 2002 *Phys. Rev.* **C 66** 054902
[46] Abelev B I *et al.* 2010 *Phys. Rev.* **C 81** 024911
Aggarwal M M *et al.* 2011 *Phys. Rev.* **C 83** 024901
[47] Alt C *et al.* 2008 *Phys. Rev.* **C 78** 034918

IOP Conf. Series: Journal of Physics: Conf. Series **878** (2017) 012019 doi:10.1088/1742-6596/878/1/012019

Bounds for lighter Higgses in extensions of the Standard Model

Aldo Deandrea

Université de Lyon, France; Université Lyon 1, CNRS/IN2P3, UMR5822 IPNL, F-69622 Villeurbanne Cedex, France.

E-mail: deandrea@ipnl.in2p3.fr

Abstract. Higgs bosons lighter than the 125 GeV one can be present in extensions of the Standard Model and exclusion or discovery of such a possibility allows to improve our knowledge of the Higgs sector. I consider Two Higgs Doublet Models, which are the minimal scalar structure of a large number of models, in order to explore the possibility of constraining a neutral scalar particle lighter than the 125 GeV Higgs boson. Such a lighter particle is not yet completely excluded by present data. I show that some new constraints can be obtained at the LHC for these light particles.

1. Introduction

The Large Hadron Collider (LHC) has confirmed the effective description of the electroweak sector given by the Standard Model (SM) Lagrangian with the discovery of the 125 GeV Higgs boson and the analysis of its properties [1, 2]. Many studies, both from the theoretical and experimental side, have considered extensions of the Standard Model (SM) with an enlarged scalar sector, but most studies have considered the possibility of new scalars heavier than the 125 GeV Higgs boson which was discovered at the LHC. It is however possible to have a spectrum in which lighter scalars are present together with an SM-like Higgs boson at 125 GeV. Among these possibilities there are detailed BSM models as well as effective descriptions including only the extended scalar sector. Two Higgs Doublet Models (2HDMs) constitute one of the simplest possibilities, where the SM Lagrangian is extended by the addition of a second scalar doublet. Previous phenomenological studies describing the lighter Higgs bosons include [3, 4, 5, 6, 7, 8, 9]. At masses below 125 GeV, the main search channel at the LHC is the di-photon decay channel [11, 12].

2. Two Higgs Doublet Models

The 2HDMs are extension of the Standard Model including two complex SU(2) doublets, ϕ_1 and ϕ_2. A $\mathbb{Z}_2$ symmetry is introduced to avoid flavour-changing currents, so that all fermions of a given electric charge couple to at most one Higgs doublet. The convention usually adopted is given in Table 1.

	Type I	Type II	Flipped (Type Y)	Lepton Specific (Type X)
Up-type quark	ϕ_2	ϕ_2	ϕ_2	ϕ_2
Down-type quark	ϕ_2	ϕ_1	ϕ_1	ϕ_2
Leptons	ϕ_2	ϕ_1	ϕ_2	ϕ_1

Table 1. The couplings between the SM fermions and the two scalar doublets in 2HDMs.

The 2HDMs potential constrained by the $\mathbb{Z}_2$ symmetry can be written as:

$$\begin{aligned} V =& m_{11}^2\phi_1^\dagger\phi_1 + m_{22}^2\phi_2^\dagger\phi_2 - m_{12}^2\left(\phi_1^\dagger\phi_2 + \phi_2^\dagger\phi_1\right) + \frac{\lambda_1}{2}\left(\phi_1^\dagger\phi_1\right)^2 + \frac{\lambda_2}{2}\left(\phi_2^\dagger\phi_2\right)^2 + \\ & \lambda_3\left(\phi_1^\dagger\phi_1\right)\left(\phi_2^\dagger\phi_2\right) + \lambda_4\left(\phi_1^\dagger\phi_2\right)\left(\phi_2^\dagger\phi_1\right) + \frac{\lambda_5}{2}\left[\left(\phi_1^\dagger\phi_2\right)^2 + \left(\phi_2^\dagger\phi_1\right)^2\right], \end{aligned} \tag{1}$$

where all the parameters are real. The parameter m_{12}^2 is a soft breaking term of the $\mathbb{Z}_2$ symmetry. The two scalar doublets get vacuum expectation values (vevs):

$$\phi_1 = \begin{pmatrix} 0 \\ \frac{v_1}{\sqrt{2}} \end{pmatrix}, \qquad \phi_2 = \begin{pmatrix} 0 \\ \frac{v_2}{\sqrt{2}} \end{pmatrix}, \tag{2}$$

with $v \equiv \sqrt{v_1^2 + v_2^2}$.

After symmetry breaking there are five physical scalars: two neutral $\mathcal{CP}$-even states h and H, one neutral $\mathcal{CP}$-odd state A and two charged ones $H^\pm$. The connection of the potential of eq. 1 to mass-eigenstates is given by two angles: α which mixes the $\mathcal{CP}$-even scalar states to give mass-eigenstates and β, with $\tan\beta = \frac{v_2}{v_1}$, which rotates the two doublets in a basis where only one of them acquires a vev. The relation between the previous parameters and the mass basis is:

$$\begin{gathered} \lambda_1,\ \lambda_2,\ \lambda_3,\ \lambda_4,\ \lambda_5,\ m_{11}^2,\ m_{22}^2,\ m_{12}^2 \\ \Updownarrow \\ m_h,\ m_H,\ m_A,\ m_{H^\pm},\ \tan\beta,\ \sin(\beta-\alpha),\ v,\ m_{12}^2 \end{gathered} \tag{3}$$

where v is the 246 GeV electroweak scale, and one of the masses of the $\mathcal{CP}$-even states is set to be equal to the measured Higgs boson mass. The masses of the two $\mathcal{CP}$-even states are ordered with $m_h < m_H$, where I will call h the light Higgs boson and H the heavy Higgs boson of the model. In the following the heavy Higgs boson H of the model is identified with the Higgs boson discovered at LHC, $m_H = 125$ GeV, while the remaining six parameters are left free.

3. Bounds on 2HDMs

The spectrum of neutral and charged scalars of the 2HDMs is a minimal extension of the scalar sector with one additional doublet and gives rise to five physical scalars: two charged $H^\pm$ and three neutral h, H and A states. If the Higgs boson discovered at the LHC is associated with the heavier H, the two other neutral states can be candidates for a lighter Higgs boson. I briefly discuss the different constraints used to impose bounds on the model: indirect constraints, LEP constraints and LHC constraints.

	Experimental values
S	0.05 ± 0.11
T	0.09 ± 0.13
U	0.01 ± 0.11

	Correlations
ST	+0.90
SU	-0.59
TU	-0.83

Table 2. Experimental values of the oblique parameters with 1σ uncertainty and correlations between them [15].

3.1. Indirect constraints

The indirect constraints include limits on the precision electroweak parameters S, T and U [13], flavour constraints and theoretical requirements to ensure stability of the potential, unitarity and perturbativity. The electroweak parameters are computed using the program `2HDMC` [14] and compared to the experimental limits [15] at 2σ (see Table 2 for the experimental values with 1σ uncertainties and the correlations between them).

The stability of the potential requires:

$$\lambda_1 \geq 0, \qquad \lambda_2 \geq 0, \qquad \lambda_3 \geq -\sqrt{\lambda_1 \lambda_2}, \\ \lambda_3 + \lambda_4 - |\lambda_5| \geq -\sqrt{\lambda_1 \lambda_2} \tag{4}$$

In addition we require to have tree-level perturbative unitarity for the scattering of Higgs bosons and the longitudinal parts of electroweak gauge bosons [16].
In order to trust perturbative calculations, a condition is added on the quartic Higgs bosons couplings $C_{h_i h_j h_k h_l}$:

$$|C_{h_i h_j h_k h_l}| \leq 4\pi \tag{5}$$

These three conditions are also computed via the `2HDMC` program.

The parameter space of the model is further tested against flavour bounds looking at the branching ratios $\mathcal{BR}(\overline{B} \to X_s\gamma)$ and $\mathcal{BR}(B_s \to \mu^+\mu^-)$, which obtain contributions from the charged Higgs bosons and the neutral ones respectively and at the isospin asymmetry $\Delta_0(B \to K^*\gamma)$ and the ΔM_d frequency oscillation which are sensitive to the presence of charged Higgs bosons. The value of each process is computed in the 2HDMs via the program `SuperIso` [17, 18] and then compared to the experimental limits at 2σ. In order to take into account the theoretical uncertainties in the 2HDMs calculation, which are not evaluated in `SuperIso`, we add to the experimental 1σ uncertainty σ_{Exp} of each process the 1σ theoretical uncertainty σ_{Th}. A summary of the results is available in Table 3.

Process	Experimental values	Theoretical computation	Combined error at 1σ
$\mathcal{BR}(\overline{B} \to X_s\gamma)$	$(3.43 \pm 0.22) \times 10^{-4}$ [19]	$(3.40 \pm 0.19) \times 10^{-4}$ [20]	0.29×10^{-4}
$\mathcal{BR}(B_s \to \mu^+\mu^-)$	$(2.9 \pm 0.7) \times 10^{-9}$ [21, 22]	$(3.54 \pm 0.27) \times 10^{-9}$ [20]	0.8×10^{-9}
$\Delta_0(B \to K^*\gamma)$	$(5.2 \pm 2.6) \times 10^{-2}$ [23]	$(5.1 \pm 1.5) \times 10^{-2}$ [20]	3.0×10^{-2}
ΔM_d	0.510 ± 0.003 ps^{-1} [19]	0.543 ± 0.091 ps^{-1} [24]	0.091 ps^{-1}

Table 3. Values of the experimental and theoretical flavour constraints.

3.2. LEP and LHC constraints

The `HiggsBounds` program [25, 26, 27, 28] is able to test a model against experimental data coming from LEP, Tevatron and the LHC. In our analysis we use `HiggsBounds` version 4.2.1 with the LEP experiment constraints only, in order to impose LHC constraints separately.

The LHC constraints are coming from the experimental results on the 125 GeV Higgs boson at Run 1, *i.e.* the 2HDM heavy Higgs boson H, in our case. We use the exclusion contours in the plane of the signal strength for each individual production mode $\mu_{VBF/VH}$ vs $\mu_{ggh/tth}$ given by the combined ATLAS and CMS experiments [29]. Assuming a Gaussian profile for the likelihood $\mathcal{L}$ at 68% C.L., we fit the ellipses for each decay channel $Y = \{WW, ZZ, \gamma\gamma, \tau\tau, b\bar{b}\}$ and hence obtain the parametrisation for each of them (see [30] for details). Combining the log-likelihood ratios:

$$\Delta\chi^2(p_j) = \sum_Y \chi^2_Y(p_j) - \sum_Y \chi^2_Y(\widehat{p_j}), \tag{6}$$

with p_j the set of free parameters on which the function depends and $\widehat{p_j}$ their value minimising the χ^2 function. In our case, we have six degrees of freedom (κ_g^2, κ_V^2, $BR^{2HDM}_{H\to WW}$, $BR^{2HDM}_{H\to\tau\tau}$, $BR^{2HDM}_{H\to\gamma\gamma}$, $BR^{2HDM}_{H\to b\bar{b}}$, as $BR^{2HDM}_{H\to ZZ}$ is linked to $BR^{2HDM}_{H\to WW}$). This choice in the free parameters implies that we assume no correlation between the kappas and the branching ratios, which is correct as long as the deviation of the branching ratios is not too large with respect to the Standard Model values. A point in the 2HDM parameter space passing the LHC constraints, therefore, has a $\Delta\chi^2$ value lower than 12.85, which is the value at 95% C.L. for 6 degrees-of-freedom.

4. Searches for a lighter scalar Higgs

A light resonance decaying into two photons was searched for by CMS [11] in the range of mass between 80 and 110 GeV. To compare with the experimental sensitivity, we need to compute the expected production cross sections in the different production modes and branching ratios into the observed final states.

4.1. Cross sections and branching ratios

The program 2HDMC [14] version 1.7.0 is used to compute the branching ratios of the different Higgs bosons of the theory. The cross sections are calculated using an approximation that we denote in the following as the "kappa trick". Defining the generic parameter κ_Y as $\kappa_Y^2 = \frac{\Gamma_Y^{2HDM}}{\Gamma_Y^{SM}}$ for a specific decay channel Y, the cross sections can be approximated as:

$$\sigma^{2HDM}_{ggh} \simeq \kappa_g^2 \times \sigma^{SM}_{ggh}, \qquad \sigma^{2HDM}_{VBF/VH} \simeq \kappa_V^2 \times \sigma^{SM}_{VBF/VH} = \sin^2(\beta-\alpha) \times \sigma^{SM}_{VBF/VH}. \tag{7}$$

As the couplings of the light scalar Higgs boson to the W and Z bosons are rescaled in the same way compared to the SM couplings, we have that $\kappa_Z = \kappa_W \equiv \kappa_V = \sin(\beta - \alpha)$. The SM cross section is taken from the LHC Higgs Cross-Section Working Group [31]. The kappas are computed using 2HDMC. A comment on the validity of this approximation is in order. The cross section production in VBF and VH mode does not cause any problem as the leading effect arises at tree level, however for the gluon fusion mode a loop induced coupling is present and thus it is important to check the validity of the "kappa trick" for this production mode. Note indeed that for loop induced vertices the use of an effective kappa factor is not always appropriate and more general parameterisations exist (see for example [32, 33]). In order to explore this issue, we performed a comparison between the cross sections in gluon fusion obtained via the program SusHi and the ones obtained with the "kappa trick". This comparison shows that the deviation is less than 3% for the whole mass range and stable upon modification of the values of the input parameters. As it stays within the range allowed by the uncertainties (theoretical, PDF and α_s) calculated by the LHC Higgs Cross-Section Working Group [31], we can safely use our method for the light Higgs boson.

4.2. Constraining the 2HDMs parameter space

In this section we study the influence of the three sets of constraints defined in Section 3 (indirect, LEP and LHC constraints) on the free parameters of the model. For this purpose we generate a set of one million points for each of the four different types of model defined in Table 1 with random values for each of the free parameters. The available ranges we use in the simulation are given in Table 4. The range of variation for m_h corresponds to the mass range in the CMS di-photon analysis. The lower bound of 80 GeV for $m_{H^\pm}$ comes from the bound obtained at the LEP experiment [34]. The ranges for m_A and m_{12}^2, although not totally general, are the result of previous quick scans which eliminate areas with a very low density of points passing the three sets of constraints (indirect, LEP and LHC constraints).

m_h (GeV)	m_H (GeV)	m_A (GeV)	$m_{H^\pm}$ (GeV)	$\sin(\beta-\alpha)$	$\tan\beta$	m_{12}^2 (GeV)^2
[80;110]	125	[60;1000]	[80;1000]	[-1;1]	[1/50;50]	[-(300)^2;+(200)^2]

Table 4. Range of variation for the free parameters used in the analysis.

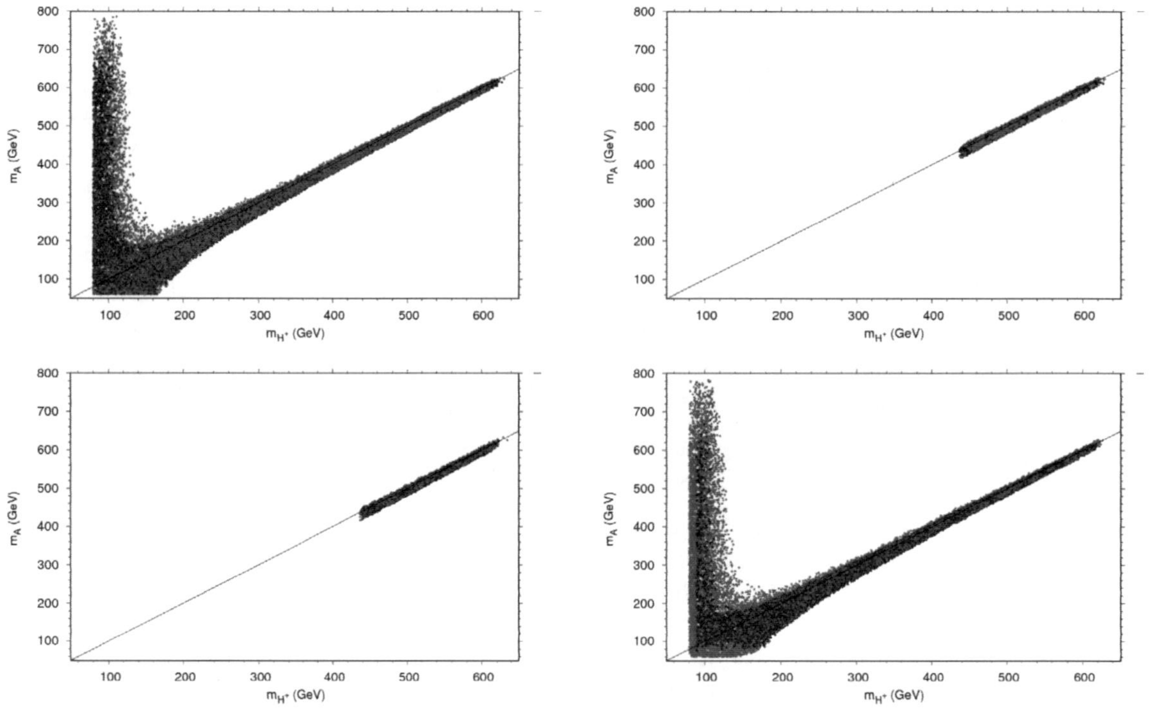

Figure 1. Constraints on the parameter space in the plane m_A vs $m_{H^\pm}$. Top left: Type I. Top right: Type II. Bottom left: Flipped. Bottom right: Lepton Specific. In green the points passing indirect constraints only. In blue the points passing indirect and LEP constraints. In red the points passing indirect, LEP and LHC constraints.

In Figure 1, the generated points are plotted in the plane m_A vs $m_{H^\pm}$. The upper left panel corresponds to Type I, the upper right to Type II, the lower left to the Flipped model and the lower right to Lepton Specific model. The points passing only the indirect constraints are plotted in green, those passing indirect and LEP constraints are in blue and those passing indirect, LEP and LHC constraints are in red. I will use these same conventions in the following.

Note that the m_A and $m_{H^\pm}$ masses are correlated: when m_A and $m_{H^\pm}$ grow, the indirect constraints force them to be near the black line corresponding to $m_A = m_{H^\pm}$. This is due to the T parameter which is very sensitive to these two masses and enforces them to be close to each other. In Type I, we find that most of the red points lie in the ranges $m_A \in$[60 GeV; 650 GeV] and $m_{H^\pm} \in$[80 GeV; 630 GeV]. In Type II and Flipped, the two masses are much more constrained $m_A \in$[400 GeV; 650 GeV] and $m_{H^\pm} \in$[430 GeV; 630 GeV]: this is due to the fact that the down-type quarks couple now to the ϕ_1 doublet instead of the ϕ_2 doublet as in Type I, thus the $\mathcal{BR}(B \to X_s\gamma)$ flavour limit imposes a very strong constraint on the mass of the charged Higgs bosons. Note that the most recent BELLE data [35] further constrain Type II and Flipped models (see for details [36]). These recent constraints are not included in the previous plots, but for Type I (on which I will focus in the following) the limit only affect the very low $\tan\beta < 2$ values. When the bounds are combined with the T parameter constraint, they impose also limits on the pseudo-scalar mass. The Lepton Specific case is very similar to Type I, as the couplings of the down-type quark are the same. We find $m_A \in$[80 GeV; 630 GeV] and $m_{H^\pm} \in$[80 GeV; 630 GeV] to be the preferred regions. For $\sin(\beta - \alpha)$, the allowed range is close to zero, which is consistent with the choice of $m_H = 125$ GeV: as $\sin(\beta - \alpha) \simeq 0$, we have $\cos(\beta - \alpha) \simeq 1$ so that the couplings of the heavy Higgs boson H to gauge bosons are similar to the SM ones (this is the alignment limit described in [8]). We find that the preferred ranges are $\sin(\beta - \alpha) \in$[-0.4; 0.3] for Type I, $\sin(\beta - \alpha) \in$[-0.5; 0.05] for Type II and Flipped model and $\sin(\beta - \alpha) \in$[-0.3; 0.2] for Lepton Specific model.

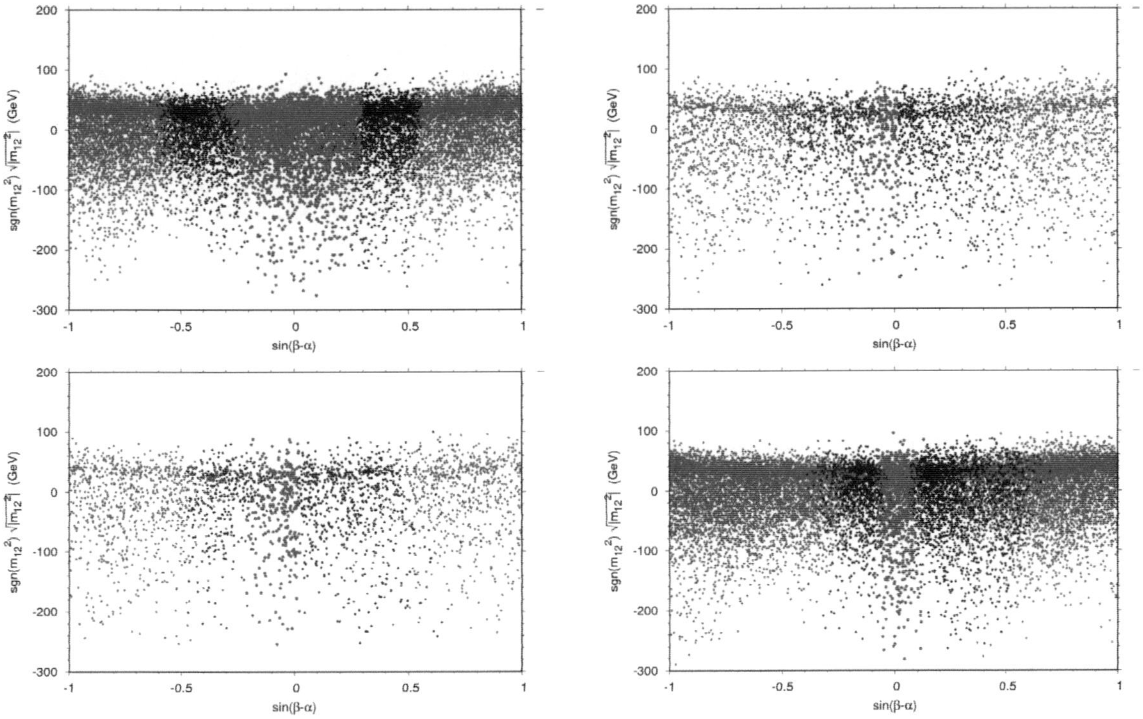

Figure 2. Constraints on the free parameters in the plane m_{12} vs $\sin(\beta - \alpha)$. Top left: Type I. Top right: Type II. Bottom left: Flipped. Bottom right: Lepton Specific. Same colour code as in Figure 1.

Inspecting the plane m_{12} vs $\sin(\beta - \alpha)$ we can see the constrains on m_{12}^2 (see Figure 2): $m_{12}^2 <(100\ \text{GeV})^2$ in all the four types.

The parameter space can be limited in order to increase the statistics of the allowed points using the domains indicated by the previous plots. In addition to this, as we are interested in checking the sensitivity to a lighter Higgs boson at LHC Run 1 in the di-photon decay channel, we can further restrict to the areas of relatively high values of cross section times branching ratio into two photons. The minimum value of the CMS observed upper limit [11] is 0.032 pb in the gluon fusion channel, obtained for $m_h = 103$ GeV and 0.019 pb in the VBF/VH channel, obtained for $m_h = 100.5$ GeV. We checked that with the results for the gluon fusion production mode, CMS is not sensitive to a lighter Higgs boson in this particular channel at LHC Run 1. On the contrary for the VBF channel there is some sensitivity for model type I at the LHC, as shown in figure 3. One can deduce from these results that in the Type II, Flipped and Lepton

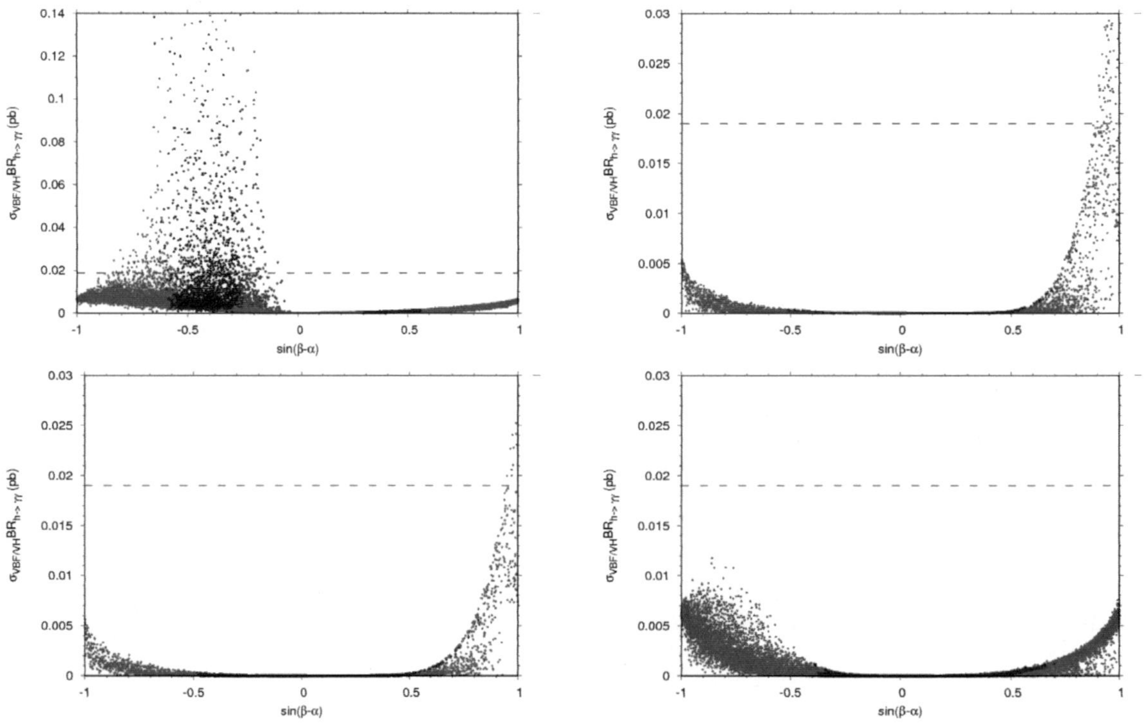

Figure 3. 2HDM generated points in the plane $\sigma \times BR_{h\to\gamma\gamma}$ vs $\sin(\beta-\alpha)$ in the VBF/VH production mode. Top left: Type I. Top right: Type II. Bottom left: Flipped. Bottom right: Lepton Specific. Same colour code as in Figure 1. The dashed line corresponds to the minimum value of the CMS observed upper limit in the VBF/VH production mode.

Specific models, CMS had no sensitivity to a lighter Higgs boson at LHC Run 1 in the $h \to \gamma\gamma$ decay channel, neither in the gluon fusion nor in the VBF/VH production mode. For Type I, there is no sensitivity in the gluon fusion channel, but only in the VBF/VH channel, where we find red points above the dashed line. As the value of the CMS observed upper limit depends on the mass of the light Higgs boson considered, the dashed line represented on the plots is not an absolute bound. Some of the red points above it can be in practice below the CMS observed limit, but it is a good indication of the potential capability of the VBF/VH channel for some exclusion.

We use Figure 3 even further by choosing to consider only areas where the points have relatively high values of cross section times branching ratio. We choose a lower bound at 0.01 pb to select the points, which corresponds to $\sin(\beta-\alpha) \in$[-0.3;-0.05]. We can similarly work

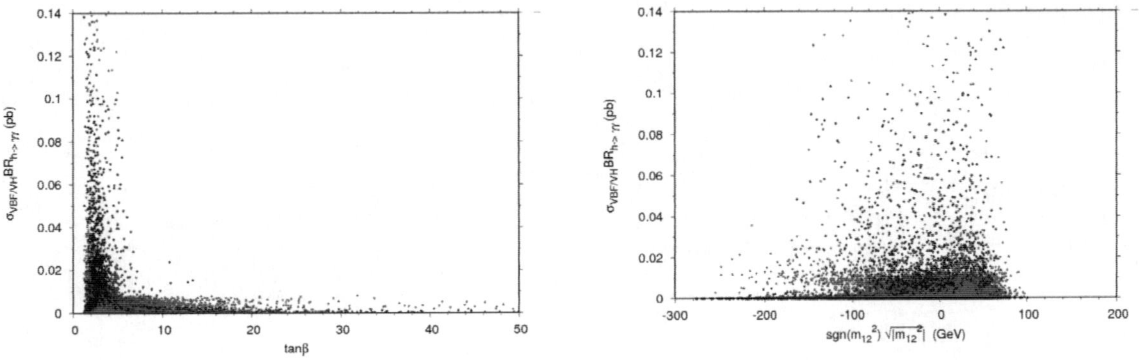

Figure 4. Value of the cross section times branching ratio in the VBF/VH production mode as a function of $\tan\beta$ (left) and $sgn(m_{12}) \times \sqrt{|m_{12}^2|}$ (right) in Type I. Same colour code as in Figure 1.

with tighter ranges for the parameters $\tan\beta$ and m_{12}^2 (see Figure 4). We choose $\tan\beta \in$[2;12] and $m_{12}^2 \in$[-(100 GeV)^2;+(100 GeV)^2].

4.3. Comparison with the CMS low mass di-photon analysis
We now perform a second "focused" scan and make a detailed comparison with the sensitivity of the CMS search at 8 TeV with one million points, this time for Type I only, using the restricted parameter ranges we found in the previous section (see Table 5).

m_h (GeV)	m_H (GeV)	m_A (GeV)	$m_{H^\pm}$ (GeV)	$\sin(\beta-\alpha)$	$\tan\beta$	$m12^2$
[80;110]	125	[60;650]	[80;630]	[-0.3;-0.05]	[2;12]	[-(100)^2;+(100)^2]

Table 5. Allowed range of variation for the free parameters.

The resulting points of this second scan are plotted in Figure 5 in the plane $\sigma \times BR_{h\to\gamma\gamma}$ in the gluon fusion production mode (left panel) and the VBF/VH production mode (right panel) vs m_h, superimposed on the public exclusion limits of CMS collaboration. For convenience only the red points, *i.e.* the points passing all of the indirect, LEP and LHC constraints, are plotted here. These results confirm the expectation that there is no sensitivity in the gluon fusion production mode but many points are above the CMS observed limit in the VBF/VH production mode for a light Higgs boson with mass below 105 GeV. As the points above the observed CMS upper limit are excluded at 95% C.L., we can expect to exclude some new region in the parameter space thanks to this analysis.

5. Conclusions
The search for an extended Higgs sector represents one of the most important avenues for probing the possible structure of physics beyond the Standard Model. In the simplified setting of Two Higgs Doublet Models, we have explored current constraints from flavour, precision electroweak tests and direct collider searches. We have studied the reach of the CMS experiment at the LHC Run 1 for a second Higgs particle lighter than the 125 GeV Higgs boson. We have explored the different production modes (gluon fusion, vector boson fusion, associated production with a gauge boson) and the decay of the Higgs boson to two photons for the light boson. There is indeed some sensitivity in these last two production modes as shown recasting an existing Run

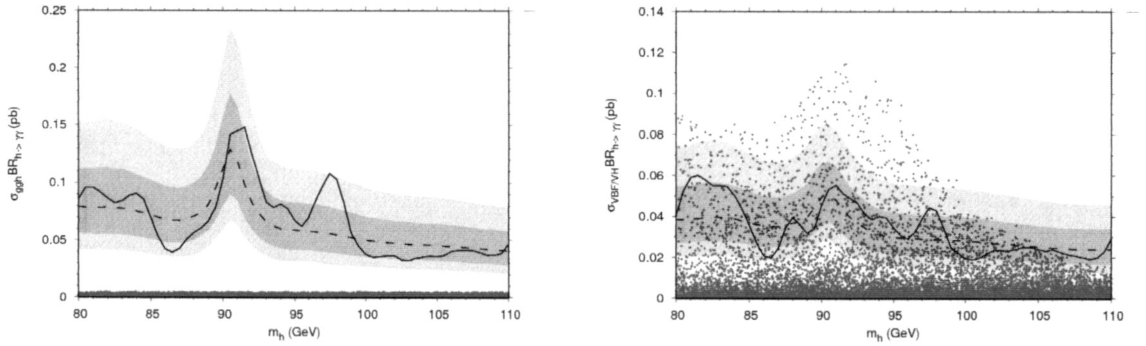

Figure 5. Points generated in the 2HDM Type I passing indirect, LEP and LHC constraints, superimposed on the results of the CMS 8 TeV low-mass di-photon analysis [11] in the gluon fusion production mode (left panel) and the combined VBF and VH production mode (right panel). The dashed line corresponds to the expected upper limit on $\sigma \times BR_{h\to\gamma\gamma}$ at 95% C.L., with 1 and 2 sigma errors in green and yellow respectively. The solid line is the observed upper limit at 95% C.L.

1 CMS analysis. A lighter neutral scalar particle is not completely excluded by present bounds and searches. Out of the four types of 2HDMs, in the low-mass region for a neutral scalar, only Type I has in its parameter space points with large enough cross section times branching ratio to allow detection or exclusion in the two photon decay channel. It is for sure interesting to perform a similar low mass search (even for lower masses than those considered at Run 1) at 13 TeV for the LHC in Run 2 as the increased sensitivity to lower cross section values allow to further explore and constrain or possibly discover new scalar neutral particles and in any case will allow a better understanding of an extended Higgs sector.

Acknowledgements
I wish to thank my collaborators on this subject: Giacomo Cacciapaglia, Suzanne Gascon-Shotkin, Solène Le Corre, Morgan Lethuillier and Junquan Tao for the many fruitful discussions and common work. AD is partially supported by Institut Universitaire de France, the Labex-LIO (Lyon Institute of Origins) under grant ANR-10-LABX-66, FRAMA (FR3127, Fédération de Recherche "André Marie Ampère") and IN2P3 Theory-LHC France funding.

References
[1] Aad G et al. (ATLAS) 2012 Phys. Lett. **B716** 1–29 (*Preprint* 1207.7214)
[2] Chatrchyan S et al. (CMS) 2012 Phys. Lett. **B716** 30–61 (*Preprint* 1207.7235)
[3] Ferreira P M, Santos R, Sher M and Silva J P 2012 Phys. Rev. **D85** 035020 (*Preprint* 1201.0019)
[4] Chang S, Kang S K, Lee J P, Lee K Y, Park S C and Song J 2013 JHEP **05** 075 (*Preprint* 1210.3439)
[5] Chang S, Kang S K, Lee J P, Lee K Y, Park S C and Song J 2014 JHEP **09** 101 (*Preprint* 1310.3374)
[6] Celis A, Ilisie V and Pich A 2013 JHEP **07** 053 (*Preprint* 1302.4022)
[7] Cacciapaglia G, Deandrea A, Drieu La Rochelle G and Flament J B 2015 Phys. Rev. **D91** 015012 (*Preprint* 1311.5132)
[8] Bernon J, Gunion J F, Haber H E, Jiang Y and Kraml S 2016 Phys. Rev. **D93** 035027 (*Preprint* 1511.03682)
[9] Bernon J, Gunion J F, Jiang Y and Kraml S 2015 Phys. Rev. **D91** 075019 (*Preprint* 1412.3385)
[10] Ellwanger U and Rodriguez-Vazquez M 2016 JHEP **02** 096 (*Preprint* 1512.04281)
[11] 2015 Search for new resonances in the diphoton final state in the mass range between 80 and 110 GeV in pp collisions at $\sqrt{s}$ = 8 TeV Tech. Rep. CMS-PAS-HIG-14-037 CERN Geneva URL http://cds.cern.ch/record/2063739
[12] Aad G et al. (ATLAS) 2014 Phys. Rev. Lett. **113** 171801 (*Preprint* 1407.6583)
[13] Peskin M E and Takeuchi T 1992 Phys. Rev. **D46** 381–409

[14] Eriksson D, Rathsman J and Stal O 2010 Comput. Phys. Commun. **181** 189–205 (*Preprint* 0902.0851)
[15] Baak M, Cth J, Haller J, Hoecker A, Kogler R, Mnig K, Schott M and Stelzer J (Gfitter Group) 2014 Eur. Phys. J. **C74** 3046 (*Preprint* 1407.3792)
[16] Arhrib A, Benbrik R and Gaur N 2012 Phys. Rev. **D85** 095021 (*Preprint* 1201.2644)
[17] Mahmoudi F 2009 Comput. Phys. Commun. **180** 1579–1613 (*Preprint* 0808.3144)
[18] Mahmoudi F 2008 Comput. Phys. Commun. **178** 745–754 (*Preprint* 0710.2067)
[19] Amhis Y et al. (Heavy Flavor Averaging Group (HFAG)) 2014 (*Preprint* 1412.7515)
[20] Hurth T, Mahmoudi F and Neshatpour S 2016 Nucl. Phys. **B909** 737–777 (*Preprint* 1603.00865)
[21] Aaij R et al. (LHCb) 2013 Phys. Rev. Lett. **111** 101805 (*Preprint* 1307.5024)
[22] Khachatryan V et al. (LHCb, CMS) 2015 Nature **522** 68–72 (*Preprint* 1411.4413)
[23] Olive K A et al. (Particle Data Group) 2014 Chin. Phys. **C38** 090001
[24] Lenz A 2014 B-mixing in and beyond the Standard model 8th International Workshop on the CKM Unitarity Triangle (CKM2014 (*Preprint* 1409.6963) URL https://inspirehep.net/record/1318833/files/arXiv:1409.6963.pdf
[25] Bechtle P, Brein O, Heinemeyer S, Weiglein G and Williams K E 2010 Comput. Phys. Commun. **181** 138–167 (*Preprint* 0811.4169)
[26] Bechtle P, Brein O, Heinemeyer S, Weiglein G and Williams K E 2011 Comput. Phys. Commun. **182** 2605–2631 (*Preprint* 1102.1898)
[27] Bechtle P et al. 2012 PoS **CHARGED2012** 024 (*Preprint* 1301.2345)
[28] Bechtle P et al. 2014 Eur. Phys. J. **C74** 2693 (*Preprint* 1311.0055)
[29] Aad G et al. (ATLAS, CMS) 2016 (*Preprint* 1606.02266)
[30] Flament J B 2015 (*Preprint* 1504.07919)
[31] Andersen J R et al. (LHC Higgs Cross Section Working Group) 2013 (*Preprint* 1307.1347)
[32] Cacciapaglia G, Deandrea A and Llodra-Perez J 2009 JHEP **06** 054 (*Preprint* 0901.0927)
[33] Cacciapaglia G, Deandrea A, Drieu La Rochelle G and Flament J B 2014 Phys. Rev. Lett. **113** 201802 (*Preprint* 1406.1757)
[34] Abbiendi G et al. (LEP, DELPHI, OPAL, ALEPH, L3) 2013 Eur. Phys. J. **C73** 2463 (*Preprint* 1301.6065)
[35] Abdesselam A et al. (Belle) 2016 Measurement of the inclusive $B \to X_{s+d}\gamma$ branching fraction, photon energy spectrum and HQE parameters 38th International Conference on High Energy Physics (ICHEP 2016) Chicago, IL, USA, August 03-10, 2016 (*Preprint* 1608.02344) URL https://inspirehep.net/record/1479946/files/arXiv:1608.02344.pdf
[36] Misiak M and Steinhauser M 2017 (*Preprint* 1702.04571)
[37] Artoisenet P et al. 2013 JHEP **11** 043 (*Preprint* 1306.6464)
[38] Spira M, Djouadi A, Graudenz D and Zerwas P M 1995 Nucl. Phys. **B453** 17–82 (*Preprint* hep-ph/9504378)

IOP Conf. Series: Journal of Physics: Conf. Series **878** (2017) 012020 doi:10.1088/1742-6596/878/1/012020

Higgs production at NLO in the Standard Model Effective Field theory

Nicolas Deutschmann[1,2]

[1]Center for Cosmology, Particle Physics and Phenomenology (CP3), Université Catholique de Louvain, 1348 Louvain-La-Neuve, Belgium.

[2]Univ. Lyon, Université Claude Bernard Lyon 1, CNRS/IN2P3, UMR5822 IPNL, F-69622, Villeurbanne, France.

E-mail: n.deutschmann@ipnl.in2p3.fr

Abstract. The Effective Field Theory approach is a fruitful way of constraining new physics in a model independent way. As the Higgs sector is one of the most popular candidate for deviations from the Standard Model, it is particularly important that the constraints extracted from the experimental data on the Higgs boson be as meaningful as possible, which entails making accurate and precise theoretical predictions.

In this proceeding, I discuss a two loop calculation performed to improve the existing Leading Order result for the Higgs gluon-fusion cross section in the Standard Model Effective Field Theory. We present the modern multi-loop calculation techniques employed to obtain this result and highlight the unusual divergence structure of the amplitude.

1. Introduction

The Run I of the LHC has yielded the much awaited discovery of a scalar boson with properties compatible with the Standard Model (SM) Higgs boson [1, 2]. Observing the Higgs boson at ATLAS and CMS in different production channels and decay modes has opened the possibility to probe its couplings to other SM particles, showing a good agreement between predictions and observations [3].

Despite the current success of the SM in predicting Higgs boson properties and other LHC phenomena, we know that our theory for fundamental interactions is incomplete. The Standard Model Effective Field Theory (SMEFT) provides a consistent framework to parametrize possible deviations in a model independent way by encapsulating new physics effects in higher-dimensional operators involving the SM fields [4–6]. This approach is very powerful as it can be systematically related to any renormalizable theory beyond the SM (BSM) in which all the new fields are much heavier than LHC energies.

Recent years have seen a rapid progress in the development of precise predictions in the SMEFT, in particular in the form of a number of next-to-leading-order (NLO) results in top physics [8–11] and in Higgs boson physics [12–15].

Our work is part of this effort to provide precise SMEFT predictions for key LHC processes. We wish to improve the existing leading-order (LO) prediction for Higgs-boson production through gluon fusion [16] by including the NLO QCD radiative corrections. We focus on operators that modify the couplings of the top quark and the Higgs boson.

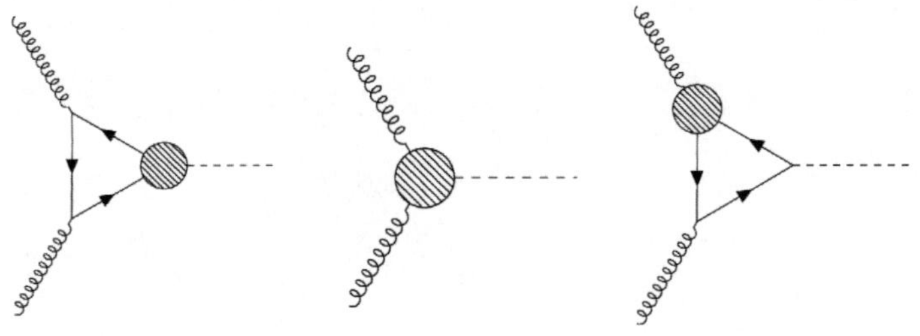

(a) O_1 (b) O_2 (c) O_3

Figure 1: Sample diagrams in the leading order contribution to gluon fusion for O_1, O_2, and O_3. The blobs indicate the dimension-6 operators.

This proceeding focuses on the calculation of the virtual corrections to gluon fusion in the SMEFT and is organized as follows: section 2 details how we set up the calculation in the SMEFT and discusses the relevant operators. section 3 discusses the structure of the amplitude. Finally, section 4 describes the multi-loop techniques used for the evaluation of the virtual corrections.

2. EFT Setup

The SMEFT is the extension of the SM by the tower of higher dimensional operators involving SM fields:

$$\mathcal{L}_{\text{SMEFT}} = \mathcal{L}_{\text{SM}} + \sum_{i=1}^{\infty}\sum_{j=0}^{N_i} \frac{C_j^{(i)}}{\Lambda^i} O_j^{(i)}, \tag{1}$$

where the operators $O_j^{(i)}$ have an energy dimension of $4+i$, Λ is an energy scale larger than typical LHC energies and $C_j^{(i)}$ are dimensionless couplings. While non-renormalizable in general, the SMEFT Lagrangian can be safely truncated at some order $1/\Lambda^n$ and the resulting set of operators is stable under renormalization as long as observables are also truncated at that order. For most practical purposes it is sufficient to work at the order $1/\Lambda^2$: it is the first non-trivial order which features operators that affect virtually all aspects of LHC physics. A dimension 5 operator exists but is related to neutrino physics and bears little effect on collider observables [4].

Three CP-conserving operators affect gluon fusion by modifying the top loop [11]:

$$O_1 = \bar{Q}_L \phi t_R (\phi^\dagger \phi - \frac{v^2}{2}) + \text{h.c.}, \tag{2}$$

$$O_2 = g_s^2 G^a_{\mu\nu} G_a^{\mu\nu} (\phi^\dagger \phi - \frac{v^2}{2}) + \text{h.c.}, \tag{3}$$

$$O_3 = g_s \bar{Q}_L T_a \sigma^{\mu\nu} t_R G^a_{\mu\nu} \phi + \text{h.c.}. \tag{4}$$

Their leading-order contribution to this process is shown in figure 1. The one-loop renormalization of the SMEFT has been studied in detail [7] and we can use these results to obtain the required counterterms. We work in dimensional regularization with a number of dimensions $d = 4 - 2\epsilon$. We use the same conventions as [11], adding to each operator its Hermitian conjugate even if it is already Hermitian, and subtracting vacuum expectation values from operators to avoid shifting SM definitions of fields and couplings. This leads to including counterterms for the strong coupling and the top quark mass proportional to EFT couplings.

The renormalization of the relevant parameters is the following:

$$
\begin{aligned}
Z_g &= \mu^{-2\epsilon}\left(1+\alpha_s\tfrac{\delta Z_g}{\epsilon}+\alpha_s C_3\tfrac{\delta Z_g^C}{\epsilon}\right), & \alpha_s^B &= \mu^{2\epsilon}S_\epsilon^{-1}\alpha_s(1+\alpha_s\tfrac{\delta Z_\alpha}{\epsilon}+\alpha_s C_3\tfrac{\delta Z_\alpha^C}{\epsilon}),\\
m_t^B &= m_t(1+\alpha_s\tfrac{\delta Z_m}{\epsilon})+\alpha_s C_3\tfrac{\delta Z_m^C}{\epsilon}, & C_1^B &= \mu^{3\epsilon}\left(C_1(1+\alpha_s\tfrac{\delta Z_{11}}{\epsilon})+C_3\alpha_s\tfrac{\delta Z_{13}}{\epsilon}\right),\\
C_3^B &= \mu^\epsilon C_3(1+\alpha_s\tfrac{\delta Z_{33}}{\epsilon}), & C_2^B &= C_2+C_3\tfrac{\delta Z_{23}}{\epsilon}+\alpha_s C_3\left(\tfrac{\delta Z_{23}^2}{\epsilon^2}+\tfrac{\delta Z_{23}^1}{\epsilon}\right).
\end{aligned} \tag{5}
$$

We denote bare parameters with a superscript B and renormalized parameters without a superscript. The renormalization scale is μ, Z_g is the gluon field renormalization constant and the δZ_X are the counterterms (the superscript C indicates counterterms to SM parameters generated by the EFT). Finally, $S_\epsilon = (4\pi)^\epsilon \exp(-\gamma\epsilon)$ and γ is the Euler-Mascheroni constant.

These counterterms highlight a recurrent theme in EFT renormalization: operator mixing. The bare SMEFT couplings are expressed not only in terms of their renormalized counterpart, but also have contributions coming from other SMEFT couplings. This behavior does not only impact on counterterms, but also the running of these operators, which mix together when the renormalization scale changes. All the counterterms required for the calculation exist in the literature except for δZ_{23}^1, which is a two-loop counterterm that will be fixed by our calculation.

3. Structure of the amplitude

The SM gluon fusion process, as well as the contributions generated by O_1 and O_3 are loop-induced processes and their NLO virtual corrections will therefore consist of two-loop diagrams. We will first focus on the LO calculation, which has unusual features, and then move on to discussing the NLO corrections.

3.1. Leading-order amplitude

The amplitude $\mathcal{A}$ at leading order is the sum of four contributions, which are exemplified in figure 1:

$$\mathcal{A} = \alpha_s^B \mathcal{A}_{\text{SM}}^{(0)}(m_t^B) + \alpha_s^B C_1^B \mathcal{A}_1^{(0)}(m_t^B) + \alpha_s^B C_2^B \alpha_s^B \mathcal{A}_2^{(0)}(m_t^B) + C_3^B \mathcal{A}_3^{(0)}(m_t^B), \tag{6}$$

where $\mathcal{A}_{\text{SM}}^{(0)}$ is the sum of LO SM diagrams and the $\mathcal{A}_i^{(0)}$ are the sums of diagrams containing operators O_i. It is interesting to note that the diagrams in $\mathcal{A}_3^{(0)}$ have UV divergences already at leading order. This divergence has already been noted in previous work [16, 17] and the LO amplitude is renormalized by the leading-order mixing of C_2 with C_3: $C_2^B = C_2 + C_3(\delta Z_{23}/\epsilon) + \mathcal{O}(\alpha_s)$. As a result the expression of the amplitude in terms of renormalized quantities is

$$
\begin{aligned}
\mathcal{A} = {} & \alpha_s \mathcal{A}_{\text{SM}}^{(0)}(m_t) + \alpha_s C_1 \mathcal{A}_1^{(0)}(m_t) + \alpha_s C_2 \mathcal{A}_2^{(0)}(m_t) \\
& + \alpha_s C_3 \left(\mathcal{A}_3^{(0)}(m_t) + S_\epsilon^{-1}\frac{\delta Z_{23}}{\epsilon}\mathcal{A}_2^{(0)}(m_t)\right) + \mathcal{O}(\alpha_s^2).
\end{aligned} \tag{7}
$$

3.2. Next-to-leading-order amplitude

The next-to-leading order correction to the amplitude is the sum of two contributions: virtual corrections, which are corrections of the LO arising through the exchange of a virtual gluon, and real emissions, in which an extra parton is emitted. In the final cross section the square of the real emissions and the interference between the Born-level diagrams and the renormalized virtual

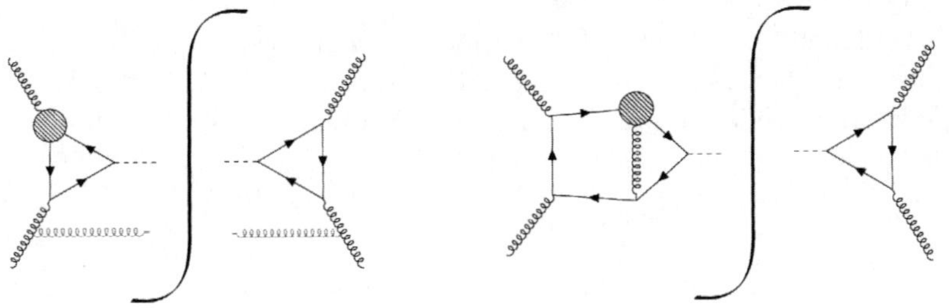

(a) Reals (b) Virtuals

Figure 2: Real and virtual contributions to the square amplitude. The two terms around the $\int$ symbols represent the product of a piece of the amplitude and another of the complex-conjugate amplitude.

corrections contribute to the same order and constitute the NLO corrections to the amplitude as shown in figure 2. The real emissions are at most one-loop diagrams and can therefore be handled by `MadGraph5_aMC@NLO` [18]. We will focus on discussing the virtual corrections, which include two-loop diagrams and exhibit more unusual behavior related to the leading-order renormalization. In terms of bare quantities, the amplitude at NLO is expressed as

$$\begin{aligned} \mathcal{A} = Z_g \Big(&\alpha_s^B \mathcal{A}_{\rm SM}^{(0)}(m_t^B) + \alpha_s^B C_1^B \mathcal{A}_1^{(0)}(m_t^B) + \alpha_s^B C_2^B \alpha_s^B \mathcal{A}_2^{(0)}(m_t^B) + C_3^B \mathcal{A}_3^{(0)}(m_t^B) \\ &+(\alpha_s^B)^2 \mathcal{A}_{\rm SM}^{(1)}(m_t^B) + (\alpha_s^B)^2 C_1^B \mathcal{A}_1^{(1)}(m_t^B) + (\alpha_s^B)^2 C_2^B \mathcal{A}_2^{(1)}(m_t^B) + (\alpha_s^B)^2 C_3^B \mathcal{A}_3^{(1)}(m_t^B) \Big), \end{aligned} \tag{8}$$

where $\mathcal{A}_{\rm SM}^{(1)}$ is the sum of NLO SM diagrams and the $\mathcal{A}_i^{(1)}$ are the sum of NLO diagrams generated by operator O_i. We can then express our amplitude in terms of renormalized quantities

$$\begin{aligned} \mathcal{A} = {} & \alpha_s \mathcal{A}_{\rm SM}^{(0)}(m_t) + \alpha_s^2 S_\epsilon^{-1} \left(\mathcal{A}_{\rm SM}^{(1)}(m_t) + \frac{\delta Z_\alpha + \delta Z_g}{\epsilon} \mathcal{A}_{\rm SM}^{(0)}(m_t) + m_t \frac{\delta Z_m}{\epsilon} \frac{\partial \mathcal{A}_{\rm SM}^{(0)}(m_t)}{\partial m_t} \right) \\ & + \alpha_s C_1 \mathcal{A}_1^{(0)}(m_t) + \alpha_s^2 C_1 S_\epsilon^{-1} \left(\mathcal{A}_1^{(1)}(m_t) + \frac{\delta Z_\alpha + \delta Z_g + \delta Z_{11}}{\epsilon} \mathcal{A}_1^{(0)}(m_t) + m_t \frac{\delta Z_m}{\epsilon} \frac{\partial \mathcal{A}_1^{(0)}(m_t)}{\partial m_t} \right) \\ & + \alpha_s C_2 \mathcal{A}_2^{(0)}(m_t) + \alpha_s^2 C_2 S_\epsilon^{-1} \left(\mathcal{A}_2^{(1)}(m_t) + \frac{\delta Z_\alpha + \delta Z_g}{\epsilon} \mathcal{A}_2^{(0)}(m_t) + m_t \frac{\delta Z_m}{\epsilon} \frac{\partial \mathcal{A}_2^{(0)}(m_t)}{\partial m_t} \right) \\ & + \alpha_s C_3 \left(\mathcal{A}_3^{(0)}(m_t) + \frac{\delta Z_{23}}{\epsilon} \mathcal{A}_2^{(0)}(m_t) \right) + \alpha_s^2 C_3 S_\epsilon^{-1} \Bigg[\mathcal{A}_3^{(1)}(m_t) + \frac{\delta Z_\alpha + \delta Z_g + \delta Z_{33}}{\epsilon} \mathcal{A}_3^{(0)}(m_t) \\ & + \frac{\delta Z_{13}}{\epsilon} \mathcal{A}_1^{(0)}(m_t) + \frac{\delta Z_{23}}{\epsilon} \mathcal{A}_2^{(1)}(m_t) + \left(\frac{\delta Z_{23}^1}{\epsilon} + \frac{\delta Z_{23}^2}{\epsilon^2} \right) \mathcal{A}_2^{(0)}(m_t) + \frac{\delta Z_\alpha^C}{\epsilon} \mathcal{A}_{\rm SM}^{(0)}(m_t) \\ & + m_t \frac{\delta Z_m}{\epsilon} \frac{\partial \mathcal{A}_3^{(0)}(m_t)}{\partial m_t} + m_t \frac{\delta Z_m^C}{\epsilon} \frac{\partial \mathcal{A}_{\rm SM}^{(0)}(m_t)}{\partial m_t} \Bigg], \end{aligned} \tag{9}$$

which makes apparent the UV divergences of the amplitude.

The renormalized NLO virtual amplitude still has divergences that will be canceled against divergences that appear in the infrared and collinear regions of the phase space of the extra parton in the real emission contributions [19]. The structure of infrared and collinear divergences is universal: for the virtual contributions, we know that the divergences of the renormalized amplitude $\mathcal{A} = \alpha_s/(2\pi)\mathcal{A}^{(0)} + (\alpha_s/(2\pi))^2\mathcal{A}^{(1)}$ appear in a factorized form: $\mathcal{A}^{(1)} = \mathcal{A}^{(1)}_{\text{finite}} + I_1\mathcal{A}^{(0)}$, where I_1 is a universal operator acting on the colors of the external particles. In the case of $gg \to H$, the color structure is trivial, which makes I_1 take the simple form

$$I_1^{gg\to H} = -\frac{e^{\epsilon\gamma}}{\Gamma(1-\epsilon)}\left(\frac{C_A}{\epsilon^2} + \frac{\beta_0}{\epsilon}\right)\left(\frac{\mu^2}{-s}\right)^{\epsilon}. \tag{10}$$

For example, if we look at the part of our amplitude proportional to C_2, we therefore know that

$$\mathcal{A}_2^{(1)}(m_t) + \frac{\delta Z_\alpha + \delta Z_g}{\epsilon}\mathcal{A}_2^{(0)}(m_t) + m_t\frac{\delta Z_m}{\epsilon}\frac{\partial \mathcal{A}_2^{(0)}(m_t)}{\partial m_t} = \mathcal{A}^{(1)}_{2,\text{finite}} + I_1\mathcal{A}_2^{(0)}. \tag{11}$$

By subtracting its infrared poles from the complete amplitude, we obtain an expression that we know to be finite. We can therefore use this constraint to fix the value of δZ^1_{23} by imposing that all the poles in the amplitude cancel.

4. Multi-loop calculation techniques

The virtual corrections to the gluon fusion process consist of a combination of one- and two-loop diagrams. In this section we discuss the techniques we used to evaluate them.

The first step in such a calculation is to obtain expressions for the diagrams. To this end we use `QGRAF` to generate diagrams that are then translated to `FORM` expressions using a private `Python` program. These expressions are manipulated in `FORM` to evaluate Dirac matrix traces and index contractions and then imported using `Mathematica`. It is much easier to work with scalar objects, which we can do by exploiting the simple Lorenz structure of the amplitude:

$$\mathcal{A} = \varepsilon_\mu(p_1)\varepsilon_\nu(p_2)\delta_{ab}\left(p_1\cdot p_2 g^{\mu\nu} - p_1^\mu p_2^\nu - p_1^\nu p_2^\mu\right)\mathcal{M}, \tag{12}$$

where a, b are the color indices. The two-loop part of this form factor $\mathcal{M}$ is a linear combination of scalar integrals:

$$\mathcal{M}\big|_{2L} = \sum a_i \int \frac{d^dk_1}{(2\pi)^d}\frac{d^dk_2}{(2\pi)^d}\frac{\mathcal{N}_i}{D_{i1}\dots D_{iq}}, \tag{13}$$

where the numerators $\mathcal{N}$ are products of contracted momenta and the denominators are the usual $D_{ij} = (Q_{ij}^2 - m_{ij}^2)$ where Q_{ij} is some linear combination of momenta and m_{ij} some internal mass (here the top mass or 0). These integrals number in the hundreds and it would not be wise to evaluate them one by one; an organizational principle is required. Integrals are regrouped into "families", also called "topologies", which are sets of integrals that share a common list of denominators with arbitrary powers. In our case we defined three families, described in figure 3, which can account for all integrals in our amplitude. Numerators are also accounted for using negative powers of the denominators. For example,

$$\begin{aligned}\int \frac{d^dk_1}{(2\pi)^d}\frac{d^dk_2}{(2\pi)^d}\frac{k_2\cdot k_2}{D_{11}D_{12}D_{17}} &= \int \frac{d^dk_1}{(2\pi)^d}\frac{d^dk_2}{(2\pi)^d}\frac{m_t^2}{D_{11}D_{12}D_{17}} \\ &\quad + \int \frac{d^dk_1}{(2\pi)^d}\frac{d^dk_2}{(2\pi)^d}\frac{1}{D_{11}D_{12}\left(D_{16}\right)^{-1}D_{17}} \\ &= m_t^2 I_1(1,1,0,0,0,0,1) + I_1(1,1,0,0,0,-1,1).\end{aligned} \tag{14}$$

As a result we can specify any integral in a given family by giving the exponent of each integral, and each can be seen as a point in $\mathbb{Z}^n$ ($n = 7$ here).

IOP Conf. Series: Journal of Physics: Conf. Series **878** (2017) 012020 doi:10.1088/1742-6596/878/1/012020

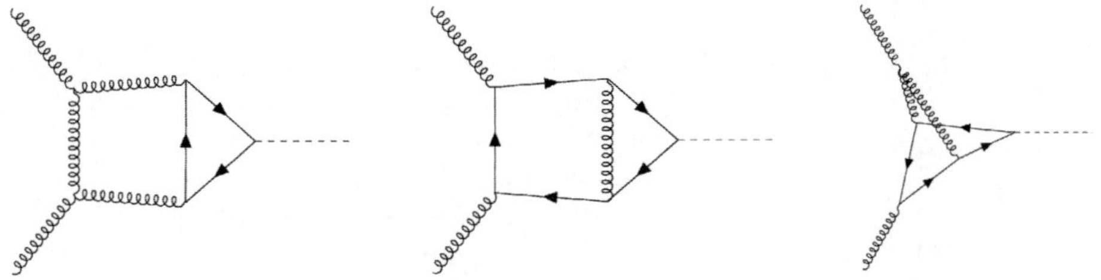

Topology 1	Topology 2	Topology 3
$D_{11} = k_1^2$	$D_{21} = k_1^2 - m_t^2$	$D_{31} = k_1^2 - m_t^2$
$D_{12} = (k_1 + p_1)^2$	$D_{22} = (k_1 + p_2)^2 - m_t^2$	$D_{23} = (k_1 - k_2 - p_1)^2$
$D_{13} = (k_1 + p_1 + p_2)^2$	$D_{23} = (k_1 + p_1 + p_2)^2 - m_t^2$	$D_{33} = (k_1 + p_1 + p_2)^2 - m_t^2$
$D_{14} = (k_2 + p_1 + p_2)^2 - m_t^2$	$D_{24} = (k_2 + p_1 + p_2)^2 - m_t^2$	$D_{34} = (k_2 + p_1 + p_2)^2 - m_t^2$
$D_{15} = (k_2 + p_1)^2 - m_t^2$	$D_{25} = (k_2 + p_2)^2 - m_t^2$	$D_{35} = (k_2 + p_1)^2 - m_t^2$
$D_{16} = k_2^2 - m_t^2$	$D_{26} = k_2^2 - m_t^2$	$D_{36} = (k_1 + p_1)^2 - m_t^2$
$D_{17} = (k_1 - k_2)^2 - m_t^2$	$D_{27} = (k_1 - k_2)^2$	$D_{37} = (k_1 - k_2)^2$,

Figure 3: The three families used for our calculations. The set of denominators for each family is the same as the denominators appearing in each diagram.

Integral families are not only a bookkeeping device, we know that the infinite number of integrals within a family can always be expressed as a linear combination of a finite basis of integrals [29], often called *master integrals*. Not only do we know a finite basis exists, but we also know how to obtain it using a method called the Laporta algorithm [30]. This method relies on Integration-By-Parts Identities (IBP), which exploit the fact that integrals of total derivatives are 0 in dimensional regularization and that integral families are stable under differentiation,

$$\int \frac{d^d k_1}{(2\pi)^d} \frac{d^d k_2}{(2\pi)^d} \frac{\partial}{\partial k_i^\mu} \frac{v^\mu}{D_{i1}^{n_1} \dots D_{i7}^{n_7}} = 0. \tag{15}$$

The derivative shifts powers in the denominators and generates numerators, meaning that the left hand side of equation 15 is a linear combination of integrals in the topology. Let us see in a simple example how this can be used to find a basis for a family of master integrals. Our example will be the family of one-loop bubble integrals: $I_b(a_1, a_2) = \int d^d k \frac{1}{(k^2)^{a_1}((k+p)^2)^{a_2}}$.

We can generate two independent IBP relations (differentiating with respect to k_1 or k_2) for this family:

$$I_b(a_1, a_2) = \frac{a_1 + a_2 - d - 1}{p^2(a_2 - 1)} I(a_1, a_2 - 1) + \frac{1}{p^2} I(a_1 - 1, a_2) \tag{16}$$

$$= \frac{a_1 + a_2 - d - 1}{p^2(a_1 - 1)} I(a_1 - 1, a_2) + \frac{1}{p^2} I(a_1, a_2 - 1). \tag{17}$$

These have been set up to highlight how we use these relations: to reduce the powers of the denominators. If we need an integral with a given set of (a_1, a_2), we can express it using these relations recursively in terms of integrals with lower powers. Fortunately, the recursion ends when either power is 0 because the resulting integral is scaleless: $I_b(a_1, 0) = I_b(0, a_2) = 0$. As shown in figure 4, this results in any integral being expressed in terms of $I(1,1)$. In more complicated cases, not all integrals can be expressed in terms of a single master, but a linear

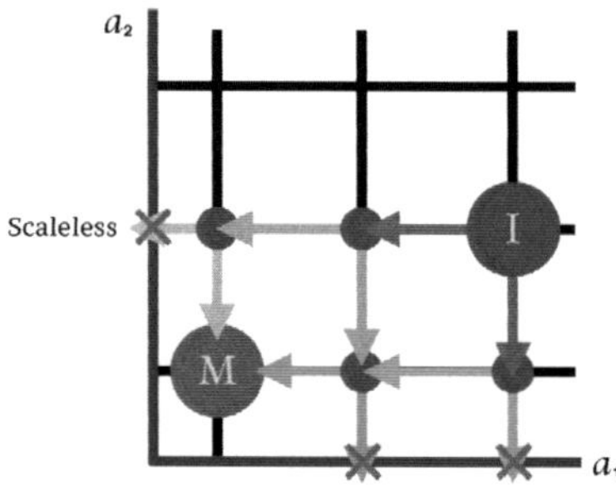

Figure 4: Reduction of $I_b(3, 2)$ to the master integral $I_b(1, 1)$ by recursive application of the IBP identities for the bubble topology illustrated in the plane (a_1, a_2). The recursion ends on the lines $a_1 = 0$ and $a_2 = 0$ because all integrals on these are scaleless.

combination of several integrals, which can be obtained from others by recursively applying IBP relations.

In our case the 1969 integrals we need to evaluate can be expressed as linear combinations of only 17 master integrals, which are the same as those required for the SM calculation of Higgs production at two loops [31, 32]. We performed this reduction with `LiteRed` [24] and cross-checked it with `FIRE` [25]. All 17 of them have been calculated in previous work [31–35] and we could use these results to evaluate $\mathcal{M}$.

The final amplitude is a linear combination of multiple polylogarithms (MPL) depending on the variable

$$x = \frac{\sqrt{4m_t^2 - s} - \sqrt{-s}}{\sqrt{4m_t^2 - s} + \sqrt{-s}}. \tag{18}$$

This class of function appears in a large number of loop calculations. MPLs are expressed as iterated integrals over a logarithmic kernel:

$$G(a_1, \ldots, a_n; x) = \int_0^x \frac{dt}{t - a_1} G(a_2, \ldots, a_n, t) \tag{19}$$

$$G(a; x) = \log(1 - x/a). \tag{20}$$

Using these expressions for the scalar integrals, we can evaluate the amplitude in a closed form. We checked our implementation of the published integrals by numerically evaluating their expressions in terms of polylogarithms with `Ginac` [28], and comparing with the direct numerical evaluation of the integrals in `FIESTA` [26] and `SecDec` [27]. The explicit form of the amplitude is of little interest for this proceeding, but its evaluation allowed us to determine the value of the new counterterm that appears in the calculation, which we will report in an upcoming paper.

5. Summary and conclusion

We have reported on the progress achieved in the calculation of the NLO virtual amplitude for the Higgs boson production through gluon fusion. We have shown that several dimension-six operators can modify this process in the SMEFT and discussed how modern multi-loop techniques have enabled us to compute the amplitude for virtual corrections to this process. We are now poised to combine this result with the automatic calculation of the real emission contribution in `MadGraph5_aMC@NLO` and to provide a more detailed discussion of our result as well as to study the NLO phenomenology of this process in a forthcoming paper.

References

[1] S. Chatrchyan *et al.* [CMS Collaboration], Phys. Lett. B **716** (2012) 30 [arXiv:1207.7235 [hep-ex]].

[2] G. Aad *et al.* [ATLAS Collaboration], Phys. Lett. B **716** (2012) 1 [arXiv:1207.7214 [hep-ex]].
[3] G. Aad *et al.* [ATLAS and CMS Collaborations], JHEP **1608** (2016) 045 [arXiv:1606.02266 [hep-ex]].
[4] S. Weinberg, Physica A **96** (1979) 327.
[5] W. Buchmuller and D. Wyler, Nucl. Phys. B **268** (1986) 621.
[6] C. N. Leung, S. T. Love and S. Rao, Z. Phys. C **31** (1986) 433.
[7] E. E. Jenkins, A. V. Manohar and M. Trott, JHEP **1310** (2013) 087 [arXiv:1308.2627 [hep-ph]].
[8] D. Buarque Franzosi and C. Zhang, Phys. Rev. D **91** (2015) no.11, 114010 [arXiv:1503.08841 [hep-ph]].
[9] C. Zhang, Phys. Rev. Lett. **116** (2016) no.16, 162002 [arXiv:1601.06163 [hep-ph]].
[10] O. Bessidskaia Bylund, F. Maltoni, I. Tsinikos, E. Vryonidou and C. Zhang, JHEP **1605** (2016) 052 [arXiv:1601.08193 [hep-ph]].
[11] F. Maltoni, E. Vryonidou and C. Zhang, JHEP **1610** (2016) 123 [arXiv:1607.05330 [hep-ph]].
[12] C. Hartmann and M. Trott, JHEP **1507** (2015) 151 [arXiv:1505.02646 [hep-ph]].
[13] M. Ghezzi, R. Gomez-Ambrosio, G. Passarino and S. Uccirati, JHEP **1507** (2015) 175 [arXiv:1505.03706 [hep-ph]].
[14] C. Hartmann and M. Trott, Phys. Rev. Lett. **115** (2015) no.19, 191801 [arXiv:1507.03568 [hep-ph]].
[15] R. Gauld, B. D. Pecjak and D. J. Scott, JHEP **1605** (2016) 080 [arXiv:1512.02508 [hep-ph]].
[16] C. Degrande, J. M. Gerard, C. Grojean, F. Maltoni and G. Servant, JHEP **1207** (2012) 036 Erratum: [JHEP **1303** (2013) 032] [arXiv:1205.1065 [hep-ph]].
[17] M. Grazzini, A. Ilnicka, M. Spira and M. Wiesemann, PoS EPS **-HEP2015** (2015) 144 [arXiv:1511.08059 [hep-ph]].
[18] J. Alwall *et al.*, JHEP **1407** (2014) 079 [arXiv:1405.0301 [hep-ph]].
[19] S. Catani and M. H. Seymour, Nucl. Phys. B **485** (1997) 291 Erratum: [Nucl. Phys. B **510** (1998) 503] [hep-ph/9605323].
[20] S. Frixione, Z. Kunszt and A. Signer, Nucl. Phys. B **467** (1996) 399 [hep-ph/9512328].
[21] R. Frederix, S. Frixione, F. Maltoni and T. Stelzer, JHEP **0910** (2009) 003 [arXiv:0908.4272 [hep-ph]].
[22] P. Nogueira, J. Comput. Phys. **105** (1993) 279.
[23] J. Kuipers, T. Ueda, J. A. M. Vermaseren and J. Vollinga, Comput. Phys. Commun. **184** (2013) 1453 [arXiv:1203.6543 [cs.SC]].
[24] R. N. Lee, J. Phys. Conf. Ser. **523** (2014) 012059 [arXiv:1310.1145 [hep-ph]].
[25] A. V. Smirnov, Comput. Phys. Commun. **189** (2015) 182 [arXiv:1408.2372 [hep-ph]].
[26] A. V. Smirnov, Comput. Phys. Commun. **204** (2016) 189 [arXiv:1511.03614 [hep-ph]].
[27] S. Borowka, G. Heinrich, S. P. Jones, M. Kerner, J. Schlenk and T. Zirke, Comput. Phys. Commun. **196** (2015) 470 [arXiv:1502.06595 [hep-ph]].
[28] J. Vollinga, Nucl. Instrum. Meth. A **559** (2006) 282 [hep-ph/0510057].
[29] K. G. Chetyrkin and F. V. Tkachov, Nucl. Phys. B **192** (1981) 159.
[30] S. Laporta, Int. J. Mod. Phys. A **15** (2000) 5087 [hep-ph/0102033].
[31] U. Aglietti, R. Bonciani, G. Degrassi and A. Vicini, JHEP **0701** (2007) 021 [hep-ph/0611266].
[32] C. Anastasiou, S. Beerli, S. Bucherer, A. Daleo and Z. Kunszt, JHEP **0701** (2007) 082 [hep-ph/0611236].
[33] R. Bonciani, P. Mastrolia and E. Remiddi, Nucl. Phys. B **661** (2003) 289 Erratum: [Nucl. Phys. B **702** (2004) 359] [hep-ph/0301170].
[34] R. Bonciani, P. Mastrolia and E. Remiddi, Nucl. Phys. B **690** (2004) 138 [hep-ph/0311145].
[35] R. Bonciani, V. Del Duca, H. Frellesvig, J. M. Henn, F. Moriello and V. A. Smirnov, JHEP **1612** (2016) 096 [arXiv:1609.06685 [hep-ph]].

IOP Conf. Series: Journal of Physics: Conf. Series **878** (2017) 012021 doi:10.1088/1742-6596/878/1/012021

Time Dependent $\hat{q}$ from AdS/CFT

W. A. Horowitz

Department of Physics, University of Cape Town, Private Bag X3, Rondebosch 7701, South Africa

E-mail: wa.horowitz@uct.ac.za

Abstract. We present the first ever AdS/CFT calculation of $\hat{q}$ for a light quark jet as a function of position or, equivalently, time. Our result does not suffer from the gamma factor blow up of the usual time-independent AdS/CFT heavy quark setup and is qualitatively similar to, but about a factor of 2/3 larger than, the light flavor result from Liu, Rajagopal, and Wiedemann. Our findings can be immediately implemented into any $\hat{q}$-based energy loss model.

Our $\hat{q}$ derivation relies on our calculation of the average distance squared, $s^2(t)$, travelled by the endpoint of a string falling in an AdS_3-Schwarzschild spacetime. The early time behavior is ballistic, $s^2(t) \sim t^2$, but the late time behavior is the usual diffusive Brownian motion, $s^2(t) \sim t$. These late time dynamics are universal and depend only on the near-horizon physics, which allows us to generalize our results to arbitrary dimensions and thus make contact with the physics explored by RHIC and LHC.

Additionally, we find that AdS/CFT predicts angular ordering for radiation in a medium, just as in vacuum, and in contradistinction to weak-coupling. Finally, our results also imply, sensibly, that AdS/CFT predicts a smooth interpolation between the angular correlations of open heavy flavor and light flavor observables.

1. Introduction

The remarkable success of nearly inviscid relativistic hydrodynamics in high multiplicity pp, pA, and AA collisions at RHIC and LHC [1–4] suggests the medium created in these collisions is strongly coupled and thus best described using the techniques of AdS/CFT [5]. A quantitative understanding of high momentum particles interacting with the quark-gluon plasma produced in these colliders provides a crucial test of this understanding of the dynamics of many-body quantum chromodynamics (QCD) at temperatures $T \gtrsim T_c$, where T_c is the temperature at which a phase transition from normal nuclear matter occurs [6–8].

Previous work [9–11] has demonstrated that AdS/CFT [12, 13] can provide a qualitatively correct description of the suppression pattern of decay products associated with open heavy flavor production at RHIC and LHC. Crucially, one must include the momentum fluctuations due to the thermal nature of the QGP medium [14–16]. Additional work has shown that, after a renormalization procedure, AdS/CFT also provides a qualitatively similar suppression for light flavor jets when compared to data [17–19].

It is natural then to ask: what is the impact of momentum fluctuations on the propagation of high momentum heavy quarks and, equivalently, light quarks?

We report in this proceedings the first attempt to answer this question.

IOP Conf. Series: Journal of Physics: Conf. Series **878** (2017) 012021 doi:10.1088/1742-6596/878/1/012021

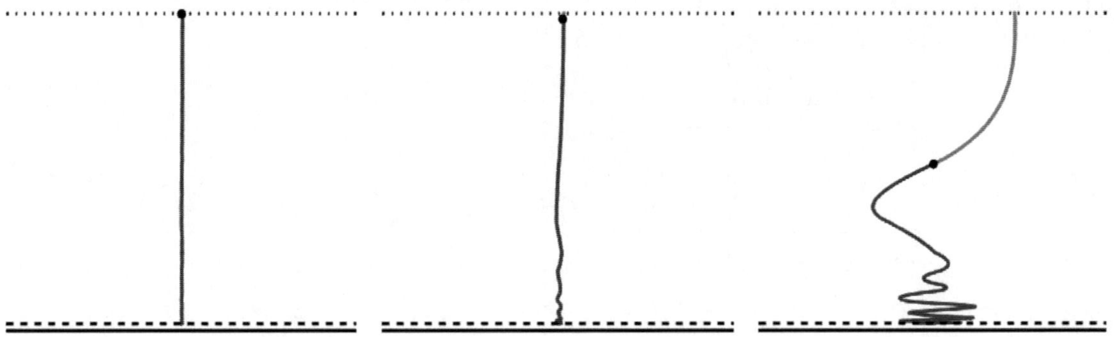

Figure 1: Snapshots in time (with time increasing from left to right) of a falling limp noodle (blue) equilibrating with a black hole (whose horizon is given by the dashed black line). The free endpoint of the falling limp noodle can be interpreted as an observer travelling down the stretched string of de Boer et al. [20] (red curve) at the local speed of light.

2. Method

Computing the thermal momentum spectrum for a light quark propagating with some velocity through a strongly coupled plasma is a very difficult problem. We thus first answer the question: what is the Brownian diffusion of the endpoint of a falling string (wobbly limp noodle) initially at rest in a strongly coupled plasma? See Fig. 1. Again, the problem is very difficult to solve in an AdS space in general dimension, so we will focus on AdS_3.

The outline of the solution is to 1) derive the classical solution; 2) quantize the perpendicular directions of motion on the string given the classical solution; 3) populate the quanta in these perpendicular directions according to Bose statistics, which is equivalent to the semi-classical approximation; and 4) compute the relevant correlators. For full details, one may read [21].

We will find it useful to interpolate between the heavy flavor result for the distance squared traversed by the string endpoint and that for a light quark. For a heavy quark, the free string endpoint stays at a fixed height in the direction orthogonal to the black brane horizon. For a light quark, the free string endpoint falls at the local speed of light. We will take a to be the fraction of the local speed of light at which we allow the free string endpoint to fall. Then we have that

$$s^2(t;a) = \langle :(\hat{X}_{\text{End}}(t;a)) - \hat{X}_{\text{End}}(0;a))^2 : \rangle \tag{1}$$

$$= \frac{\beta^2}{4\pi^2\sqrt{\lambda}} \int_0^\infty \frac{d\omega}{\omega} \frac{1}{e^{\beta\omega} - 1} |f_\omega(\sigma_f - at) - f_\omega(\sigma_f) e^{i\omega t}|^2. \tag{2}$$

3. Results

One is able to solve Eq. 2 analytically in several limits: for small virtualities at all times, for arbitrary virtualities at early times, and for arbitrary virtualities at late times.

IOP Conf. Series: Journal of Physics: Conf. Series **878** (2017) 012021 doi:10.1088/1742-6596/878/1/012021

3.1. Small Virtualities

For small virtualities we have that

$$s^2_{\text{small}}(t;a) = \frac{\beta^2}{\pi^2\sqrt{\lambda}}\int_0^\infty \frac{d\omega}{\omega}\frac{1}{e^{\beta\omega}-1}\Big(1+\cos^2(a\omega t) - 2\cos(a\omega t)\cos(\omega t)\Big)$$
$$= \frac{\beta^2}{4\pi^2\sqrt{\lambda}}\ln\left(\frac{2a\beta^3\sinh^2\left(\frac{\pi(a+1)t}{\beta}\right)\sinh^2\left(\frac{\pi(a-1)t}{\beta}\right)\text{csch}\left(\frac{2\pi a t}{\beta}\right)}{\pi^3\left(a^2-1\right)^2 t^3}\right). \quad (3)$$

Albeit fully analytic, the above formula is not the most transparent. One may expand Eq. 3 for early times, $t \ll 1/T$, to find

$$s^2_{\text{small}}(t;a) \xrightarrow{t\ll\beta} \frac{t^2}{6\sqrt{\lambda}} + \mathcal{O}((t/\beta)^4). \quad (4)$$

The $s^2 \sim t^2$ dependence is characteristic of ballistic motion.

One may also expand Eq. 3 for late times, $t \gg 1/T$, to find

$$s^2_{\text{small}}(t;a) \xrightarrow{\beta\ll t} \frac{\beta t}{\pi\sqrt{\lambda}}\left(1-\frac{a}{2}\right) + \frac{\beta^2}{4\pi^2\sqrt{\lambda}}\left\{\begin{array}{ll} 4\ln\left(\frac{\beta}{2\pi t}\right), & \text{if } a=0 \\ \ln\left(\frac{a\beta^3}{4\pi^3(a^2-1)^2t^3}\right), & \text{if } 0<a<1 \\ \ln\left(\frac{\beta}{4\pi t}\right), & \text{if } a=1 \end{array}\right\} + \mathcal{O}(1). \quad (5)$$

The $s^2 \sim t$ dependence is characteristic of Brownian motion. Note the subdiffusive log corrections.

3.2. Arbitrary Virtualities

Unfortunately one cannot solve analytically the arbitrary virtualities result for arbitrary times. One can, however, find the early and late time expansions. For $t \ll 1/T$

$$s^2(t;a)\Big|_{t\ll\beta} = \frac{4t^2}{\sqrt{\lambda}}\int_0^\infty d\nu \frac{\nu}{e^{2\pi\nu}-1}\frac{1+\nu^2}{1+\nu^2\tilde{r}_0^2}$$
$$= \frac{t^2}{6\tilde{r}_0^4\sqrt{\lambda}}\left(\tilde{r}_0^2 - 6(\tilde{r}_0^2-1)\left(-2\gamma_E - \pi\cot\left(\frac{\pi}{\tilde{r}_0}\right) + H_{-1/\tilde{r}_0} + H_{1/\tilde{r}_0} + 2\ln(\tilde{r}_0)\right)\right), \quad (6)$$

where γ_E is the Euler-Mascheroni constant and $H_{\mp 1/\tilde{r}_0}$ are harmonic numbers. Again, one finds ballistic $s^2 \sim t^2$ behavior.

For $t \gg 1/T$ one finds

$$s^2(t;a)\Big|_{t\gg\beta} = s^2_{\text{small}}(t;a), \quad (7)$$

where $s^2_{\text{small}}(t;a)$ is exactly the expression of Eq. 3.

Critically, notice that the late time dynamics are *universal* in the sense that 1) there is only one scale that determines the difference between early and late times, $1/T$, (in particular, there is no dependence on a) and 2) the late time behavior is independent of the initial string length (virtuality of the quark). Thus the late time behavior is set completely by the near horizon dynamics of the string.

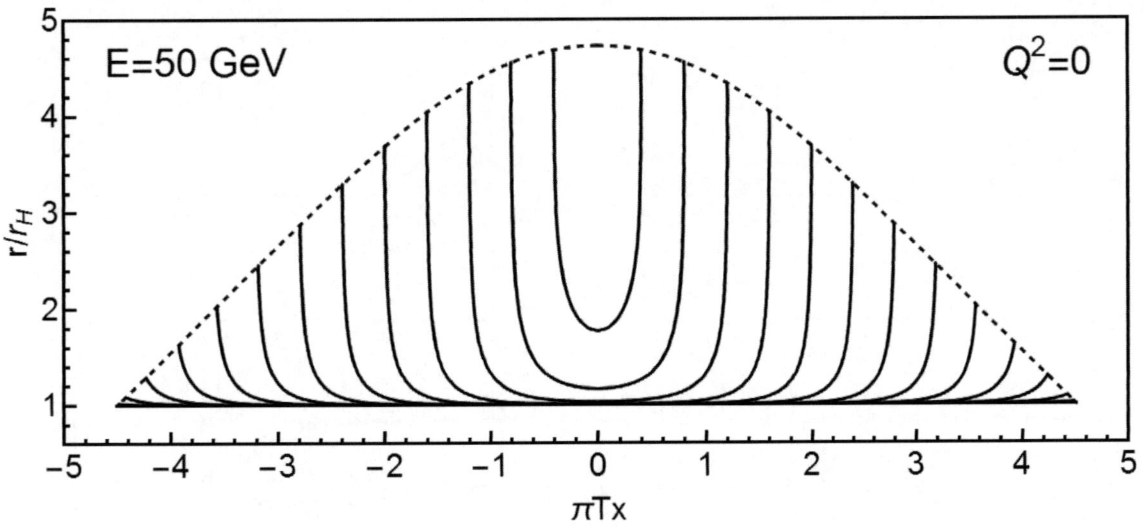

Figure 2: Snapshots in time of a typical numerical falling string profile (black curve) of Chesler et al. [22] representing a hard light quark–anti-quark pair production event in the dual field theory. The falling string nucleates at $x = 0$ as a point and evolves to an extended object with the endpoints falling towards the stretched horizon at $r \sim r_H$ along trajectories (dashed black curve) approximated by null geodesics. Initially the transverse fluctuations are characterized by $a = 0$ while close to the stretched horizon $a = 1$.

3.3. Arbitrary Dimensions

Since the late time dynamics is determined solely by the near horizon physics, we may extend our derivation of the late time dynamics to arbitrary numbers of spatial dimensions. From the metric for AdS_d space and repeating the computation that led to Eq. 3 in AdS_3, we find that

$$s^2_{\text{small}}(t; a, d) = \frac{1}{\sqrt{\lambda}} \left(\frac{(d-1)\beta}{4\pi} \right)^2 \ln \left(\frac{2a\beta^3 \sinh^2\left(\frac{\pi(a+1)t}{\beta}\right) \sinh^2\left(\frac{\pi(a-1)t}{\beta}\right) \operatorname{csch}\left(\frac{2\pi a t}{\beta}\right)}{\pi^3 \left(a^2-1\right)^2 t^3} \right), \quad (8)$$

which reduces to our previous result for $d = 3$, Eq. 3.

For processes that exhibit Brownian $s^2 \sim t$ motion, one may define the diffusion coefficient,

$$s^2(a, d) \equiv 2D(a, d)t. \quad (9)$$

Expanding Eq. 8 in powers of β/t, we can again extract the diffusion coefficient by comparing to Eq. 9:

$$D(a, d) := \frac{(d-1)^2 \beta}{8\pi\sqrt{\lambda}} \left(1 - \frac{a}{2}\right). \quad (10)$$

Using this very different approach to computing the diffusion coefficient, we exactly reproduce the known low velocity heavy quark, $a = 0$, $d = 5$ results of [14, 15].

3.4. $\hat{q}(t)$

Let's now make a non-trivial conjecture that we can apply the above results to compute the dynamical $\hat{q}(t)$ for the first time in AdS/CFT. Recall that $\hat{q}$ is the average transverse momentum

IOP Conf. Series: Journal of Physics: Conf. Series **878** (2017) 012021 doi:10.1088/1742-6596/878/1/012021

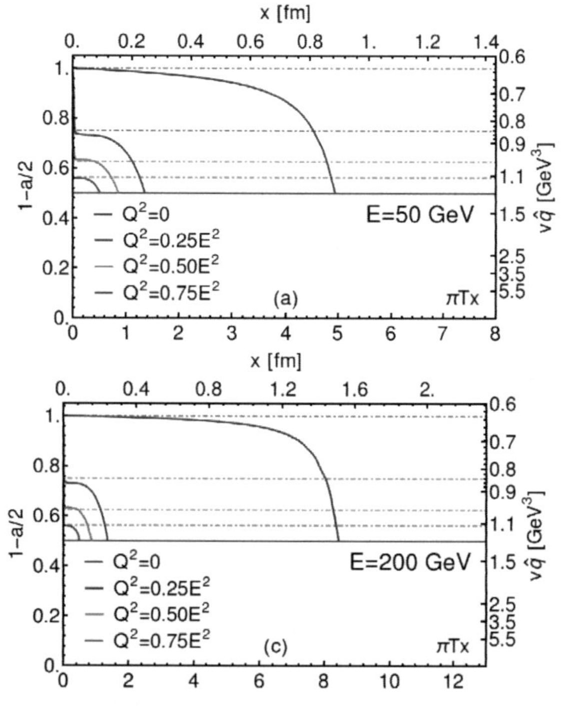

Figure 3: $1 - a/2 \propto 1/(v \times \hat{q})$ plotted as a function of x (left column) and t (right column) for an $E = 50$ GeV quark (top row) and $E = 200$ GeV quark (bottom row) for various virtualities in a $T = 350$ MeV plasma with $\lambda = 5.5$.

squared transferred from the plasma to the probe quark per unit distance travelled by the probe quark [7, 23]. $\hat{q}$ is an important transport coefficient, which can be used to determine the rate of radiative energy loss of a probe quark in pQCD [7, 23].

From the above definition, one may relate $\hat{q}$ to the diffusion coefficient [14],

$$\hat{q} = \frac{\langle p_T^2 \rangle}{\lambda_{mfp}} = \frac{2\kappa_T}{v} = \frac{4T^2}{vD}$$
$$= \frac{32\pi\sqrt{\lambda}T^3}{(d-1)^2(1-a/2)v}, \tag{11}$$

where in the second line we used Eq. 10 applicable for times large compared to $1/T$.

We thus find that for heavy quarks ($a = 0$) in the usual 4 spacetime dimensions, we have for a high momentum heavy quark moving at approximately constant velocity v but for which we do not input energy to maintain a constant velocity,

$$\hat{q}_{MH} = \frac{2\pi\sqrt{\lambda}T^3}{v}, \tag{12}$$

which should be compared to the previous calculation in which the heavy quark is explicitly kept at a constant velocity [14],

$$\hat{q}_{Gubser} = \frac{2\pi\sqrt{\lambda}T^3}{v}\sqrt{\gamma}. \tag{13}$$

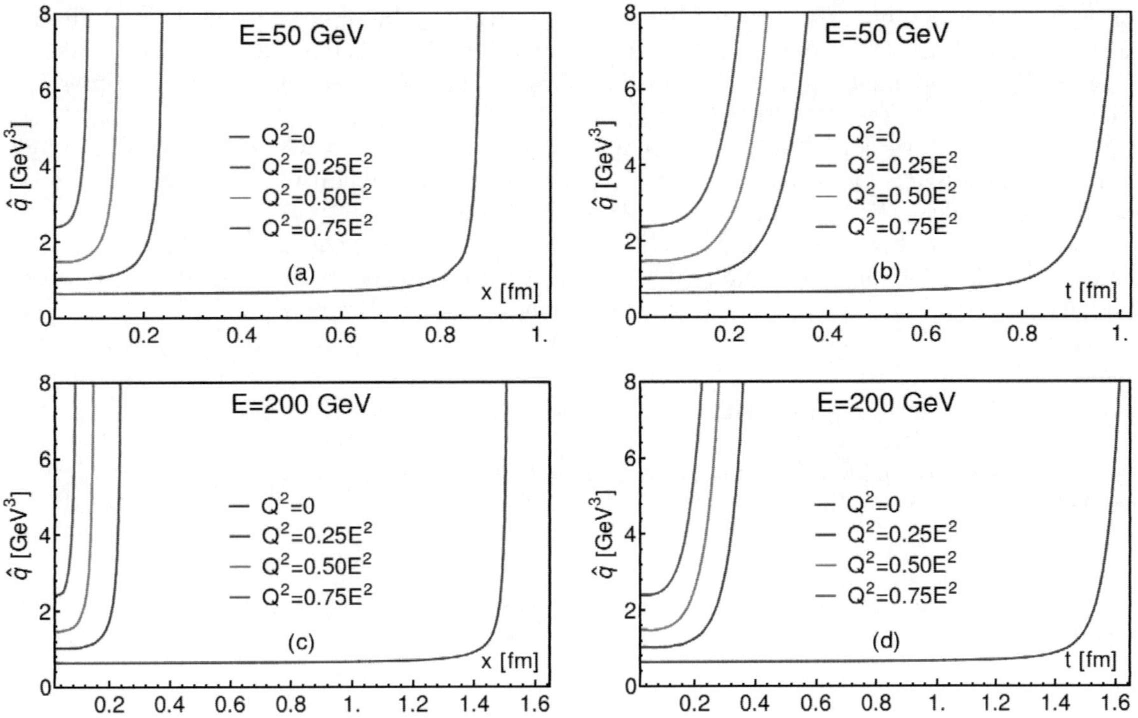

Figure 4: (Colour online) $\hat{q}$ plotted as a function of x (left column) and t (right column) for an $E = 50$ GeV quark (top row) and $E = 200$ GeV quark (bottom row) for various virtualities in a $T = 350$ MeV plasma with $\lambda = 5.5$.

For high momentum light quarks, we have that

$$\hat{q}_{MH} = \frac{2\pi\sqrt{\lambda}T^3}{(1-a/2)v} \simeq (2-4)\pi\sqrt{\lambda}T^3, \tag{14}$$

which should be compared to the LRW result [24]

$$\hat{q}_{LRW} = \frac{\pi^{3/2}\Gamma(3/4)}{\Gamma(5/4)}\sqrt{\lambda}T^3 \simeq 7.5\sqrt{\lambda}T^3. \tag{15}$$

We can, however, go even further for the light quark case by numerically solving the falling string equations of motion for the usual light flavor initial conditions. We show snapshots in time of the shape of the falling string in Fig. 2. From the numerical solution, we may extract explicitly $a(t)$, as shown in Fig. 3. One may thus determine $\hat{q}(t)$, which we show in Fig. 4. Our derivation for the diffusion coefficient in AdS_5 required the assumption of examining the probe system at times late compared to $1/T$. We nevertheless show $\hat{q}$ for all times for phenomenologically relevant values of energy and virtuality for purposes of illustration. At very late times, the endpoint of the falling Chesler string has essentially stopped moving in the horizontal direction, and $\hat{q}$ blows up trivially.

4. Conclusions

We have shown here the first results for $\hat{q}(t)$ for light and heavy quarks from AdS/CFT. For heavy quarks, we find, crucially, that the momentum diffusion picked up per unit distance does not grow with heavy quark velocity. This lack of growth with the Lorentz boost factor implies that the application of Langevin energy loss model to heavy quarks does not break down due to unphysically large momentum fluctuations and leads to a prediction of much greater suppression of heavy quarks at high momenta [25].

For light flavor, we found $\hat{q}(t)$ by mapping the rate of descent of the endpoint of a falling numerical open string in AdS_5/Schwarzschild space to the Brownian motion of a falling limp noodle as derived in [21]. We show the result for $\hat{q}(t)$ in Fig. 4.

One can see from Fig. 3 that a is a monotonically increasing function of time (or, equivalently, distance travelled by the string endpoint). Thus the quark experiences its greatest transverse fluctuations at the initial time, and the fluctuations decrease monotonically with time. This decrease in transverse fluctuations with time/distance is consistent with the pQCD result of angularly ordered gluon emissions from an off-shell parton in vacuum [26, 27] and in contradistinction to the anti-angular ordering observed in medium [28].

Finally, since the transverse momentum fluctuations smoothly interpolate between light flavor and heavy flavors through a, one naturally expects a smooth interpolation between the angular correlations of $q\bar{q}$ pairs at high-p_T.

5. Acknowledgments

The author wishes to thank the South African National Research Foundation and SA-CERN for support.

References

[1] Schenke B, Jeon S and Gale C 2011 *Phys. Rev. Lett.* **106** 042301 (*Preprint* 1009.3244)

[2] Bozek P 2012 *Phys. Rev.* **C85** 014911 (*Preprint* 1112.0915)

[3] Bozek P and Broniowski W 2013 *Phys. Rev.* **C88** 014903 (*Preprint* 1304.3044)

[4] Weller R D and Romatschke P 2017 (*Preprint* 1701.07145)

[5] Casalderrey-Solana J, Liu H, Mateos D, Rajagopal K and Wiedemann U A 2011 (*Preprint* 1101.0618)

[6] Horowitz W A and Gyulassy M 2008 *Phys. Lett.* **B666** 320–323 (*Preprint* 0706.2336)

[7] Majumder A and Van Leeuwen M 2011 *Prog. Part. Nucl. Phys.* **66** 41–92 (*Preprint* 1002.2206)

[8] Horowitz W A 2013 *Nucl. Phys.* **A904-905** 186c–193c (*Preprint* 1210.8330)

[9] Akamatsu Y, Hatsuda T and Hirano T 2009 *Phys. Rev.* **C79** 054907 (*Preprint* 0809.1499)

[10] Horowitz W A 2010 *Probing the Frontiers of QCD* Ph.D. thesis Columbia U. (*Preprint* 1011.4316) URL https://inspirehep.net/record/878251/files/arXiv:1011.4316.pdf

[11] Horowitz W A 2015 *Phys. Rev.* **D91** 085019 (*Preprint* 1501.04693)

[12] Herzog C P, Karch A, Kovtun P, Kozcaz C and Yaffe L G 2006 *JHEP* **07** 013 (*Preprint* hep-th/0605158)

[13] Gubser S S 2006 *Phys. Rev.* **D74** 126005 (*Preprint* hep-th/0605182)

[14] Gubser S S 2008 *Nucl. Phys.* **B790** 175–199 (*Preprint* hep-th/0612143)

[15] Casalderrey-Solana J and Teaney D 2006 *Phys. Rev.* **D74** 085012 (*Preprint* hep-ph/0605199)

[16] Casalderrey-Solana J and Teaney D 2007 *JHEP* **04** 039 (*Preprint* hep-th/0701123)

[17] Horowitz W A and Gyulassy M 2011 *J. Phys.* **G38** 124114 (*Preprint* 1107.2136)

[18] Morad R and Horowitz W A 2014 *JHEP* **11** 017 (*Preprint* 1409.7545)

[19] Casalderrey-Solana J, Gulhan D C, Milhano J G, Pablos D and Rajagopal K 2014 *JHEP* **10** 019 [Erratum: JHEP09,175(2015)] (*Preprint* 1405.3864)

[20] de Boer J, Hubeny V E, Rangamani M and Shigemori M 2009 *JHEP* **07** 094 (*Preprint* 0812.5112)

[21] Moerman R W and Horowitz W A 2016 (*Preprint* 1605.09285)

[22] Chesler P M, Jensen K, Karch A and Yaffe L G 2009 *Phys. Rev.* **D79** 125015 (*Preprint* 0810.1985)

[23] Burke K M *et al.* (JET) 2014 *Phys. Rev.* **C90** 014909 (*Preprint* 1312.5003)

[24] Liu H, Rajagopal K and Wiedemann U A 2006 *Phys. Rev. Lett.* **97** 182301 (*Preprint* hep-ph/0605178)

[25] Hambrock R and Horowitz W A 2017 AdS/CFT predictions for azimuthal and momentum correlations of $b\bar{b}$ pairs in heavy ion collisions *8th International Conference on Hard and Electromagnetic Probes of High-energy Nuclear Collisions: Hard Probes 2016 (HP2016) Wuhan, Hubei, China, September 23-27, 2016* (*Preprint* 1703.05845) URL https://inspirehep.net/record/1518153/files/arXiv:1703.05845.pdf

[26] Mueller A H 1981 *Phys. Lett.* **B104** 161–164

[27] Ermolaev B I and Fadin V S 1981 *JETP Lett.* **33** 269–272 [Pisma Zh. Eksp. Teor. Fiz.33,285(1981)]

[28] Mehtar-Tani Y, Salgado C A and Tywoniuk K 2011 *Phys. Rev. Lett.* **106** 122002 (*Preprint* 1009.2965)

IOP Conf. Series: Journal of Physics: Conf. Series **878** (2017) 012022 doi:10.1088/1742-6596/878/1/012022

Leading the non-cancelling IR divergences in massless gauge theories: Abelian case

Abdullah Khalil[1] **and W. A. Horowitz**[1]

[1]Department of Physics, University of Cape Town, Rondebosch 7701, South Africa

E-mail: abdullah@aims.ac.za

Abstract. We study the cancellation of both soft and collinear infrared divergences at next-to-leading order correction in a process where a massless electron is scattered off of a static point charge. We show that a self-consistent application of the KLN theorem requires including all initial and final degenerate states, which require including diagrams with an arbitrary number of disconnected photons forming a divergent series. We improve on previous work by rearranging the series in a controlled way and exploit the Monotone Convergence Theorem to prove the uniqueness of our rearranged result. This rearrangement yields a factorization of the infinite contribution from the initial state soft photons, that then cancels in the physically observable cross section.

1. Introduction

The infra-red (IR) catastrophe in purely massless gauge theories are known in two forms: soft, which is due to the massless nature of the radiation and appear when a soft (the word soft means that the photon energy is less than some experimental energy resolution Δ) massless particle is absorbed or emitted from the radiator, and collinear, which comes when the radiating particle is considered to be massless and usually happens when a particle is emitted or absorbed or emitted collinearly with the radiator.

The first approach towards eliminating the IR divergences was given by Bloch and Nordsieck (BN) [1, 2]. Their idea is, briefly, that it is impossible to specify a final state of a scattering process involving a charged particle because an arbitrary number of soft photons can be emitted without being detected. The BN cancellation requires adding both the virtual and the real emission corrections. The BN theorem has long been known for the success in eliminating the soft IR divergences. However, the appearance of quantum chromodynamics (QCD) introduces the collinear IR divergences where one should deal with light quarks. A more general theorem has been introduced to solve the mass singularities in gauge theories by Kinoshita [3], Lee and Nauenberg [4] (KLN). The KLN theorem states, in a modern language, that one should sum over both the initial and final degenerate indistinguishable states to get rid of the mass singularities. A combination of the BN theorem and the KLN theorem to eliminate the soft and collinear IR divergences respectively produce an IR finite formula for the cross section. However, as is shown by M. Lavelle and D. McMullan (LM) [5], such a treatment is inconsistent, where it breaks the time reversal symmetry and one should stick with the more general KLN theorem. A naive attempt to solve the problem by including the initial state absorption of a soft photon will spoil

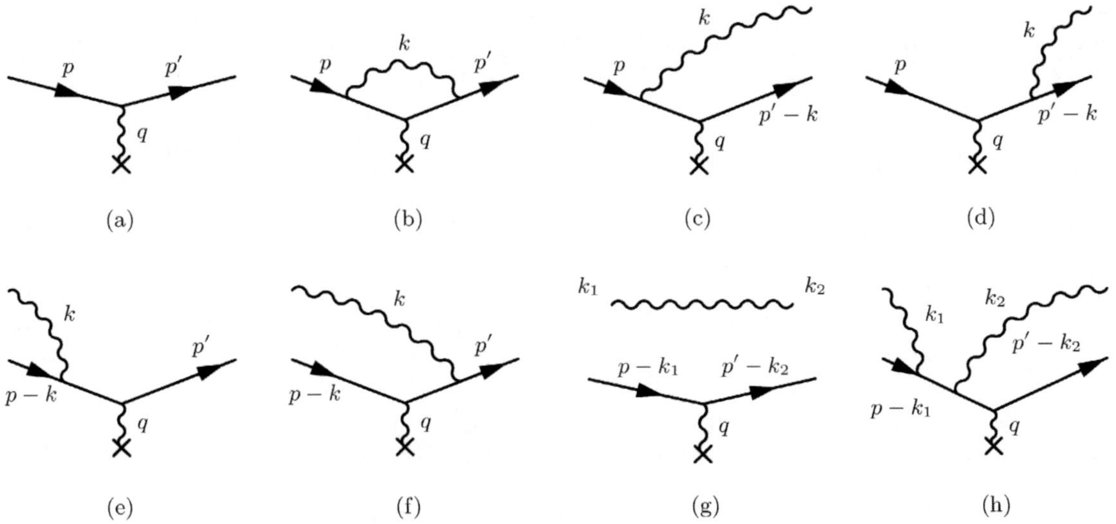

Figure 1. The (a) leading order (LO) and (b) next-to-leading order (NLO) vertex contributions to the $1 \to 1$ process in Rutherford scattering. The (c) and (d) potentially degenerate final state radiative processes. The (e) and (f) potentially degenerate initial state radiative processes. A disconnected photon amplitude such as (g) might interfere with an amplitude with a photon absorption and emission process such as (h).

the cancellation of the soft IR divergences. The same problem exists for the scalar ϕ^3 theory in six dimensions [6].

Many attempts have been made to eliminate the non-cancelling IR divergences from the initial state absorption of a soft photon by invoking the role of the disconnected diagrams, where the disconnected amplitudes can interfere with connected ones in which there are both emission and absorption of a soft photon, forming a connected cut diagram [5–9]. However, the former subsequent works summed over some, but not all, possible degenerate initial and final states, a full discussion on the incompleteness of their calculations can be found in Ref. [5]. LM concluded with a claim that no formal and consistent way exist so far to describe the correct way of applying the KLN theorem and urgent work needs to be done [5]. Ito [10] and Akhoury, Sotiropoulos, and Zakharov [11] (IASZ) included all possible initial and final degenerate states with an infinite number of disconnected soft photons forming a divergent infinite series at next-to-leading order (NLO). IASZ showed that by rearranging the series through formal manipulations, it will be sufficient to sum over a finite number of diagrams, but, as was pointed out in [5], the IASZ rearrangement yields a NLO result that is exactly 0 at tree level which makes their argument unsafe.

In this proceeding we consider a simple QED example, where an electron is scattered off of a static point charge and we check the IR cancellation at the NLO correction. We show that the IASZ treatment describes the consistent way of applying the KLN theorem. However, by providing a different rearrangement one may be able to seek the finiteness of the cross section. We also give a rigorous mathematical proof of uniqueness of our rearrangement.

2. Lee-Nauenberg treatment

We consider a massless electron scattering by a static point charge with initial and final four momenta $p = (E, \vec{p})$ and $p' = (E, \vec{p}\,')$ respectively, where the momentum transfer would be $q = p' - p$. We use the photon mass m_γ and the electron mass m_e to regularize the soft and

IOP Conf. Series: Journal of Physics: Conf. Series **878** (2017) 012022 doi:10.1088/1742-6596/878/1/012022

collinear divergences. The BN theorem requires first including the tree level Fig. 1 (a) and NLO vertex (b) as well as the emission of a soft photon from either the incoming or the outgoing electrons (c) and (d). The contribution from these processes to the differential cross section is given by

$$d\sigma_{BN} = d\sigma_0 \left\{ 1 + \frac{\alpha_e}{\pi^2} \left[\log\left(\frac{E^2}{\Delta^2}\right) \left(1 - \log\left(\frac{-q^2}{m_e^2}\right)\right) + \frac{3}{2}\log\left(\frac{-q^2}{m_e^2}\right) + \mathcal{O}(1) \right] \right\} , \qquad (1)$$

where α_e is the QED coupling constant and $\mathcal{O}(1)$ are terms that are finite at zero photon and electron masses. We note that Eq. (1) is free of soft IR divergences, but it diverges logarithmically as we set $m_e \to 0$. We now use the KLN theorem to remove the potentially large logs, $\log(-q^2/m_e^2)$, in the high energy limit $(-q^2 \gg m^2)$ by summing over the indistinguishable hard collinear final emission Fig. 1 (d) and initial absorption (e) processes. We note that the remaining amplitudes where a hard photon is taking part in the scattering process can be distinguished from either the incoming or the outgoing electrons [12]. One can easily see that the LN result can be obtained by replacing the m_e^2 by $\delta^2 E^2$ [4], and the result is free of any IR divergences. Here we point out that we used a combination of the BN and KLN theorems for such a cancellation, which is clearly inconsistent [5], and to stay in the spirit of the KLN theorem one should also add the absorption of a soft photon as a degenerate initial state, which spoils the finiteness of the cross section at NLO.

3. The role of the disconnected diagrams

The disconnected diagrams were first introduced in [4] to provide the cancellation of the mass singularities from the diagrams neglected above when the photon is hard and collinear in such a way that it can be distinguished, and have been used later by De. Calan [7] for his simplified model and T. Muta in QCD [8]. However, they seemed to include an asymmetric number of diagrams from the initial and final states, where we will follow their lead to check the inconsistency of their treatment. We first include the contributions from the interference between the disconnected diagram shown in Fig. 1 (g) and the absorption-emission diagrams such as the one in (h). We note that the Feynman rule for the disconnected photon is $(2\pi)^3\, 2\omega_{k_1}\delta^{(3)}(\vec{k}_1 - \vec{k}_2)\, \delta_{\epsilon_1\epsilon_2}$ where ω_k is the energy of the photon of momentum k, ϵ_1 and ϵ_2 are the polarization vectors for the incoming photon of momentum k_1 and the outgoing photon of momentum k_2 respectively.

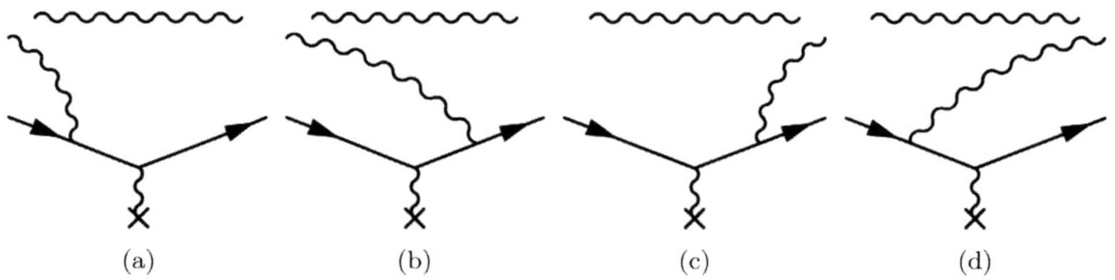

Figure 2. The absorption and emission with a disconnected photon.

It can be easily shown that including the contributions from the interference of the disconnected diagram Fig. 1 (g) with the absorption-emission diagrams Fig. 1 (h) is not sufficient for the cancellation, even adding the contributions from the absorption or emission with one disconnected photon, such as the ones in Fig. 2, keeps the cross section IR divergent and raises the question of adding even more disconnected photons to the former processes, while the most important question becomes; is there a full systematic method to apply the KLN

IOP Conf. Series: Journal of Physics: Conf. Series **878** (2017) 012022 doi:10.1088/1742-6596/878/1/012022

theorem properly? We show here that the resolution is to include all possible contributions from degenerate initial and final states, including the contributions from the disconnected diagrams with an arbitrary number of disconnected soft photons.

4. IASZ treatment

Following [10, 11], we consider a general form of our process with m incoming soft photons and n outgoing soft photons:

$$e^- + m\,\gamma\ (\text{soft}) \ \to e^- + n\,\gamma\ (\text{soft}) \ , \tag{2}$$

with an amplitude $\mathcal{M}_{mn}$. The transition probability for the process becomes

$$\mathcal{P}_{mn} = \frac{1}{m!}\frac{1}{n!}\sum_{i,f} |\mathcal{M}_{mn}|^2 \ , \tag{3}$$

where $\mathcal{P}_{mn}$ contains contributions from both fully and partially connected cut diagrams and the sum over initial (i) and final (f) states exist. The total LN probability will be

$$\mathcal{P} = \sum_{m,n=0}^{\infty} \mathcal{P}_{mn} \ , \tag{4}$$

where the KLN theorem ensures that the quantity $\mathcal{P}$ is free of the IR (soft or collinear) divergences.

It was shown in [10, 11] that any cut diagram from $\mathcal{P}_{mn}$ at NLO correction can be constructed from four probabilities: $\mathcal{P}_{00}$, which is the cut diagram with no real emission or absorption of soft photons, this may include the leading term, the vertex correction, the vacuum polarization, etc.; $\mathcal{P}_{10}$, and $\mathcal{P}_{01}$, which includes all cut diagrams with one soft photon in the initial or the final states respectively; $\mathcal{P}_{11}$, which includes all cut diagrams from the interference between the disconnected diagram Fig. 1 (g) and the absorption-emission diagrams like Fig. 1 (h). The fully connected cut diagrams are given by any of the previous basic probabilities, while the partially connected ones are those probabilities multiplied by a number of δ functions according to the number of disconnected photons. So we can construct $\mathcal{P}_{mn}$ by splitting each cut diagram up into connected and disconnected parts.

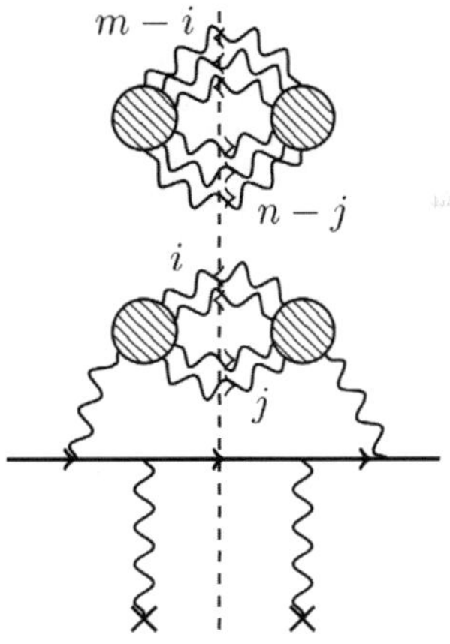

Figure 3. A generic cut diagram with m incoming and n outgoing soft photons that is partially disconnected. (Note that the blobs include the possibility of connecting incoming photon lines from the left with conjugate incoming photon lines from the right.)

We define the disconnected part by the function $\mathcal{D}(m-i, n-j)$ that describes the number of $m-i$ incoming and $n-j$ outgoing soft photons that can be joined together and become disconnected from the electron line. By definition $\mathcal{D}(0,0) = 1$ and $\mathcal{D}(a,b) = 0$ for $a \neq b$. One may show that the transition probability for the general process at NLO is given by [10, 11]

$$\mathcal{P}_{mn} = \frac{\mathcal{D}(m,n)}{m!\,n!}\mathcal{P}_{00} + \sum_{i=0} \frac{\mathcal{D}(m-i,n-i-1)}{(m-i)!\,(n-i-1)!}\mathcal{P}_{01} + \sum_{i=0} \frac{\mathcal{D}(m-i-1,n-i)}{(m-i-1)!\,(n-i)!}\mathcal{P}_{10} + \sum_{i=0} \frac{\mathcal{D}(m-i-1,n-i-1)}{(m-i-1)!\,(n-i-1)!}\widetilde{\mathcal{P}}_{11}. \tag{5}$$

IASZ were able to rearrange the full series by using the following identity

$$\frac{\mathcal{D}(m,n)}{m!\,n!} = \sum_{i=0} \frac{\mathcal{D}(m-i,n-i)}{(m-i)!\,(n-i)!} - \sum_{i=0} \frac{\mathcal{D}(m-i-1,n-i-1)}{(m-i-1)!\,(n-i-1)!}, \tag{6}$$

to give the following IR finite probability

$$\mathcal{P} = \sum_{m,n}\sum_{i} \frac{\mathcal{D}(m-i,n-i)}{(m-i)!\,(n-i)!} \left[\mathcal{P}_{00} + \mathcal{P}_{01} + \mathcal{P}_{10} + (\widetilde{\mathcal{P}}_{11} - \mathcal{P}_{00})\right]. \tag{7}$$

However, one can easily see that the probability $\mathcal{P}$ tends to zero once we send it back to tree level.

One may show that the series can be rearranged in different ways. However, we are looking for a rearrangement with special features: to be IR safe and to keep the tree level contribution finite. We give here the rearrangement that satisfies these criteria and rigorously prove that our result is unique. In order to give such a proof we manipulate Eq. (5) in a controlled way by introducing a convergence factor that becomes small for large i: we take

$$\mathcal{D}(m-i,n-j) \;\rightarrow\; \mathcal{D}(m-i,n-j)e^{-(i+j)/\Lambda} \tag{8}$$

with $\Lambda \gg 1$. The convergent factor Λ allows us to sum up to a finite number of disconnected soft photons N. Since $\mathcal{D}(m,n) = 0$ for $m \neq n$, we may simplify our manipulations by replacing the double sum over m and n with a single sum over n. With the above convergence factor we are guaranteed that $\mathcal{P} = \lim_{\Lambda\to\infty}\lim_{N\to\infty}\mathcal{P}_N(\Lambda)$ converges. Now we rearrange the partial sum $\mathcal{P}_N(\Lambda)$ up to N soft photons to find

$$\begin{aligned}\mathcal{P}_N(\Lambda) = \sum_{n=0}^{N} \frac{\mathcal{D}(n,n)}{(n!)^2}\left[\mathcal{P}_{00} + e^{-\frac{1}{\Lambda}}\mathcal{P}_{01}\right] \\ + \sum_{n=1}^{N}\sum_{i=1}^{n} \frac{\mathcal{D}(n-i,n-i)}{[(n-i)!]^2}\left[e^{-\frac{2i+1}{\Lambda}}\mathcal{P}_{01} + e^{-\frac{2i-1}{\Lambda}}\mathcal{P}_{10} + e^{-\frac{2i}{\Lambda}}\widetilde{\mathcal{P}}_{11}\right].\end{aligned} \tag{9}$$

One can easily show, using the same convergent factor, that the IASZ result is very different from Eq. (9) after swapping the limit in the full sum. So we check the possibility of switching the limits in order to prove the uniqueness of our result. We exploit the Monotone Convergence Theorem to justify that our result converges to the same value under the interchanging limits procedure by proving that the partial sum $\mathcal{P}_N(\Lambda)$ 1) monotonically increase in N for each Λ and 2) monotonically increase in Λ for each N [13]. We use the fact that $\mathcal{P}_{01} + \mathcal{P}_{10} = -\widetilde{\mathcal{P}}_{11}$ [8] to simplify Eq. (9). Then we have

$$\mathcal{P}_N(\Lambda) = \sum_{n=0}^{N} \frac{\mathcal{D}(n,n)}{(n!)^2}\left[\mathcal{P}_{00} + e^{-\frac{1}{\Lambda}}\mathcal{P}_{01}\right] + \sum_{n=1}^{N}\sum_{i=1}^{n} \frac{\mathcal{D}(n-i,n-i)}{[(n-i)!]^2}\, 2\mathcal{P}_{01}\, e^{-\frac{2i}{\Lambda}}\left[\cosh\left(\frac{1}{\Lambda}\right) - 1\right]. \tag{10}$$

Since $\mathcal{P}_{00}$, $\mathcal{P}_{01}$, Λ, and $\mathcal{D}(n,n)$ are all strictly positive, Eq. (10) clearly increases monotonically in N for fixed Λ. To show that $\mathcal{P}_N(\Lambda)$ increases monotonically in Λ for fixed N, we take the derivative with respect to Λ:

$$\frac{d\mathcal{P}_N(\Lambda)}{d\Lambda} = \sum_{n=0}^{N} \frac{\mathcal{D}(n,n)}{(n!)^2}\left[\frac{1}{\Lambda^2}e^{-\frac{1}{\Lambda}}\mathcal{P}_{01}\right] + \mathcal{O}\left(\frac{1}{\Lambda^3}\right). \tag{11}$$

Although one finds that the higher order in $1/\Lambda$ correction term is negative, for any N we can find a Λ large enough such that the first term, which is strictly positive, dominates. We have thus proved that we may exchange limits for our rearranged formula Eq. (10), and we may evaluate the $\Lambda \to \infty$ limit first, yielding our main result:

$$\mathcal{P} = (\mathcal{P}_{00} + \mathcal{P}_{01}) \sum_{n=0}^{\infty} \frac{\mathcal{D}(n,n)}{(n!)^2}. \tag{12}$$

We note that the contribution from the emission or absorption of a hard photon may also have an arbitrary number of disconnected soft photons and remains degenerate with our state, in such a way that none of the disconnected photons can be attached to the hard photon. Let us call the probabilities from the contribution of the absorption of hard and collinear photons $\mathcal{P}_{10}^{h,c}$ and the contribution from the emission of hard and collinear photons $\mathcal{P}_{01}^{hc}$, then the contribution from the diagrams where a hard photon is taking part with an infinite number of disconnected photons becomes

$$\mathcal{P}^{hc} = (\mathcal{P}_{10}^{hc} + \mathcal{P}_{01}^{hc}) \sum_{n=0}^{\infty} \frac{\mathcal{D}(n,n)}{(n!)^2}. \tag{13}$$

Finally, the total probability contributions from *all* the initial and final degenerate states will be

$$\mathcal{P}_{tot} = (\mathcal{P}_{00} + \mathcal{P}_{01} + \mathcal{P}_{10}^{hc} + \mathcal{P}_{01}^{hc}) \sum_{n=0}^{\infty} \frac{\mathcal{D}(n,n)}{(n!)^2}. \tag{14}$$

We find that all the soft initial state physics of the infinite number of undetectably soft photons completely factorizes. When the cross section is computed, one simply divides out by this unobserved infinity. We have thus rendered all soft and collinear IR divergences harmless. One can now apply Eq. (14) to find the complete NLO differential cross section to the Rutherford scattering [14]

$$d\sigma_{NLO} = d\sigma_0 \Bigg\{ 1 + \frac{\alpha}{\pi}\Bigg[\log\left(\frac{-q^2}{\delta^2 E^2}\right)\left(\log\left(\frac{E^2}{\Delta^2}\right) + \frac{3}{2}\right) + \frac{2}{3}\log\left(\frac{-q^2}{\mu_{\overline{\text{MS}}}^2}\right) + \frac{\pi^2}{2}\left(\left(\frac{4E^2}{-q^2} - 1\right)^{-\frac{1}{2}} - \frac{8}{3}\right) + \frac{5}{36} + \mathcal{O}(m^2, \delta^2)\Bigg]\Bigg\}. \tag{15}$$

5. Conclusions

The application of the KLN theorem requires including *all* degenerate initial and final states to achieve the IR finiteness of the cross section. These degenerate states include disconnected diagrams with an arbitrary number of disconnected photons which form a divergent series of diagrams at the transition probability level. We performed a rearrangement of this formally divergent series under the control of a convergent factor, which helped us to check the validity and the uniqueness of our rearrangement through the application of the Monotone Convergence Theorem. Our rearrangement shows that we only need to sum up to a finite number of diagrams, where the disconnected part remains factorized, and more important that the sum of these diagrams is IR safe and leads to the tree level contribution when we take the limit of the coupling constant to 0. Therefore, we arrived at the extremely non-trivial result that for NLO Rutherford scattering the summation over *all* indistinguishable initial and final states is equivalent to the summation over *only* the initial hard collinear and final soft, hard collinear, and soft and collinear degenerate states.

Acknowledgments
We thank the South African National Research Foundation and the SA-CERN consortium for their support. AK also thanks the African Institute for Mathematical Sciences (AIMS) and the University of Cape Town for their support.

References

[1] F. Bloch and A. Nordsieck, "Note on the Radiation Field of the electron," *Phys. Rev.*, vol. 52, pp. 54–59, 1937.

[2] D. R. Yennie, S. C. Frautschi, and H. Suura, "The infrared divergence phenomena and high-energy processes," *Annals Phys.*, vol. 13, pp. 379–452, 1961.

[3] T. Kinoshita, "Mass singularities of Feynman amplitudes," *J. Math. Phys.*, vol. 3, pp. 650–677, 1962.

[4] T. D. Lee and M. Nauenberg, "Degenerate Systems and Mass Singularities," *Phys. Rev.*, vol. 133, B1549–B1562, 1964, [,25(1964)].

[5] M. Lavelle and D. McMullan, "Collinearity, convergence and cancelling infrared divergences," *JHEP*, vol. 03, p. 026, 2006. arXiv: `hep-ph/0511314 [hep-ph]`.

[6] M. Lavelle, D. McMullan, and T. Steele, "Soft Collinear Degeneracies in an Asymptotically Free Theory," *Adv. High Energy Phys.*, vol. 2012, p. 379 736, 2012. arXiv: `1008.3949 [hep-ph]`.

[7] C. De Calan and G. Valent, "Infra-red divergence and incident photons," *Nucl. Phys.*, vol. B42, pp. 268–280, 1972.

[8] T. Muta and C. A. Nelson, "Role of Quark - Gluon Degenerate States in Perturbative QCD," *Phys. Rev.*, vol. D25, p. 2222, 1982.

[9] B. Mirza and M. Zarei, "Cancellation of soft and collinear divergences in noncommutative QED," *Phys. Rev.*, vol. D74, p. 065 019, 2006. arXiv: `hep-th/0609181 [hep-th]`.

[10] I. Ito, "INFRARED DIVERGENCES IN QCD AND KINOSHITA-LEE-NAUENBERG THEOREM," *Prog. Theor. Phys.*, vol. 67, p. 1216, 1982.

[11] R. Akhoury, M. G. Sotiropoulos, and V. I. Zakharov, "The KLN theorem and soft radiation in gauge theories: Abelian case," *Phys. Rev.*, vol. D56, pp. 377–387, 1997. arXiv: `hep-ph/9702270 [hep-ph]`.

[12] R. Gastmans, "Radiative corrections in gauge theories," Tech. Rep., 1981.

[13] A. W. Knapp, *Basic real analysis*. Springer Science & Business Media, 2005.

[14] A. Khalil and W. A. Horowitz, in preperation.

IOP Conf. Series: Journal of Physics: Conf. Series **878** (2017) 012023 doi:10.1088/1742-6596/878/1/012023

Holography at QCD-T_c

R. Zöllner, B. Kämpfer
Helmholtz-Zentrum Dresden-Rossendorf, PF 510119, D-01314 Dresden, Germany
and
Institut für Theoretische Physik, TU Dresden, D-01062 Dresden, Germany

Abstract. Within an extended soft wall model, we study the temperature dependence of the lowest vector meson states. Scales are adjusted by using as input the ρ meson mass in vacuum and the velocity of sound from lattice QCD which displays a minimum at T_c. The melting of the ρ meson occurs at temperatures $\mathcal{O}(T_c)$.

1. Introduction

The AdS/CFT correspondence [1, 2, 3] offers the possibility of mapping strongly coupled quantum field theories – such as QCD – to weakly coupled gravity. While a gravity dual of QCD is not (yet) at our disposal, one can construct models which share selected phenomena with real-world QCD. Such holographic approaches may be anchored in string theory as top-down set-up or may belong to bottom-up settings, where in some minimalistic manner QCD features serve as input. Among the latter class is the soft wall model [4] (cf. [5] for a critical assessment) which sparked a series of applications by describing hadrons within the AdS/CFT correspondence. Interesting are, for instance, extensions to non-zero temperatures up to meson melting, as initiated in [6, 7, 8].[1] A particular formulation of the soft wall model has been put forward in [13], based on [14], where some relations to QCD have been considered on a qualitative level. Simple parameterizations have been proposed to accommodate QCD thermodynamic features together with a proper temperature dependence of the light-quark vector meson spectrum (for non-zero net baryon effects, cf. [15]). Here, we explore quantitatively the compatibility of the setting in [13] with (i) QCD thermodynamics, in particular the velocity of sound (Section 2), and (ii) the melting of the ρ meson (Section 3) within the extended soft wall model (Appendix A). Our conclusion (Section 4) is that a semi-quantitative compatibility of items (i) and (ii) can be accomplished which leaves space for further fine tuning (Appendix B).

2. Thermodynamics

In a self-consistent 'dynamical' model, e.g. the setting pioneered in [16, 17, 18] and further developed in [19, 20] to accommodate the case of 2+1 flavor QCD thermodynamics with physical quark masses, the ansatz for the metric of a 5dimensional Riemann space

$$\mathrm{d}s^2 = e^{A(z)} \left(f(z)\mathrm{d}t^2 - \mathrm{d}\vec{x}^2 - \frac{\mathrm{d}z^2}{f(z)} \right) \tag{1}$$

[1] For meson melting, cf. [5] which includes a profound discussion of the basics; [9, 10, 11, 12] among others include various particular aspects, e.g. spectra functions, light- and heavy-quark mesons etc. Our approach is more restricted as it uses less ingredients.

is employed to extract the metric functions $A(z)$, $f(z)$ as well as the dilaton profile $\Phi(z)$ by solving the Einstein equations and equation of motion resulting from the action

$$S = \frac{1}{16\pi G_5} \int \mathrm{d}^4x \, \mathrm{d}z \sqrt{g} \left(R - \frac{1}{2}(\partial\Phi)^2 - \tilde{V}(\Phi) \right). \tag{2}$$

Instead of tuning the dilaton potential $\tilde{V}(\Phi)$ appropriately, as done in [20] for instance (cf. also [19, 21, 22, 23]; for the Yang-Mills case, cf. [24]), we postpone that step for follow-up investigations and consider suitable ansätze. Such ansätze must be conform with the AdS behavior near the boundary at $z = 0$ and, for non-zero temperatures, must display a simple zero of the blackness function $f(z)$ at the horizon z_H. As demonstrated in [13] the parameterizations of Hawking temperature T and Bekenstein entropy density s

$$T(z_H) = \frac{T_x}{\theta} \left(\frac{1}{x} + \theta - 3b + 3bx - x^2 \right), \tag{3}$$

$$s(z_H) = \frac{16\pi G_5}{z_H^3} \tag{4}$$

with $x = z_H/z_x$ and parameters T_x, $\theta = \pi z_x T_x$ and b generate a phase structure in agreement with 2+1 flavor QCD with varying quark masses: $b > 1$ delivers a first-order phase transition, $b = 1$ is for a second-order phase transition and $b < 1$ implies a cross over.[2] This scans through the Columbia plot [25] on the line of degenerate quark masses. Here, we restrict ourselves to $b < 1$ and compare with the lattice QCD results [26, 27]. The velocity of sound

$$c_s^2 = \frac{s}{T} \frac{\mathrm{d}T}{\mathrm{d}s} \tag{5}$$

is a preferred quantity since it requires a minimum of scale settings.
Figure 1 exhibits two fits of the sound velocity squared, separately for the data [26] (left) and [27] (right). The solid curve is for a minimized χ^2, while the dashed curve is adjusted at the minimum of c_s^2 (softest point) fixed by the parameters $c_s^2(T_c) = \min\{c_s^2\} = 0.147$ (left, 0.14 right) at the temperature $T_c = 146$ MeV (left, 150 MeV right) and a ρ meson melting (disappearance) temperature of $T_{\text{dis}}^{\rho} = 145$ MeV (left, 150 MeV right).[3]
Alternatively, one can use the relation

$$z_H(T) = z_H^{(0)} \exp\left\{ -\frac{1}{3} \int_{T_0}^{T} \frac{\mathrm{d}\tilde{T}}{\tilde{T} c_s^2(\tilde{T})} \right\} \tag{6}$$

[2] An inflection point of $T(z_H)$ in conjunction with (4) is a sufficient condition for a local minimum of c_s^2. In fact, (3) has such an inflection at $x = 1$ and $T(x = 1) = T_x$. The example $T(z_H) = 1/(\pi z_H) + T_x - T_y^2 z_H/\pi$ proves that an inflection point is not necessary to generate a minimum of $c_s^2(T)$. We use here, however, (3) as in [13].

[3] The equation of state has several distinctive points: the minimum of sound velocity at $T_c^{c_s^2} = 145\ldots150$ MeV, the maximum of scaled interaction measure at $T_c^{(e-3p)T^4} = 205\ldots210$ MeV, as well as the inflection points of scaled pressure at $T_c^{p/T^4} = 180\ldots190$ MeV and scaled entropy density at $T_c^{s/T^3} = 155\ldots170$ MeV according to the tables in [26, 27]. Often one quotes $T_{\text{pc}} = (155 \pm 9)$ MeV as pseudo-critical temperature determined by the chiral susceptibility, for instance, cf. [27, 28, 29]. The pseudo-critical curve is parametrized as $T_c(\mu) = T_c(0)(1 + \kappa\mu^2 + \ldots)$ with displaying $T_c(0)$ =145, 155 and 165 MeV, e.g. in [30]. This is to be complemented by the chemical freeze-out temperature $T_{\text{f.o.}} = 156$ MeV [31, 32] or $164\ldots168$ MeV [33] not to be mixed with the kinetic freeze-out temperature; other recent claims are $T_{\text{f.o.}} = 165$ MeV (with afterburner) vs. $T_{\text{f.o.}} = 155$ MeV (without afterburner) [34]. [26] mentions also a hadronization temperature of 164 MeV. Here, we identify $T_c^{c_s^2} = T_c$.

IOP Conf. Series: Journal of Physics: Conf. Series **878** (2017) 012023 doi:10.1088/1742-6596/878/1/012023

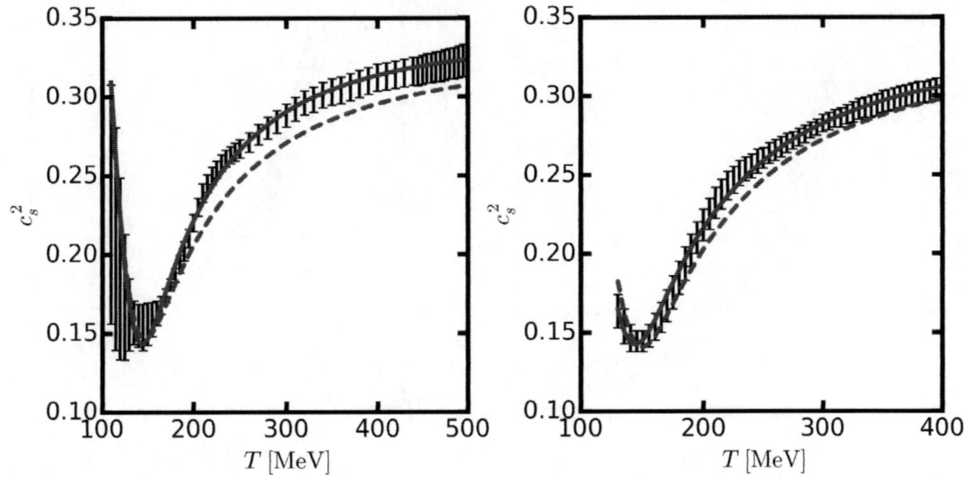

Figure 1. Speed of sound squared as a function of the temperature. Data from [26] (left panel) and [27] (right panel) are for the 1σ error range. Solid curves: minimization of χ^2 delivering T_x =130 (129) MeV, θ =1.75 (1.90) and b =0.71 (0.68) for the left (right) panel. Dashed curves: model adjustment to the minimum with coordinates T_c =146 (150) MeV and $c_s^2(T_c) = 0.147$ (0.14) and the side condition $T_{\rm dis}^{\rho}$ =145 (150) MeV; parameters are T_x =129 (129) MeV, θ =2.07 (2.02) and b =0.62 (0.66) for the left (right) panel.

to get the relation $T(z_H)$ from the lattice data $c_s^2(T)$ by exploiting the model input $\frac{1}{s}\frac{\mathrm{d}s}{\mathrm{d}z_H} = -3/z_H$ from (4). The result is displayed in Fig. 2 by the gray band which arises from using the lower and upper error bars of the data. In addition, the curves of Fig. 1 are displayed too, proving that the ansatz (3) is in fact meaningful, supposed (4) holds true. A key ingredient here is the ansatz (4) which is as in the extended soft wall model, cf. [7], while (3) is already beyond the model [7] to lift the melting temperature of vector mesons, $T_{\rm dis}$, in the order of T_c.

3. Melting of the vector mesons

There is a quantitative tension of the above thermodynamics and the melting of vector mesons. The extended soft wall model (see Appendix A for a reminder) requires specific parameters T_x, θ and b to ensure that vector mesons exist at temperatures $T \le T_{\rm dis} = \mathcal{O}(150\,\mathrm{MeV})$. The latter value comes from the thermo-statistical model analyses [35, 32, 36, 37, 38, 39] of hadron yields in ultra-relativistic heavy-ion collisions and is supported by analyses of fluctuation measures [40, 41, 42, 43, 44]. Above $T_{\rm dis}$, hadrons are 'molten', at least these ones built up by light quarks, such as the ρ mesons and its first excitations. At $T_{\rm dis}$ hadrons form and persist at temperatures below $T_{\rm dis}$. To calculate this temperature dependence one has to fix additionally

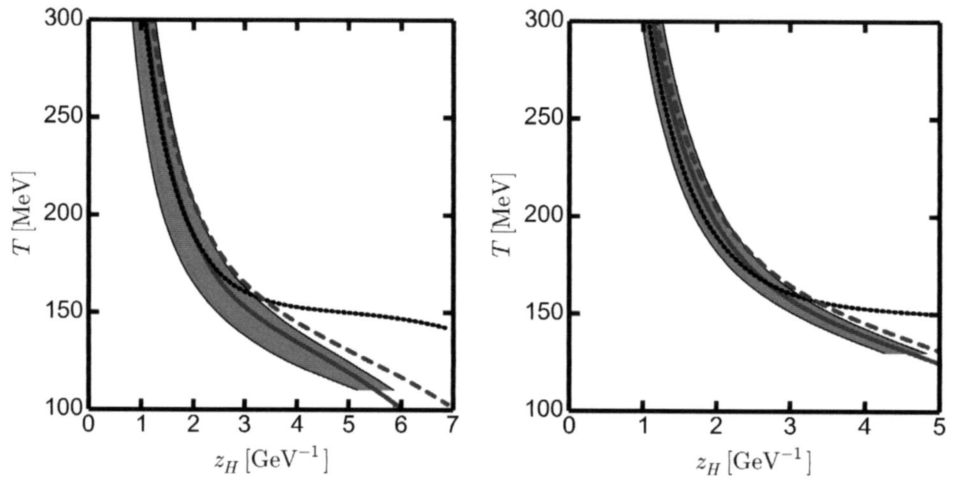

Figure 2. The temperature as a function of the horizon position z_H obtained by (6) with the data displayed in Fig. 1 (gray band). The solid and dashed curves correspond to those in Fig. 1. The dotted curve belongs to the parameter set $T_x = 150$ MeV, $\theta = 2.89$ and $b = 0.94$ which accomplishes $T^\rho_{\rm dis} = 155$ MeV.

$A(z)$ and $f(z)$. We employ here[4]

$$A(z) = -2\ln(z/L),\ Lc = 1, \tag{7}$$

$$f(x = z/z_H; z_H T(z_H)) = 1 - x^4 \left\{1 + \frac{\ln \hat{\varphi}}{e\hat{\varphi}} \left[x^{2\hat{\varphi}} - 1\right]\right\}, \tag{8}$$

$$\hat{\varphi}(z_H T(z_H)) = \exp\left\{\frac{2}{e}(\pi z_H T(z_H) - 1)\right\} \tag{9}$$

which ensure $s(z_H) = \exp\{3/2A(z_H)\} \propto L^3/z_H^3$ and $T(z_H) = -(\partial_z f)|_{z_H} /4\pi$ as well as the proper asymptotic AdS behavior. Note that the dimensional scale parameter c in the dilaton profile enters the calculation of $T^\rho_{\rm dis}$ (see Appendix A).

The χ^2 minimization for the sound velocity squared (see Fig. 1, solid curves) constrains the parameters T_x, θ and b such to deliver $T^\rho_{\rm dis} \approx 110$ MeV, while the adjustment to the softest point (see Fig. 1, dashed curves) result in $T^\rho_{\rm dis} = 145$ MeV, i.e. quite close to the wanted value of $\mathcal{O}(150$ MeV). However, even exploiting the variation of the dilaton profile, which causes a dependence $c(p)$ (see Appendix A and Fig. A1-left), does not allow to reach exactly that value. Another way to visualize such a tension is to exhibit in Fig. 2 a curve $T(z_H)$ for parameters which give $T^\rho_{\rm dis} = 155$ MeV (see dotted curve): This curve deviates significantly from the gray band in the low-temperature regime.

A remaining issue is whether the ground-state vector meson mass near and slightly below T_c suffer strong medium modifications. Figure 3 shows in fact that, despite of a minor temperature dependence at the melting point, the vacuum mass is recovered in our model. This

[4] We emphasize that the present approach is aimed at a quantification of the set-up in [13], i.e. a combination of suitable elements thought to appear in a consistent holographic model. For instance, (7) is an approximation of the proper solution of a field equation, emerging from (2), which relates the warp factor $A(z)$ and the dilaton profile $\Phi(z)$. Analogously, (8, 9) feature the general behavior of the blackness function $f(z)$ and may be considered as parametrized approximation of the solution of field equations and equations of motion, also emerging from (2).

IOP Conf. Series: Journal of Physics: Conf. Series **878** (2017) 012023 doi:10.1088/1742-6596/878/1/012023

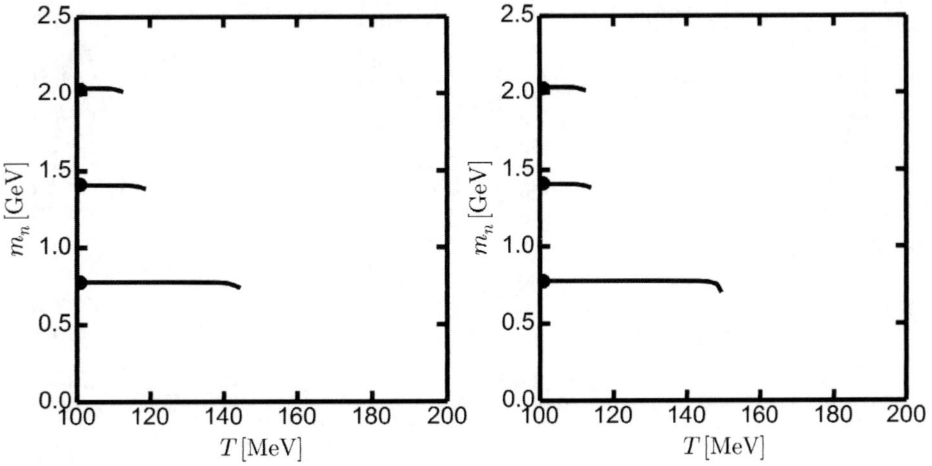

Figure 3. Vector meson masses m_n for $n = 0$ ($\rho(770)$), $n = 1$ ($\rho(1450)$) and $n = 2$ ($\rho(2030)$) as a function of the temperature. Left (right) panel: parameter set $\{T_x, \theta, b\}$ as for the dashed curves in Fig. 1-left (right) with the additional parameters $p = 4$ and $c = 310$ MeV, which cause vacuum masses as indicated by the bullets.

is remarkably consistent with the thermo-statistical model analyses of data [35, 32, 36, 37, 38, 39]. We emphasize that the first and the second excitation appears only at significantly lower temperature, and higher excitations are not found a $T > 100$ MeV. (For a parameter selection, where a few first excitations on a Regge trajectory extend to T_c, cf. [15]). Note that the effect of an increasing value of p implying an increasing altitude of the r.h.s. potential wall prevails over the influence of T_c on a decreasing scale parameter c up to $p \approx 4$. Thus by the variation of p one is able to lift $T_{\rm dis}$ a little bit. Appendix B shows a more efficient manner to lift $T_{\rm dis}$.

4. Summary

In summary we employ the extended soft wall model and show the compatibility of some useful optional model parameterizations with QCD thermodynamics near T_c. The quantity T_c is here understood as the temperature of the softest point (minimum of sound velocity) and thus related to an observable. Within the hitherto employed model ansätze, the QCD lattice data can be perfectly reproduced. At the same time, the such fixed parameters propagate into another branch of the soft wall model where the temperature dependence of light-quark vector meson states is evaluated. Keeping the ansätze for the AdS-black hole metric functions and the dilaton profile as in previous work, we find that the ρ meson melting at the temperature of $T^{\rho}_{\rm dis} = T_c$ is still compatible with lattice QCD thermodynamics. It is the low value of $T_c \cong 145$ MeV which points to a slight quantitative tension with our imagination of a hadronization temperature of $T^{\rho}_{\rm dis} \approx 150 \ldots 155$ MeV (or slightly above) to apply the thermo-statistical model for hadrons with vacuum masses. Given the simple structures of our ansätze it seems not too surprising that the available parameter space is rapidly exhausted in accommodating such vastly different aspects as the thermodynamics of the strongly interacting medium and the fate of selected individual constituents thereof. This calls for a self-consistent holographic model where the profiles of the metric functions and the dilaton as conformal symmetry breaker are solutions of field equations and equations of motion derived from a the corresponding gravity dual.

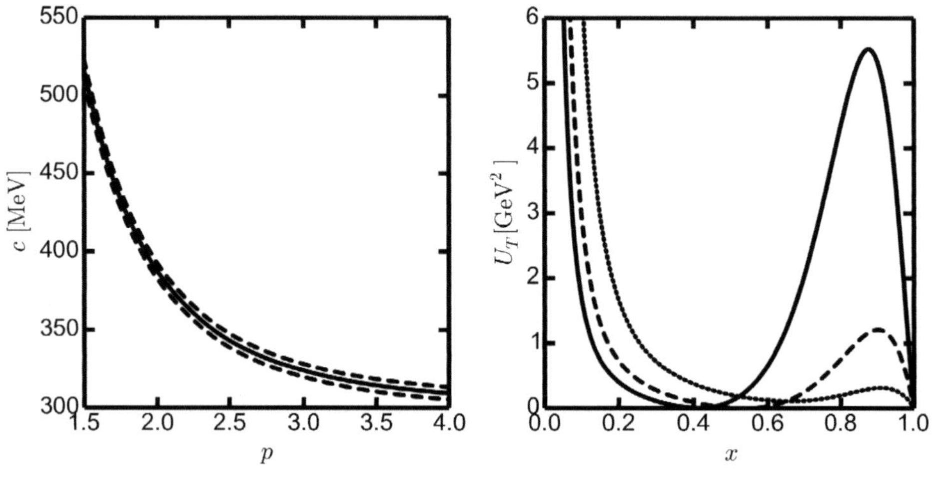

Figure A1. Left panel: The dependence $c(p)$ arising from the adjustment $m_0 = m_\rho^{(\mathrm{exp})} = 770$ MeV with ± 10 MeV variation. Right panel: The Schrödinger equivalent potential U_T (A.3) as a function of $x = z/z_H$ for the parameter set $T_x = 130$ MeV, $\theta = 1.8$, $b = 0.7$, $c = 310$ MeV and $p = 4$ for various temperatures ($T = 105$ MeV (solid), 130 MeV (dashed) and 150 MeV (dotted)). Note that for solving (A.2), U_T as a function of the tortoise coordinate $\xi(z)$ is needed.

Appendix A. Recalling the extended soft wall model

In the probe limit and focusing on a vector-like gauge field V_M the action of the soft wall model reads [4]

$$S_V = \frac{1}{k}\int \mathrm{d}^4x\,\mathrm{d}z\,\sqrt{g}e^{-\Phi(z)}F^2, \tag{A.1}$$

where $F^2 = g^{MM'}g^{NN'}F_{MN}F_{M'N'}$ with $F_{MN} = \partial_M V_N - \partial_N V_M$ $(M, N = 0, \ldots, 4)$ denoting the field strength tensor of the $U(1)$ vector field V, where the gauges $V_4 = 0$ and $\partial^\mu V_\mu = 0$ can be applied by Klein-Kaluza decomposition. V_μ is then sourced by a current operator $\bar{q}\gamma_\mu q$ in the boundary theory in the spirit of the field-operator correspondence; the determinant of the metric g_{MN} is again g, and k stands for the gauge coupling with the AdS radius L but drops out in the sequel. Greek indices run in the range $0\ldots3$. The vector field $V_\mu = \epsilon_\mu\varphi(z)\exp(ip_\nu x^\nu)$ has the polarization described by ϵ_μ; the important part of the wave function is the amplitude $\varphi(z)$ beyond the phase $p_\nu x^\nu$.

The resulting equation of motion with metric (1) can be cast into the form of a one-dimensional Schrödinger equation

$$\left(\partial_\xi^2 - (U_T - m_n^2)\right)\psi = 0 \tag{A.2}$$

by a coordinate transformation $z \to \xi$ with $\mathrm{d}\xi(z)/\mathrm{d}z = 1/f(z)$ and the transformation $\psi = \varphi\exp\{-\frac{1}{2}(A - \Phi)\}$. The potential reads

$$U_T = \left(\frac{1}{2}(\frac{1}{2}\partial_z^2 A - \partial_z^2\Phi) + \frac{1}{4}(\frac{1}{2}\partial_z A - \partial_z\Phi)^2\right)f^2 + \frac{1}{4}(\frac{1}{2}\partial_z A - \partial_z\Phi)\partial_z f^2. \tag{A.3}$$

The vector meson states correspond to normalizable solutions of (A.2) with masses squared $m_n^2 = p_\mu p^\mu$. The case $T = 0$ is equivalent to setting $f = 1$. The remaining warp factor $A(z)$ and the dilaton profile $\Phi(z)$ are to be adjusted suitably. In line with the original soft wall model [4], an easy ansatz is (7) and $\Phi(z) = (cz)^p$. We obtain the dependence $c(p)$ by the requirement

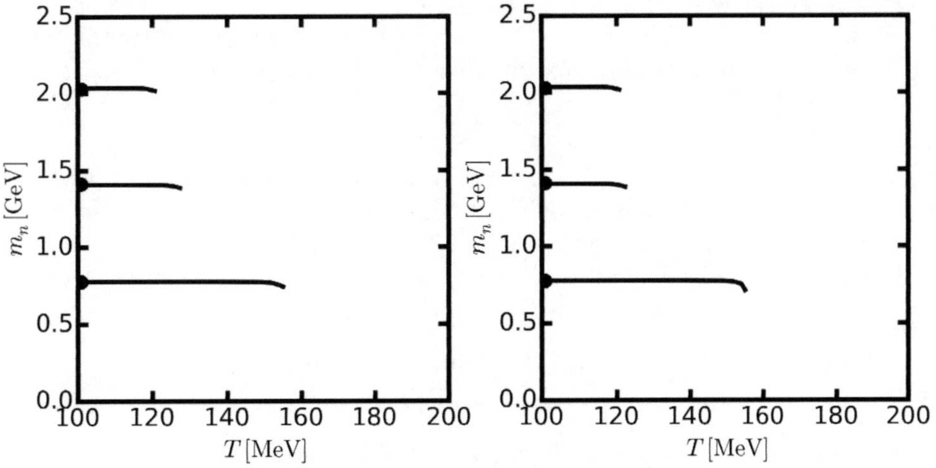

Figure A2. As in Fig. 3, but now for the entropy (B.1); $c = 310$ MeV and $p = 4$. For further parameters cf. Fig. A3.

$m_0 = m_\rho^{(\rm exp)} = 770$ MeV, see Fig. A1-left panel. In such a way, the hadron energy scale c is fixed. At non-zero temperature, we use the blackness function (8) and keep the above profiles $A(z)$ and $\Phi(z)$. An example of the potential U_T (A.3) as a function of the scaled bulk coordinate $x = z/z_H$ is exhibited in Fig. A1-right. With increasing temperature the r.h.s. potential wall diminishes rapidly so that no bound state can be accommodated at $T > T_{\rm dis}$. The temperature dependence of the lowest vector meson masses is exhibited in Fig. 3.

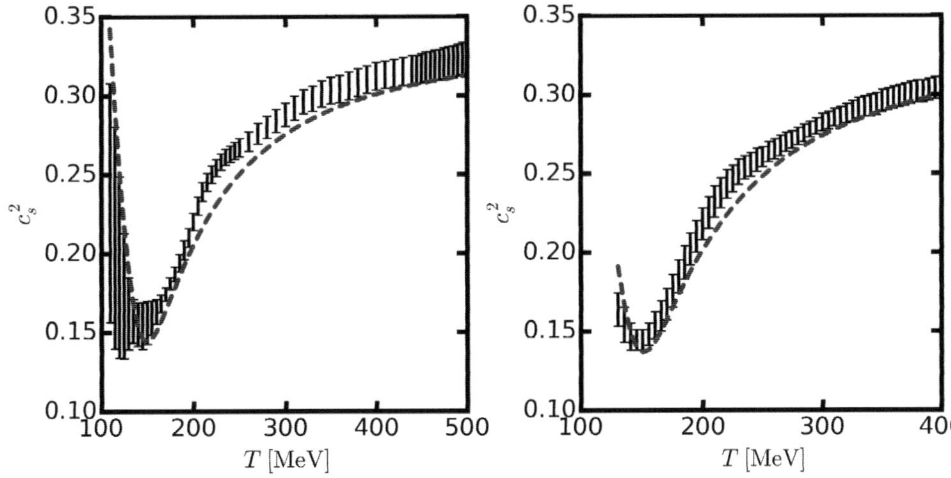

Figure A3. As in Fig. 1, but for the entropy (B.1). Parameters are T_x =134 (136) MeV, θ =1.98 (2.04), b =0.67 (0.68) and ε =12.3 (17.4) MeV for the left (right) panel and $T_{\rm c} = 150$ MeV.

Appendix B. Modified warp factor

Solving the field equations emerging from (2) points to a more involved profile of the warp function $A(z)$ than that given in (7). This in turn modifies the entropy (4). Let us consider, instead of (4), the trial ansatz

$$s(z_H) = 16\pi G_5 e^{\varepsilon z_H} z_H^{-3} \tag{B.1}$$

with the small positive dimensional parameter ε which results from an additional polynomial term in $A(z)$. The impact on the potential U_T (A.3) is very tiny. However, (B.1) allows to select values of b, θ and T_x such that the disappearance temperature $T_{\rm dis}^{\rho}$ can be tuned to 155 MeV (see Fig. A2), while the consistency with the speed of sound data is preserved (see Fig. A3).

Acknowledgments

The work is supported by Studienstiftung des deutschen Volkes.

References

[1] Maldacena J M, *Adv. Theor. Math. Phys.* **2**, 231 (1998).
[2] Gubser S S, Klebanov I R, Polyakov A M, *Phys. Lett.* B **428**, 105 (1998).
[3] Witten E, *Adv. Theor. Math. Phys.* **2**, 505 (1998).
[4] Karch A, Katz E, Son D T, Stephanov M A, *Phys. Rev.* D **74**, 015005 (2006).
[5] Iatrakis I, Kiritsis E, Paredes A, *JHEP* **1011**, 123 (2010).
[6] Colangelo P, Giannuzzi F, Nicotri S, *Phys. Rev.* D **80**, 094019 (2009).
[7] Colangelo P, Giannuzzi F, Nicotri S, *JHEP* **1205**, 076 (2012).
[8] Miranda A, Bayona B, Boschi-Filho H, Braga N, *JHEP* **0911**, 119 (2009).
[9] Fujita M, Kikuchi T, Fukushima K, Misumi T, Murata M, *Phys. Rev.* D **81**, 065024 (2010).
[10] Ishii T, Kinoshita S, Murata K, Tanahashi N, *JHEP* **1404**, 099 (2014).
[11] Braga N R F, Martin Contreras M A, Diles S, *Eur. Phys. J.* C **76**, 598 (2016).
[12] Bartz S P, Jacobson T, *Phys. Rev.* D **94**, 075022 (2016).
[13] Zöllner R, Kämpfer B, arXiv:1701.01398 [hep-ph] (2017).
[14] Zöllner R, Kämpfer B, *Phys. Rev.* C **94**, 045205 (2016).
[15] Zöllner R, Wunderlich F, Kämpfer B, arXiv:1611.04124 [hep-th] (2016).
[16] Gürsoy U, Kiritsis E, *JHEP* **0802**, 032 (2008).
[17] Gürsoy U, Kiritsis E, Nitti F, *JHEP* **0802**, 019 (2008).
[18] Gubser S, Nellore A, *Phys. Rev.* D **78**, 086007 (2008).
[19] Finazzo S, Noronha J, *Phys. Rev.* D **90**, 115028 (2014).
[20] Yaresko R, Knaute J, Kämpfer B, *Eur. Phys. J.* C **75**, 295 (2015).
[21] Finazzo S, Rougemont R, Marrochio H, Noronha J, *JHEP* **1502**, 051 (2015).
[22] Rougemont R, Ficnar A, Finazzo S, Noronha J, *JHEP* **04**, 102 (2016).
[23] Rougemont R, Noronha J, Noronha-Hostler J, *Phys. Rev. Lett.* **115**, 202301 (2015).
[24] Yaresko R, Kämpfer B, *Phys. Lett.* B **747**, 36 (2015).
[25] Pinke C, Philipsen O, PoS LATTICE2015, 149 (2016).
[26] Borsányi S et al., *Phys. Lett.* B **370**, 99 (2014).
[27] Bazavov A et al., *Phys. Rev.* D **90**, 094503 (2014).
[28] Bhattacharya T et al., *Phys. Rev. Lett.* **113**, 082001 (2014).
[29] Borsányi S et al., *JHEP* **1009**, 073 (2010).
[30] Bazavov A et al., arXiv:1701.04325 [hep-lat] (2017).
[31] Stachel J et al., *J. Phys. Conf. Ser.* **569**, 012019 (2014).
[32] Andronic A, Braun-Munzinger P, Redlich K, Stachel J, arXiv:1611.01347 [nucl-th] (2016).
[33] Adamczyk L et al. (STAR Collaboration), arXiv:1701.07065 [nucl-ex] (2017).
[34] Becattini F, Steinheimer J, Stock R, Bleicher M, *Phys. Lett.* B **764**, 241 (2017).
[35] Andronic A, Braun-Munzinger P, Stachel J, *Nucl. Phys.* A **834**, 237 (2010).
[36] Cleymans J, Kämpfer B, Kaneta M, Wheaton S, Xu N, *Phys. Rev.* C **71**, 054901 (2005).
[37] Becattini F et al., Phys. Rev. C **90**, 054907 (2014).
[38] Agakishiev G et al. (HADES Collaboration), *Eur. Phys. J.* A **52**, 178 (2016).
[39] Ferroni L, Becattini F, *Acta Phys. Polon.* B **43**, 571 (2012).
[40] Ratti C et al., *Nucl. Phys.* A **931**, 802 (2014).
[41] Borsányi S et al., *Phys. Rev. Lett.* **113**, 052301 (2014).
[42] Alba P et al., *Phys. Rev.* C **92**, 064910 (2015).
[43] Bellwied R et al., *Phys. Rev.* D **92**, 114505 (2015).
[44] Karsch F et al., *Nucl. Phys.* A **956**, 352 (2016).

Evolution of the gauge couplings and Weinberg angle in 5-dimensions for an SU(5) and flipped SU(5) gauge group

Mohammed Omer Khojali[a,1], **A. S. Cornell**[a,2], **Aldo Deandrea**[b,3] **and Giacomo Cacciapaglia**[b,4]

[a] National Institute for Theoretical Physics; School of Physics, University of the Witwatersrand, Wits 2050, South Africa
[b] Universite de Lyon, F-69622 Lyon, France; Universite Lyon 1, CNRS/IN2P3, UMR5822 IPNL, F-69622 Villeurbanne Cedex, France

E-mail: [1]khogali11@gmail.com, [2]alan.cornell@wits.ac.za, [3]deandrea@ipnl.in2p3.fr, [4] g.cacciapaglia@ipnl.in2p3.fr

Abstract. We explicitly test, in a simplified 5-dimensional model with $SU(5)$ and $SU(5) \times U(1)'$ gauge symmetries, the evolution of the gauge couplings. We assume that all the matter fields are propagating in the bulk, and consider orbifolds based on Abelian discrete groups which lead to 5-dimensional gauge theories compactified on an S^1/Z_2. The gauge couplings evolution is derived at one-loop level and used to test the impact on lower energy observables, in particular the Weinberg angle.

1. Introduction

The greatest achievement of the Large Hadron Collider (LHC) has been the discovery of the missing building block of the Standard Model (SM) [1, 2], the Higgs particle (or, at least, a particle which most likely is the SM Higgs particle). This discovery gave good support to the idea of the unification of electromagnetic and weak forces, as described by the electroweak gauge group $SU(2)_L \times U(1)_Y$, which is spontaneously broken to $U(1)_{EM}$ via the vacuum expectation value (VEV) of the Higgs fields [3].

The idea of the Grand Unification Theory (GUT) is to embed the SM gauge group ($G_{SM} \equiv SU(3)_C \times SU(2)_L \times U(1)_Y$) into a large group G. The SM group is rank 4, which means that the gauge group G must be at least rank 4. In this work, we shall study the non-supersymmetric extensions of the SM based on the gauge group $SU(5)$ and flipped $SU(5)$. In particular, we will study higher-dimensional non-supersymmetric orbifold models [4].

We consider orbifolds based on Abelian discrete groups which lead to 5-dimensional gauge theory compactified on an S^1/Z_2, where we assume that all matter fields are propagating in the bulk. The extra dimension is compactified on a circle of radius R with Z_2 orbifold. The 5-dimensional Kaluza-Klein (KK) modes of the weak doublet (Q) and singlet (q), as well as

Higgs and gauge fields (A), are given by [5, 6]

$$A(x,y) = \frac{1}{\sqrt{2\pi R}} A_0(x) + \frac{1}{2\sqrt{\pi R}} \sum_{n=1}^{\infty} A_n(x) \cos\left(\frac{ny}{R}\right), \tag{1}$$

$$Q(x,y) = \frac{1}{\sqrt{2\pi R}} Q_L^0(x) + \frac{1}{2\sqrt{\pi R}} \sum_{n=1}^{\infty} \left[Q_L^n(x) \cos\left(\frac{ny}{R}\right) + Q_R^n(x) \sin\left(\frac{ny}{R}\right)\right], \tag{2}$$

$$q(x,y) = \frac{1}{\sqrt{2\pi R}} q_R^0(x) + \frac{1}{2\sqrt{\pi R}} \sum_{n=1}^{\infty} \left[q_R^n(x) \cos\left(\frac{ny}{R}\right) + q_L^n(x) \sin\left(\frac{ny}{R}\right)\right], \tag{3}$$

where the zero modes are the 4-dimensional SM fields and there is a left- and a right-handed KK mode for each SM chiral fermion, whilst the Higgs and the gauge fields are Z_2 even fields [7].

The paper is structured as follows: In section (2) we discuss the evolution of the gauge couplings and Weinberg angle in 5-dimensions for an $SU(5)$ gauge group, whilst in section (3) we discuss the evolution of the gauge couplings and Weinberg angle in 5-dimensions for an $SU(5) \times U(1)'$ gauge group. In section (4) we conclude.

2. The gauge coupling evolution equations for an $SU(5)$ gauge group

In this section shall explore the evolution of the gauge couplings and Weinberg angle in five dimension for an $SU(5)$ gauge group, in order to have a unified theory above some energy scale M_X, with n_g generation of fermions, where we need at least 12 new gauge bosons, an $SU(2)_L$ doublet, colour triplet and their antiparticles, also a colour triplet; $SU(2)_L$ singlet of the Higgs scalars h_α [8]. We put the electro-weak doublet and the colour triplet in the 5-dimensional fundamental representation

$$5_H = \begin{pmatrix} h^r \\ h^g \\ h^b \\ \Phi^+ \\ \Phi^0 \end{pmatrix}, \tag{4}$$

where $SU(3)_C$ acts on the first 3 components, and $SU(2)_L$ acts on the last two. The $SU(5)$ gauge group breaks into the SM gauge group, $SU(3)_c \times SU(2)_L \times U(1)_Y$, when a scalar field 24_H such as the Higgs field acquires VEV, and this VEV is proportional to hypercharge generator [8].

The Higgs sector is made up of an adjoint 24_H, which acquires a VEV from the spontaneous breaking $SU(5) \to SU(3)_C \times SU(2)_W \times U(1)_Y$:

$$< 24_H >= \frac{v}{\sqrt{30}} \begin{pmatrix} 2 & 0 & 0 & 0 & 0 \\ 0 & 2 & 0 & 0 & 0 \\ 0 & 0 & 2 & 0 & 0 \\ 0 & 0 & 0 & -3 & 0 \\ 0 & 0 & 0 & 0 & -3 \end{pmatrix}. \tag{5}$$

We can write the full $SU(3)_C \times SU(2)_L \times U(1)_Y$ right handed representation of the creation operators as follows [9]:

$$\begin{aligned} u^\dagger \oplus d^\dagger \oplus e^\dagger \oplus \bar{Q}^\dagger \oplus \bar{l}^\dagger &= (3,1,2/3) \oplus (3,1,-1/3) \oplus (1,1,-1) \\ &\oplus(\bar{3},2,-1/6) \oplus (1,2,1/2). \end{aligned} \tag{6}$$

The new gauge bosons are called X and Y and they violate baryon and lepton number and carry flavour and colour. The gauge bosons are given by the adjoint representation of the $SU(5)$ gauge group:

$$24 \to (8,1,0) \oplus (1,3,0) \oplus (1,1,0) \oplus (3,2,-5/6) \oplus (\bar{3},2,5/6), \tag{7}$$

where (8,1,0) is identified as the $SU(3)_C$ gauge bosons G^α_β, (1,3,0) is identified as the $W^\pm$ and W^0 gauge bosons, (1,1,0) is identified as the B gauge boson, (3,2,-5/6) is identified as the $A^\tau_\alpha = (X_\alpha, Y_\alpha)$ gauge boson and ($\bar{3}$,2,5/6) identified as $A^\alpha_\beta = (X_\alpha, Y_\alpha)^T$ gauge boson. The covariant derivative for a fundamental representation is given by

$$D_\mu = \partial_\mu - ig\sum_{a=1}^{24}\frac{\lambda_a}{2}A^a_\mu \equiv \partial_\mu - ig\mathcal{A}_\mu, \tag{8}$$

and the matrix of the gauge bosons becomes

$$\mathcal{A}_\mu = \frac{1}{\sqrt{2}}\begin{pmatrix} G^r_r & G^r_g & G^r_b & \bar{X}_r & \bar{Y}_r \\ G^g_r & G^g_g & G^g_b & \bar{X}_g & \bar{Y}_g \\ G^b_r & G^b_g & G^b_b & \bar{X}_g & \bar{Y}_g \\ X^r & X^g & X^b & \frac{W_3}{\sqrt{2}} & W^+ \\ Y^r & Y^g & Y^b & W^- & -\frac{W_3}{\sqrt{2}} \end{pmatrix} + \sqrt{\frac{3}{5}}\begin{pmatrix} -\frac{B_\mu}{3} & 0 & 0 & 0 & 0 \\ 0 & -\frac{B_\mu}{3} & 0 & 0 & 0 \\ 0 & 0 & \frac{B_\mu}{2} & 0 & 0 \\ 0 & 0 & 0 & \frac{B_\mu}{2} & 0 \\ 0 & 0 & 0 & 0 & \frac{B_\mu}{2} \end{pmatrix}. \tag{9}$$

The new gauge bosons X and Y are clearly carrying colour and they have electric charge 4/3 and 1/3 respectively.

The one-loop beta functions for the gauge couplings in 4-dimensions for $SU(5)$ are given by:

$$16\pi^2 g_1^{-3}\beta_{g_1} = \frac{81}{20}, \tag{10}$$

$$16\pi^2 g_2^{-3}\beta_{g_2} = -\frac{19}{6}, \tag{11}$$

$$16\pi^2 g_3^{-3}\beta_{g_3} = -\frac{41}{6}. \tag{12}$$

The one-loop beta functions for the gauge couplings in 5-dimension for $SU(5)$ are given by:

$$16\pi^2 g_1^{-3}\beta_{g_1} = (S(t)-1)\left(\frac{81}{10}\right), \tag{13}$$

$$16\pi^2 g_2^{-3}\beta_{g_2} = (S(t)-1)\left(\frac{7}{6}\right), \tag{14}$$

$$16\pi^2 g_3^{-3}\beta_{g_3} = (S(t)-1)\left(-\frac{5}{2}\right), \tag{15}$$

where $t = \ln(S(t)/M_Z R)$, $S(t) = \mu R$ for $M_Z < \mu < \ln(1/M_Z R)$. For our numerical calculation we choose the initial values for the gauge couplings based on the renormalisation point M_Z scale as: $\alpha_1(M_Z) = 0.01696$, $\alpha_2(M_Z) = 0.03377$ and $\alpha_3(M_Z) = 0.1184$ [10]. In Figure 1, left panel, we present the evolution of the inverse fine structure constants in five dimensions for the one-loop beta-function, by assuming that all the matter fields are in the bulk. We see that α_1^{-1} and α_2^{-1} approximately meet at $\log(E/GeV) \sim 3.67$, α_1^{-1} and α_3^{-1} approximately meet at $\log(E/GeV) \sim 3.72$ and α_3^{-1} and α_2^{-1} approximately meet at $\log(E/GeV) \sim 3.78$. In Figure 1, right panel, we present the evolution of the Weinberg angle $\sin^2\theta_W$ for the one-loop

Scenario	$t(R_1)$	$t(R_2)$	$t(R_3)$
5D $SU(5)$	7.51	8.19	8.86

Table 1. The cut-offs in 5 dimension for an $SU(5)$ gauge group for three different compactification radii $R^{-1} = 5, 10$ and 15 TeV, where $t = \ln(\mu/M_Z)$.

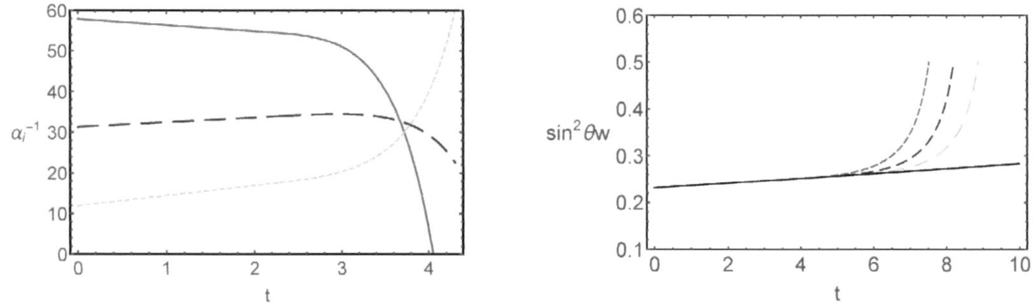

Figure 1. Left panel: The evolution of the inverse fine structure constants $\alpha_i^{-1}(\mu)$ in five dimensions as a function of t, for compactification scale $R^{-1} = 5$ TeV, where α_1^{-1} is the (green) dotted line, α_2^{-1} is the (blue) dashed line and α_3^{-1} is the (red) solid line. Right panel: Evolution of the Weinberg angle $\sin^2\theta_W$ for all matter fields in the bulk, for three different compactification scales; 5 TeV (solid line), 10 TeV (dot-dashed line), 20 TeV (dashed line) as a function of t. The solid (black) line is the SM result.

beta-function, for different values of compactification radius, for the model discussed above. Notice that the prediction that the Weinberg angle for an SU(5) gauge group is $\sin^2\theta_W \sim 3/8$ at M_X is over-shot by approximately 20%. We have summarised our results in Table 1, the cut-offs for the case of an $SU(5)$ gauge group in 5 dimensions, these values correspond to the point where $g_1 = \sqrt{\frac{5}{3}}g_2$.

3. The gauge coupling evolution equations for an $SU(5)\times U(1)'$ gauge group

In this section we shall explore the evolution of the gauge couplings and Weinberg angle in five dimensions for an $SU(5)\times U(1)'$ gauge group. The flipped $SU(5)$ gauge group is different from $SU(5)$ gauge group in the electric charge generators, it does not lie completely in $SU(5)$ gauge group. The flipped $SU(5)$ is just embeding of $SU(5)\times U(1)$ into $SO(10)$ gauge group, flipped $SU(5)$ model is very special GUT [11]. The flipped $SU(5)$ gauge group containing three generations of the quark and lepton, $(10, \bar{10})$ Higgs boson, they have the following presentation [12]:

$$F_{(10)} = [Q, d^c, \nu^c]; \qquad \bar{f}_{(\bar{5})} = [L, u^c]; \qquad l_1 = e^c, \tag{16}$$

and

$$H_{(10)} = [Q_H, d^c_H, \Phi_H] \qquad \bar{H}_{(\bar{10})} = [Q_{\bar{H}}, d^c_{\bar{H}}, \Phi_{\bar{H}}], \tag{17}$$

where the components Φ_H, $\Phi_{\bar{H}}$ breaking the flipped $SU(5)$ gauge group to the SM gauge group, once Φ_H, $\Phi_{\bar{H}}$ acquires a VEV:

$$SU(5)\times U(1)' \rightarrow SU(3)_C\times SU(2)_L\times U(1)_Y\times U(1)', \tag{18}$$

and then the gauge group $SU(3)_C\times SU(2)_L\times U(1)_Y\times U(1)'$, is spontaneous breaking to $SU(3)_C\times U(1)_{em}$, once Φ_5 acquires a VEV:

$$SU(3)_C\times SU(2)_L\times U(1)_Y\times U(1)' \rightarrow SU(3)_C\times U(1)_{em}. \tag{19}$$

Fields	SU(3)$_C$	SU(2)$_L$	$Y_5/2$
Q	3	2	$\frac{1}{6}$
L	1	2	$-\frac{1}{2}$
u^c	$\bar{3}$	1	$-\frac{2}{3}$
d^c	$\bar{3}$	1	$\frac{1}{3}$
e^c	1	1	1
Φ_5	1	2	$\frac{1}{2}$
$\bar{H}_{\bar{10}}$	$\bar{3}$	1	$-\frac{1}{3}$

Table 2. Summary the quark, lepton and Higgs field content in the flipped $SU(5)$ model and their quantum numbers.

The flipped $SU(5)$ gauge group includes $SU(5)$ gauge bosons $W^{\pm}$, W_3, B, X, Y and $\tilde{B}$, which is a $U(1)'$ gauge boson. The electric charge generator Q in flipped $SU(5)$ model is given as:

$$Q = T_3 - \frac{1}{5}Y' + \frac{2}{5}\tilde{Y}, \tag{20}$$

where Y' is the $U(1)$ inside $SU(5)$ and $\tilde{Y}$ is the one outside $SU(5)$.

The hypercharges of the known quark, lepton and Higgs field in the flipped $SU(5)$ gauge group are given as shown in Tabel 2 [11, 12]:

The one-loop beta functions for the gauge couplings in 4-dimensions for $SU(5) \times U(1)'$ are given by:

$$16\pi^2 g_1^{-3}\beta_{g_1} = \frac{53}{6}, \tag{21}$$

$$16\pi^2 g_2^{-3}\beta_{g_2} = -\frac{19}{6}, \tag{22}$$

$$16\pi^2 g_3^{-3}\beta_{g_3} = -\frac{41}{6}. \tag{23}$$

The one-loop beta functions for the gauge couplings in 5-dimensions for $SU(5) \times U(1)'$ are given by:

$$16\pi^2 g_1^{-3}\beta_{g_1} = (S(t) - 1)\left(\frac{105}{24}\right), \tag{24}$$

$$16\pi^2 g_2^{-3}\beta_{g_2} = (S(t) - 1)\left(-\frac{7}{24}\right), \tag{25}$$

$$16\pi^2 g_3^{-3}\beta_{g_3} = (S(t) - 1)\left(-\frac{5}{2}\right), \tag{26}$$

In Figure 2, left panel, we showing the evolution of the α_i^{-1} in 5-dimensions for the one-loop beta-function and in this case, one can see that α_1^{-1}, α_2^{-1} and α_3^{-1} approximately unified at $\log(E/GeV) \sim 4.0$ for compactification scale $R^{-1} = 5$ TeV. In the right panel, we present the evolution of Weinberg angle for the one-loop beta-function, for different values of compactification scales $R^{-1} = 1$ TeV and 5 TeV, for the flipped $SU(5)$ model, as an example, for $R^{-1} = 5$ TeV, $\sin^2\theta_W \sim 0.37$ at $t \sim 5.08$.

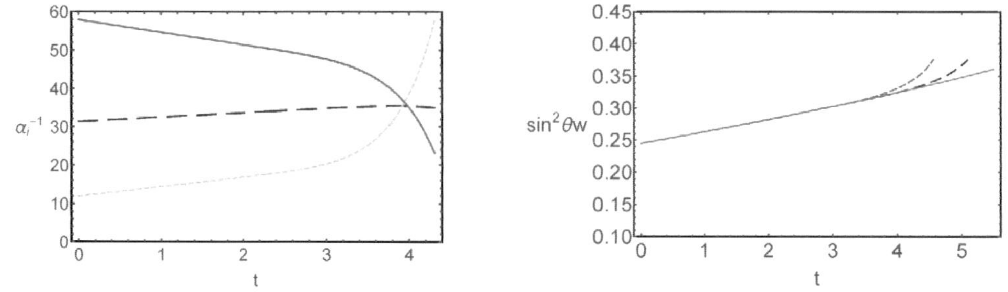

Figure 2. Left panel: The evolution of the inverse fine structure constants $\alpha_i^{-1}(\mu)$ in five dimension as a function of $\log(E/GeV)$ for compactification scale $R^{-1} = 5$ TeV. Right panel: Evolution of the Weinberg angle $\sin\theta_W$ for all matter fields in the bulk, for two different compactification scales as a function of t, for an $SU(5) \times U(1)'$ gauge group.

Scenario	$t(R_1)$	$t(R_2)$
5D flipped $SU(5)$	4.55	5.08

Table 3. The cut-offs in 5 dimensions for the flipped $SU(5)$ gauge group for two different compactification radii $R^{-1} = 1$ and 10 TeV, where $t = \ln(\mu/M_Z)$.

4. Conclusion

In this paper we derived the one-loop renormalisation group equations in five-dimensions for an $SU(5)$ and $SU(5) \times U(1)'$ gauge group compactified on an S_1/Z_2. We observed that when the fifth dimension KK-modes became kinematically accessible the evolution of the Weinberg angle rapidly increased by approximately 20% for $SU(5)$ and 7% for $SU(5) \times U(1)'$.

Acknowledgments

This work is based on the research supported by the National Research Foundation of South Africa.

References

[1] S. Chatrchyan *et al.* [CMS Collaboration], Phys. Lett. B **716** (2012) 30 [arXiv:1207.7235 [hep-ex]].
[2] G. Aad *et al.* [ATLAS Collaboration], Phys. Lett. B **716** (2012) 1 [arXiv:1207.7214 [hep-ex]].
[3] Y. Hosotani and N. Yamatsu, PTEP **2015**, 111B01 (2015) [arXiv:1504.03817 [hep-ph]].
[4] C. Csaki, C. Grojean and H. Murayama, Phys. Rev. D **67**, 085012 (2003) [hep-ph/0210133].
[5] A. S. Cornell, A. Deandrea, L. X. Liu and A. Tarhini, Mod. Phys. Lett. A **28** (2013) no.11, 1330007 [arXiv:1209.6239 [hep-ph]].
[6] A. S. Cornell and L. X. Liu, Phys. Rev. D **83**, 033005 (2011) [arXiv:1010.5522 [hep-ph]].
[7] L. X. Liu and A. S. Cornell, Phys. Rev. D **86**, 056002 (2012) [arXiv:1204.0532 [hep-ph]].
[8] G. Senjanovic, Springer Proc. Phys. **118**, 137 (2008).
[9] Marina von Steinkirch, The gauge group SU(5) as a simple GUT, State University of New York at Stony Brook, (2010).
[10] A. Abdalgabar, A. S. Cornell, A. Deandrea and A. Tarhini, Phys. Rev. D **88**, 056006 (2013) [arXiv:1306.4852 [hep-ph]].
[11] A. C. Davis and N. F. Lepora, Phys. Rev. D **52** (1995) 7265 [hep-ph/9504411].
[12] J. L. Lopez and D. V. Nanopoulos, In Erice 1995, Vacuum and vacua, 125-159 [hep-ph/9511266].

A pQCD sized problem in small systems

Isobel Kolbé[1] and W. A. Horowitz[1]

[1]Department of Physics, University of Cape Town, Private Bag X3, Rondebosch 7701, South Africa

E-mail: isobel.kolbe@gmail.com

Abstract. The Quark Gluon Plasma (QGP) has been studied extensively at the LHC, with jet quenching and particle suppression playing an important role in our ability to characterize this fundamental state of matter. A number of theoretical descriptions concerning the mechanisms whereby particle suppression occurs have been put forward with perturbative methods successfully describing suppression patterns in very central Pb-Pb collisions at the LHC. However, particle suppression is by no means the only hallmark of the existence of the QGP and many measurements at the LHC of smaller colliding systems, such as peripheral Pb-Pb and central p-Pb and p-p, have hinted at the production of a droplet of QGP in alarmingly small volumes. In stark contrast, existing perturbative Quantum Chromodynamical methods rely heavily on the assumption that the system under consideration is large, demanding an extension of pQCD methods to smaller systems. We present precisely such an extension and find corrections on the order of 100% at high energies, revealing a number of shortcomings and problematic assumptions that are present even in traditional pQCD energy loss calculations.

1. Introduction

The discovery of a number of Quark Gluon Plasma (QGP) signals (such as collective behavior [1], strangeness enhancement [2, 3], and quarkonium suppression [4]) in small colliding systems such as proton or deuteron collisions with heavy nuclei ($p/d+A$) or even proton-proton (pp) collisions has prompted a vigorous investigation into the properties of such colliding systems. Many of the observed signals are traditionally (that is, in heavy ion collisions) attributed to the presence of a QGP and have been used to characterize the QGP. In particular, jet quenching and particle suppression provided powerful femtoscopic probes of the dynamics of the degrees of freedom of the QGP.

The success of jet tomography is rooted in its successful description by a number of perturbative Quantum Chromodynamical (pQCD) energy loss models [5–8] that are able to qualitatively describe the momentum dependence and angular distribution of the suppression of high-momentum, $\sim 5-150$ GeV single particle pions [9, 10] and charged hadrons [11–13] from primordial hard light flavors and gluons and electrons [14–16] as well as D [17] and non-prompt J/ψ mesons [18] from open heavy flavor decays at mid rapidity in A+A systems from $\sqrt{s} = 0.2$ TeV to 2.76 TeV.

However, *all* of the standard pQCD energy loss models make the explicit assumption that the QGP system is large [19], calling into question their applicability to the very small systems that must be present if a QGP is created in small colliding systems. In the standard opacity expansion developed by Djordjevic, Gyulassy, Levai and Vitev (DGLV) [20,21], the large system assumption amounts to an assumption that the separation distance $\Delta z \equiv z_1 - z_0 \gg \lambda_{mfp} \gg 1/\mu_D$ between the initial production position z_0 of the hard parent parton and the position z_1 where it scatters off a QGP medium quasi-particle is large.

In order to apply DGLV energy loss to small systems, we derive the $N = 1$ opacity generalization for all separation distances. We find two curious results:

(i) Due to the large formation time assumption, only two of the relevant diagrams have non-zero corrections to the standard DGLV result.

(ii) The (negative) correction terms dominate even for relatively large system sizes at high parent parton energies.

It has been known for some time that all energy loss formalisms are sensitive to the collinear approximation [22,23], but we will demonstrate that the present sensitivity to the large formation time approximation is both new and different.

2. Setup

We follow the DGLV calculation [21], treating the high-p_T eikonal parton produced at an initial point $(t_0, z_0, \mathbf{x}_0)$ inside a finite QGP, where we have used $\mathbf{p}$ to mean transverse 2D vectors, $\vec{\mathbf{p}} = (p_z, \mathbf{p})$ for 3D vectors and $p = (p^0, \vec{\mathbf{p}}) = [p^0 + p^z, p^0 - p^z, \mathbf{p}]$ for four vectors in Minkowski and light cone coordinates respectively. The full calculation is available at [24]. As in the DGLV calculation, we consider the target to be a Gyulassy-Wang Debye screened potential [25] with Fourier and color structure given by

$$\begin{aligned} V_n &= V(\vec{\mathbf{q}}_n)e^{-i\vec{q}_n\cdot\vec{x}_n} \\ &= 2\pi\delta(q^0)v(\mathbf{q}_n, q_n^z)e^{-i\vec{q}_n\cdot\vec{x}_n}T_{a_n}(R)\otimes T_{a_n}(n). \end{aligned} \tag{1}$$

The color exchanges are handled using the applicable $SU(N_c)$ generator $T_a(n)$ in the d_n dimensional representation of the target or $T_a(R)$ in the d_R dimensional representation of the high-p_T parent parton.

In light cone coordinates the momenta (defined in Figure: 1) of the emitted gluon, the final high-p_T parton, and the exchanged medium Debye quasi-particle are

$$k = \left[xP^+, \frac{m_g^2 + \mathbf{k}^2}{xP^+}, \mathbf{k}\right], p = \left[(1-x)P^+, \frac{M^2 + \mathbf{k}^2}{(1-x)P^+}, -\mathbf{k}\right], q = [q^+, q^-, \mathbf{q}], \tag{2}$$

where the initially produced high-p_T particle of mass M has large momentum $E^+ = P^+ = 2E$ and negligible other momentum components. Notice that we include the Ter-Mikayelian plasmon effect with an effective emitted gluon mass m_g [21,26].

A shorthand for energy ratios will prove useful notationally. Following [21], we define $\omega \approx xE^+/2 = xP^+/2$, $\omega_0 \equiv \mathbf{k}^2/2\omega$, $\omega_i \equiv (\mathbf{k} - \mathbf{q}_i)^2/2\omega$, $\omega_{(ij)} \equiv (\mathbf{k} - \mathbf{q}_i - \mathbf{q}_j)^2/2\omega$, and $\tilde{\omega}_m \equiv \left(m_g^2 + M^2x^2\right)/2\omega$.

Additionally, a number of crucial assumptions are made in line with [21]: 1) the eikonal, or high energy, approximation, for which E^+ is the largest energy scale of the problem; 2) the soft (radiation) approximation $x \ll 1$; 3) collinearity, $k^+ \gg k^-$; 4) that the impact parameter varies over a large transverse area; and, most crucially for the present work, 5) the large formation time assumption $\omega_i \ll \mu_i$, where $\mu_i^2 \equiv \mu^2 + \mathbf{q}_i^2$. These assumptions allow us to 1) (eikonal) ignore the spin of the high-p_T parton; 2) (soft) assume the source current for the parent parton varies slowly with momentum $J(p - q + k) \approx J(p + k) \approx J(p)$; 3) (collinearity) complete a separation of energy scales

$$E^+ \gg k^+ \gg k^- \equiv \omega_0 \sim \omega_{(i\ldots j)} \gg \frac{(\mathbf{p} + \mathbf{k})^2}{P^+}; \tag{3}$$

and 4) take the ensemble average over0 the phase factors, which become $\langle e^{-i(\mathbf{q}-\mathbf{q}')\cdot\mathbf{b}}\rangle = ((2\pi)^2/A_\perp)\delta^2(\mathbf{q}-\mathbf{q}')$.

Within the above setup, we re-evaluated the 10 $N=1$ in opacity diagrams without the large separation distance $\Delta z \gg 1/\mu$ assumption.

3. Calculation and Results

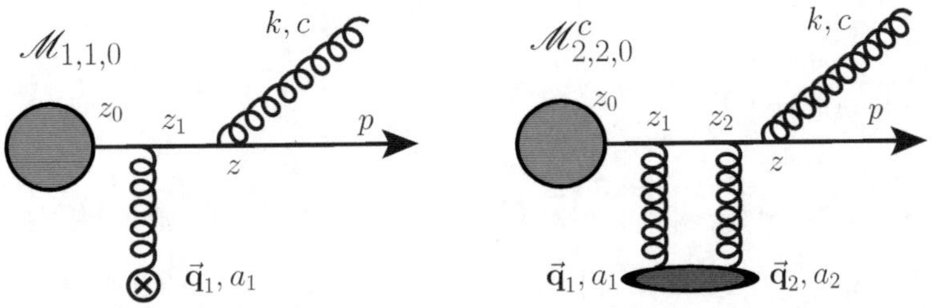

Figure 1: $\mathscr{M}_{1,1,0}$ (left panel) and $\mathscr{M}^c_{2,2,0}$ (right panel) are the only two diagrams that have non-zero short separation distance corrections in the large formation time limit. $\mathscr{M}^c_{2,2,0}$ is the double Born contact diagram, corresponding to the second term in the Dyson series in which two gluons are exchanged with the single scattering center.

Implementing the relaxation of the large separation distance assumption, $\Delta z \gg 1/\mu_D$, amounts to the retention of terms proportional to $\exp(-\mu_D\Delta z)$, that were neglected in the original evaluation of the $N=1$ in opacity energy loss derivation. Curiously, the large radiated gluon formation time approximation $\omega_i \ll \mu_i$ leads to a tremendous simplification; all but 2 of the 10 diagrams' 18 new small distance correction pole contributions are suppressed under the large formation time assumption in addition to being suppressed under the large separation distance assumption (the relaxation of which is the object of the present calculation). It should be noted that the large formation time assumption allows one to neglect terms in both the original DGLV and the current short separation distance correction. Fig. 1 shows the two diagrams that contribute non-zero small separation distance, large formation time terms to the result.

The full result for these two amplitudes under our approximation scheme is then

$$\mathscr{M}_{1,1,0} \approx -J(p)e^{ipx_0}2gT_{a_1}ca_1\int\frac{d^2\mathbf{q}_1}{(2\pi)^2}v(0,\mathbf{q}_1)e^{-i\mathbf{q}_1\cdot\mathbf{b}_1} \times\frac{\mathbf{k}\cdot\boldsymbol{\epsilon}}{\mathbf{k}^2+m_g^2+x^2M^2}\left[e^{i(\omega_0+\tilde{\omega}_m)(z_1-z_0)}-\frac{1}{2}e^{-\mu_1(z_1-z_0)}\right] \tag{4}$$

$$\mathscr{M}^c_{2,2,0} \approx J(p)e^{i(p+k)x_0}\int\frac{d^2\mathbf{q}_1}{(2\pi)^2}\int\frac{d^2\mathbf{q}_2}{(2\pi)^2}e^{-i(\mathbf{q}_1+\mathbf{q}_2)\cdot\mathbf{b}_1} \times igT_{a_2}T_{a_1}ca_2a_1v(0,\mathbf{q}_1)v(0,\mathbf{q}_1)\frac{\mathbf{k}\cdot\boldsymbol{\epsilon}}{\mathbf{k}^2+m_g^2+x^2M^2} \times\left[e^{i(\omega_0+\tilde{\omega}_m)(z_1-z_0)}+e^{-\mu_1(z_1-z_0)}\left(1-\frac{\mu_1e^{-\mu_2(z_1-z_0)}}{2(\mu_1+\mu_2)}\right)\right]. \tag{5}$$

The double differential single inclusive gluon emission distribution is given by [21]

$$d^3N_g^{(1)}d^3N_J = \frac{d^3\vec{\mathbf{p}}}{(2\pi)^32p^0}\frac{d^3\vec{\mathbf{k}}}{(2\pi)^32\omega}\left(\frac{1}{d_T}\mathrm{Tr}\langle|\mathscr{M}_1|^2\rangle+\frac{2}{d_T}\Re\,\mathrm{Tr}\langle\mathscr{M}_0^*\mathscr{M}_2\rangle\right). \tag{6}$$

From the amplitudes, the energy-weighted integral over the gluon emission distribution $\Delta E = E\int dx\, x dN_g/dx$ gives the energy loss.

Our main analytic result is then the $N = 1$ first order in opacity small distance generalization of the DGLV induced energy loss of a high-p_T parton in a QGP:

$$\Delta E^{(1)}_{ind} = \frac{C_R\alpha_s LE}{\pi\lambda_g}\int dx \int \frac{d^2\mathbf{q}_1}{\pi}\frac{\mu^2}{(\mu^2+\mathbf{q}_1^2)^2}\int\frac{d^2\mathbf{k}}{\pi}\int d\Delta z\bar{\rho}(\Delta z)\Bigg[-\frac{2\big(1-\cos\{(\omega_1+\tilde{\omega}_m)\Delta z\}\big)}{(\mathbf{k}-\mathbf{q}_1)^2+m_g^2+x^2M^2}$$
$$\times\left(\frac{(\mathbf{k}-\mathbf{q}_1)\cdot\mathbf{k}}{\mathbf{k}^2+m_g^2+x^2M^2}-\frac{(\mathbf{k}-\mathbf{q}_1)^2}{(\mathbf{k}-\mathbf{q}_1)^2+m_g^2+x^2M^2}\right)$$
$$+\frac{1}{2}e^{-\mu_1\Delta z}\Bigg\{\left(\frac{\mathbf{k}}{\mathbf{k}^2+m_g^2+x^2M^2}\right)^2\left(1-\frac{2C_R}{C_A}\right)\Big(1-\cos\{(\omega_0-\tilde{\omega}_m)\Delta z\}\Big)$$
$$+\frac{\mathbf{k}\cdot(\mathbf{k}-\mathbf{q}_1)}{(\mathbf{k}^2+m_g^2+x^2M^2)\big((\mathbf{k}-\mathbf{q}_1)^2+m_g^2+x^2M^2\big)}\big(\cos\{(\omega_0-\tilde{\omega}_m)\Delta z\}-\cos\{(\omega_0-\omega_1)\Delta z\}\big)\Bigg\}\Bigg]. \qquad (7)$$

The standard DGLV result, with characteristic color triviality, is apparent in the first two lines of 7, while the last two lines are the short separation distance correction. The correction has the expected properties that it vanishes for both large separation distances ($\Delta z\mu_D \ll 1$) and vanishing separation distances (the LPM effect). Our correction term also has a number of surprising features: the term proportional to $2C_R/C_A$ breaks color triviality to all orders in opacity and in fact, although only apparent through numerical analysis, the correction term in fact dominates the original DGLV calculation at high energies.

The numerical investigation of equation (7) is performed following [21] and results in figures 2a, 2b and 2c.The numerical analysis used the following values: $\mu = 0.5$ GeV, $\lambda_{mfp} = 1$ fm, $C_R = 4/3$, $C_A = 3$, $\alpha_s = 0.3$, $m_{charm} \equiv m_c = 1.3$ GeV and $m_{bottom} \equiv m_b = 4.75$ GeV, and the QCD analogue of the Ter-Mikayelian plasmon effect was taken into account by setting $m_{gluon} \equiv m_g = \mu/\sqrt{2}$. As in [26], kinematic upper limits were used for the momentum integrals such that $0 \le k \le 2x(1-x)E$ and $0 \le q \le \sqrt{3E\mu}$, due to finite kinematics. This choice of k_{max} guarantees that the final momentum of the parent parton is collinear to the initial momentum of the parent parton and that the momentum of the emitted gluon is collinear to the momentum of the parent parton. The fraction of momentum carried away by the radiated gluon, x, was integrated over from 0 to 1. The distribution of scattering centers was assumed to be exponential in order to account for the rapidly expanding medium, $\bar{\rho}(z) = 2\exp(-2\Delta z/L)/L$.

The fractional energy loss of charm and bottom quarks in a 4 fm long brick of static QGP as a function of parent parton energy is shown in Fig. 2a. Generally, the small distance correction is an energy *gain* in addition to which the correction term increases with energy relative to the DGLV result.

The fractional energy loss of charm and bottom quarks as a function of path length for parent parton with energy $E = 10$ GeV is shown in Fig. 2b. The non-negligible effect for large path lengths of the small separation distance correction is a well understood feature and diminishes with path length.

The most prominent result from our numerical analysis is shown in Fig. 2c. Here, the fractional energy loss of 100 GeV charm and bottom quarks propagating up to 5 fm through a QGP shows clearly that the small distance "correction" term dominates, even out to path lengths of ~ 3 fm, over the leading DGLV result.

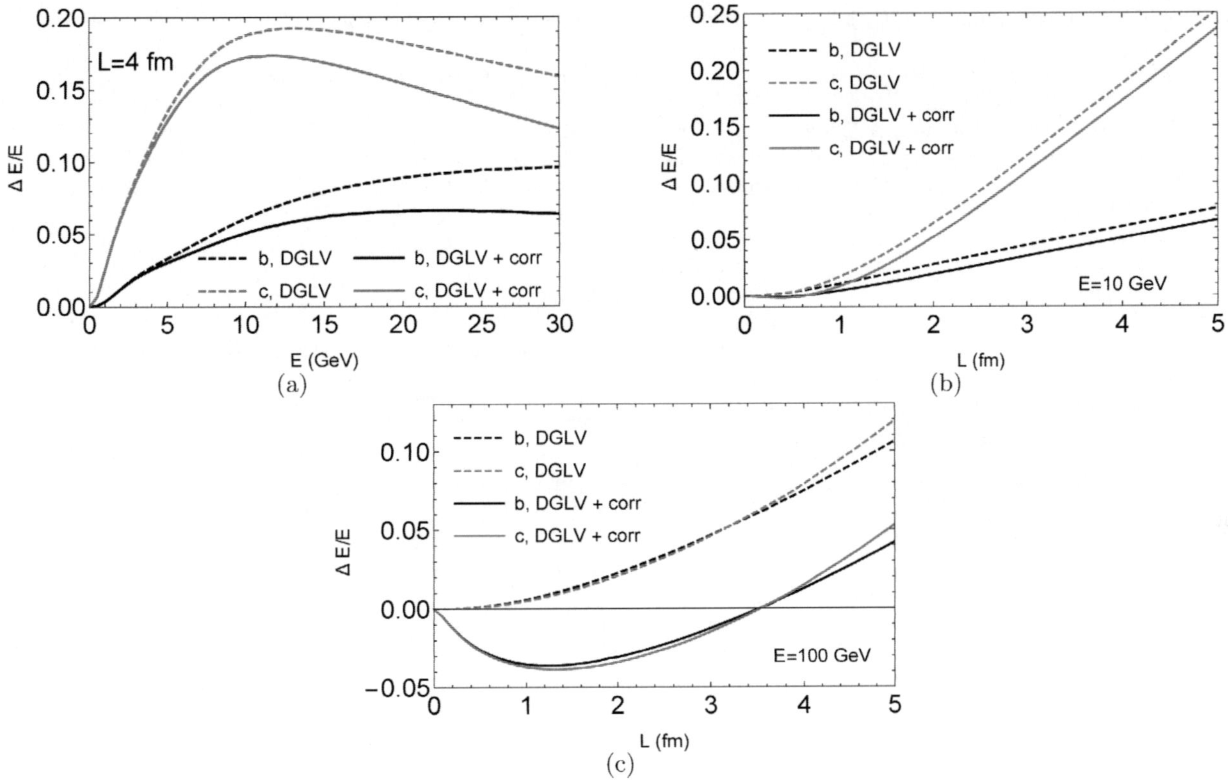

Figure 2: Fractional energy loss of charm and bottom quarks in a QGP with $\mu = 0.5$ GeV and $\lambda_{mfp} = 1$ fm for (a) fixed path length $L = 4$ fm, (b) fixed energy $E = 10$ GeV, and (c) fixed energy $E = 100$ GeV. In the figures, "DGLV" dashed curves are computed from the original $N = 1$ in opacity large separation distance DGLV formula while "DGLV + corr" solid lines are from our all separation distance generalization of the $N = 1$ DGLV result, equation (7).

4. Conclusions

We have performed a short separation distance correction to the standard DGLV calculation and have found that the assumption scheme that results from relaxing the large separation distance assumption breaks down at asymptotically large energies. As such, a more detailed interrogation of the validity of the large formation time assumption in pQCD-based energy loss calculations is necessary.

5. Acknowledgments

The authors wish to thank the SA-CERN Collaboration, the South African National Research Foundation (NRF), and the National Institute for Theoretical Physics (NITheP) for their generous support. The authors also wish to thank CERN for the hospitality extended to them during the completion of part of this work. Finally, the authors wish to thank Miklos Gyulassy, Ulrich Heinz, Sangyong Jeon, and Carlos Salgado for valuable discussions.

References

[1] V. Khachatryan *et al.*, "Evidence for Collective Multiparticle Correlations in p-Pb Collisions," *Phys. Rev. Lett.*, vol. 115, no. 1, p. 012301, 2015.

[2] B. B. Abelev *et al.*, "Multiplicity Dependence of Pion, Kaon, Proton and Lambda Production in p-Pb Collisions at $\sqrt{s_{NN}} = 5.02$ TeV," *Phys. Lett.*, vol. B728, pp. 25–38, 2014.

[3] CMS Collaboration, "Multiplicity and rapidity dependence of strange hadrons spectra in pp, pPb, and PbPb collisions at LHC energies," 2015.

[4] J. Adam *et al.*, "Centrality dependence of inclusive J/ production in p-Pb collisions at $\sqrt{s_{\rm NN}} = 5.02$ TeV," *JHEP*, vol. 11, p. 127, 2015.

[5] N. Armesto, M. Cacciari, A. Dainese, C. A. Salgado, and U. A. Wiedemann, "How sensitive are high-p(T) electron spectra at RHIC to heavy quark energy loss?," *Phys. Lett.*, vol. B637, pp. 362–366, 2006.

[6] W. A. Horowitz, "Heavy Quark Production and Energy Loss," *Nucl. Phys.*, vol. A904-905, pp. 186c–193c, 2013.

[7] K. M. Burke *et al.*, "Extracting the jet transport coefficient from jet quenching in high-energy heavy-ion collisions," *Phys. Rev.*, vol. C90, no. 1, p. 014909, 2014.

[8] M. Djordjevic, M. Djordjevic, and B. Blagojevic, "RHIC and LHC jet suppression in non-central collisions," *Phys. Lett.*, vol. B737, pp. 298–302, 2014.

[9] A. Adare *et al.*, "Azimuthal anisotropy of π^0 and η mesons in Au + Au collisions at $\sqrt{s_{NN}} = 200$ GeV," *Phys. Rev.*, vol. C88, no. 6, p. 064910, 2013.

[10] B. B. Abelev *et al.*, "Neutral pion production at midrapidity in pp and Pb-Pb collisions at $\sqrt{s_{\rm NN}} = 2.76$ TeV," *Eur. Phys. J.*, vol. C74, no. 10, p. 3108, 2014.

[11] S. Chatrchyan *et al.*, "Study of high-pT charged particle suppression in PbPb compared to *pp* collisions at $\sqrt{s_{NN}} = 2.76$ TeV," *Eur. Phys. J.*, vol. C72, p. 1945, 2012.

[12] B. Abelev *et al.*, "Centrality Dependence of Charged Particle Production at Large Transverse Momentum in Pb–Pb Collisions at $\sqrt{s_{\rm NN}} = 2.76$ TeV," *Phys. Lett.*, vol. B720, pp. 52–62, 2013.

[13] G. Aad *et al.*, "Measurement of charged-particle spectra in Pb+Pb collisions at $\sqrt{s_{\rm NN}} = 2.76$ TeV with the ATLAS detector at the LHC," *JHEP*, vol. 09, p. 050, 2015.

[14] B. I. Abelev *et al.*, "Transverse momentum and centrality dependence of high-p_T non-photonic electron suppression in Au+Au collisions at $\sqrt{s_{NN}} = 200$ GeV," *Phys. Rev. Lett.*, vol. 98, p. 192301, 2007. [Erratum: Phys. Rev. Lett.106,159902(2011)].

[15] S. Sakai, "Measurement of R_{AA} and ν_2 of electrons from heavy-flavour decays in Pb-Pb collisions at $\sqrt{s_{NN}} = 2.76$ TeV with ALICE," *Nucl. Phys.*, vol. A904-905, pp. 661c–664c, 2013.

[16] A. Adare *et al.*, "Single electron yields from semileptonic charm and bottom hadron decays in Au+Au collisions at $\sqrt{s_{NN}} = 200$ GeV," 2015.

[17] J. Adam *et al.*, "Transverse momentum dependence of D-meson production in Pb-Pb collisions at $\sqrt{s_{\rm NN}} = 2.76$ TeV," 2015.

[18] S. Chatrchyan *et al.*, "Suppression of non-prompt J/ψ, prompt J/ψ, and Y(1S) in PbPb collisions at $\sqrt{s_{NN}} = 2.76$ TeV," *JHEP*, vol. 05, p. 063, 2012.

[19] U. A. Wiedemann, "Jet Quenching in Heavy Ion Collisions," pp. 521–562, 2010. [Landolt-Bornstein23,521(2010)].

[20] M. Gyulassy, P. Levai, and I. Vitev, "Reaction operator approach to nonAbelian energy loss," *Nucl. Phys.*, vol. B594, pp. 371–419, 2001.

[21] M. Djordjevic and M. Gyulassy, "Heavy quark radiative energy loss in QCD matter," *Nucl. Phys.*, vol. A733, pp. 265–298, 2004.

[22] W. A. Horowitz and B. A. Cole, "Systematic theoretical uncertainties in jet quenching due to gluon kinematics," *Phys. Rev.*, vol. C81, p. 024909, 2010.

[23] N. Armesto *et al.*, "Comparison of Jet Quenching Formalisms for a Quark-Gluon Plasma 'Brick'," *Phys. Rev.*, vol. C86, p. 064904, 2012.

[24] I. Kolbé and W. A. Horowitz, "Short path length pQCD corrections to energy loss in the quark gluon plasma," arXiv:1509.06122.

[25] M. Gyulassy and X.-n. Wang, "Multiple collisions and induced gluon Bremsstrahlung in QCD," *Nucl. Phys.*, vol. B420, pp. 583–614, 1994.

[26] S. Wicks, W. Horowitz, M. Djordjevic, and M. Gyulassy, "Elastic, inelastic, and path length fluctuations in jet tomography," *Nucl. Phys.*, vol. A784, pp. 426–442, 2007.

[27] A. Majumder and M. Van Leeuwen, "The Theory and Phenomenology of Perturbative QCD Based Jet Quenching," *Prog. Part. Nucl. Phys.*, vol. A66, pp. 41–92, 2011.

[28] A. Adare *et al.*, "Centrality-dependent modification of jet-production rates in deuteron-gold collisions at $\sqrt{s_{NN}}$=200 GeV," 2015.

[29] G. Aad *et al.*, "Centrality and rapidity dependence of inclusive jet production in $\sqrt{s_{\rm NN}} = 5.02$ TeV proton-lead collisions with the ATLAS detector," *Phys. Lett.*, vol. B748, pp. 392–413, 2015.

IOP Conf. Series: Journal of Physics: Conf. Series **878** (2017) 012026 doi:10.1088/1742-6596/878/1/012026

Using Spin Observables and Polarizations to Probe Top-Higgs FCNC Couplings at Colliders

Blaženka Melić[1], **Monalisa Patra**[2]

Rudjer Bošković Institute, Department of Theoretical Physics, Bijenička 54, HR-10000 Zagreb, Croatia

E-mail: [1]melic@irb.hr,[2]mpatra@irb.hr

Abstract. We present how the polarized linear colliders can be used, complementary to the LHC, to fully determine the top-Higgs flavor changing neutral current (FCNC) couplings by using produced asymmetries and top spin polarizations and correlations.

1. The flavor changing top quark couplings at hadron colliders

The discovery of the Higgs boson at the LHC, has lead to a dedicated program of measuring its various properties so as to look for deviations from the standard model (SM). In the SM the presence of a single Higgs doublet leads to no FCNC transitions at tree level mediated by the Higgs boson. For e.g. the SM branching ratio of $t \to cH$ is extremely small, of the order $\mathrm{BR}(t \to cH)_{\mathrm{SM}} \approx 10^{-15}$ [1, 2], which is many orders of magnitude smaller than the value to be measured at the LHC at 14 TeV. Therefore an affirmative observation of the process $t \to qH$, well above the SM rate, will be a conclusive indication of new physics beyond the SM. Limits on the FCNC couplings in the light quark-Higgs sector can be obtained from neutral meson oscillations ($K^0 - \bar{K}^0$, $B^0 - \bar{B}^0$ and $D^0 - \bar{D}^0$) [3].

The most general FCNC tqH Lagrangian considered is of the form

$$\begin{aligned} \mathcal{L}^{tqH} &= g_{tu}\bar{t}_R u_L H + g_{ut}\bar{u}_R t_L H + g_{tc}\bar{t}_R c_L H + g_{ct}\bar{c}_R t_L H + h.c \\ &= \bar{t}(g_{qt}P_R + g^*_{tq}P_L)qH + \bar{q}(g_{tq}P_R + g^*_{qt}P_L)tH. \end{aligned} \tag{1}$$

The total decay width of the top in the presence of these FCNC couplings is then

$$\Gamma_t = \Gamma_t^{SM} + \Gamma_{t\to qH} \approx \Gamma_t^{SM} + 0.155\,(|\, g_{tq}\,|^2 + |\, g_{qt}\,|^2)\,, \tag{2}$$

where the width $\Gamma_t^{SM} = \Gamma_{t\to W^+b} = 1.35$ GeV, for $m_t = 173.3$ GeV at NLO, while the experimentally observed value of the total top-quark width is $\Gamma_t = 1.41^{+0.19}_{-0.15}$ GeV. The additional FCNC decay processes give positive contributions to Γ_t, proportional to $(|\, g_{tq}\,|^2 + |\, g_{qt}\,|^2)$ and from the experimentally observed Γ_t an upper bound on $\sqrt{|\, g_{tq}\,|^2 + |\, g_{qt}\,|^2}$ can be obtained. The ATLAS and the CMS collaborations have set upper limits on the tqH coupling through the top pair production, with one top decaying to Wb and the other top assumed to decay to qH. The 95% CL upper limit obtained by ATLAS by combining the analysis of the different Higgs decay channel, at $\sqrt{s} = 8$ TeV and an integrated luminosity of 20.3 fb^{-1} are $\mathrm{Br}(t \to cH) \leq 4.6 \times 10^{-3}$ and $\mathrm{Br}(t \to uH) \leq 4.5 \times 10^{-3}$ [4, 5]. It was shown in Ref. [6], that studying the single top

+ Higgs production in $pp \to (t \to W^+b)H$ at the LHC, will lead to the disentangling of the tuH and tcH couplings. They analysed in detail the multilepton, diphoton plus lepton and vector boson plus Higgs final state and obtained bounds on the FCNC couplings. They found the lepton rapidity distribution from the multilepton searches improve the bounds obtained on BR(t $\to$ uH) by a factor of 1.5 and that they can be used to discriminate between the tcH and tuH couplings.

2. Analysis of the tqH final state in the polarized e^-e^+ linear collider

The tqH coupling has also been studied in the context of the next generation e^-e^+ linear colliders, the International Linear Collider (ILC) and the Compact Linear Collider (CLIC) [7, 8]. The baseline machine design for both the colliders allows for up to $P_{e^-}^{L,T} = \pm 80\%$ electron (longitudinal and transversal) polarization, while provisions have been made to allow positron polarization of $P_{e^+}^{L,T} = \pm 30\%$ as an upgrade option. Both these machines are designed to operate at centre of mass energies of 350, 500 and 1000 GeV, with the possibility of CLIC to be also adapted for the 3 TeV operation.

We study the $t\bar{t}$ production in the context of the e^-e^+ linear collider as

$$\begin{aligned} &e^-(p_1) + e^+(p_2) \to t(q_1) + \bar{t}(q_2), \\ &\qquad t(q_1) \to q(p_q) + H, \qquad \bar{t}(q_2) \to \bar{b}(p_b) + l^+(p_l) + \nu(p_\nu). \end{aligned} \tag{3}$$

where $q = u, c$.

The total phase space for the process under study is split into the product of the differential cross-section for the $t\bar{t}$ production, the three-particle decay of the anti-top quark and the two-particle decay of the top quark, with the Higgs decaying to $b\bar{b}$. The analytical analysis is done first considering the decay of t to qH along with the inclusive decay of $\bar{t}$. The dominant SM background to this process is also considered, with the inclusive decay of $\bar{t}$ and the SM decay of top to W^+b, with the W decaying hadronically. The total squared matrix element $\mid \bar{\mathcal{M}}^2 \mid$ of the process in (3) is defined as

$$\mid \bar{\mathcal{M}}^2 \mid \;=\; \sum_{L,R} \sum_{(\lambda_t \lambda_t' = \pm)} \rho^{P^{t\bar{t}}}_{LR,\lambda_t\lambda_t'} \rho^{D^t}_{\lambda_t\lambda_t'} = \sum_{L,R} \sum_{(\lambda_t \lambda_t' = \pm)} \mathcal{M}^{L,R}_{\lambda_t} \mathcal{M}^{*L,R}_{\lambda_t'} \rho^{D^t}_{\lambda_t\lambda_t'}, \tag{4}$$

where $\mathcal{M}^{L,R}_{\lambda_t}$ is the production helicity amplitude of the top with a given helicity λ_t. The helicities of the anti-top are summed over. The decay matrix of the top along with the production helicity amplitudes are listed in the Appendix of Ref. [9]. Our calculations are done in the frame where the electron beam direction is in the positive z direction, with the top emitted at a polar angle θ_t and the quark emitted in the top decay makes a polar θ_q angle with the electron beam. The four-vector in the rest frame of the top are related to the c.m. frame by boosting in the z direction and rotating around the y axis.

Combining the production and the density matrices in the narrow width approximation for t, we get the polar distribution of the emitted quark q in the presence of the beam polarization, after integrating over ϕ_q, θ_t, to be

$$\frac{d\sigma}{ds\, d\cos\theta_q\, d\phi_t} \;=\; |T|^2, \tag{5}$$

where the most general formula for the matrix element square $|T_{e^-e^+}|^2$ for arbitrary polarized e^-e^+ beams is given in [10, 11] as

$$\begin{aligned} |T|^2 \;=\; &\frac{1}{4}\Big\{(1 - P_{e^-}^L)(1 + P_{e^+}^L)|T_{e_L^- e_R^+}|^2 + (1 + P_{e^-}^L)(1 - P_{e^+}^L)|T_{e_R^- e_L^+}|^2 \\ &- 2P_{e^-}^T P_{e^+}^T \,\mathrm{Re}\; e^{i(\eta - 2\phi_t)} T^*_{e_R^- e_L^+} T_{e_L^- e_R^+}\Big\}, \end{aligned} \tag{6}$$

where $\eta = \alpha_- + \alpha_+$, with $\alpha_\mp$ denoting the angle of polarization of the electron and the positron, respectively. In Eq. (6) $T_{e^-_{\lambda_1} e^+_{\lambda_2}}$ is the helicity amplitude for the process under consideration and λ_1, λ_2 are the helicities of the electron and the positron, respectively. The degree of the longitudinal and the transversal polarization for the electrons and positrons are denoted by $P^L_{e^\mp}$ and $P^T_{e^\mp}$, respectively, and particular pieces $|T_{e^-_L e^+_R}|^2$, $|T_{e^-_R e^+_L}|^2$ and $T^*_{e^-_R e^+_L} T_{e^-_L e^+_R}$ are given by

$$
\begin{aligned}
|T_{e^\mp_L e^\pm_R}|^2 &= \left(|g_{tq}|^2 + |g_{qt}|^2\right)\left(a_0 + a_1 \cos\theta_q + a_2 \cos^2\theta_q\right) \\
&\quad \pm \left(|g_{tq}|^2 - |g_{qt}|^2\right)\left(b_0 + b_1 \cos\theta_q + b_2 \cos^2\theta_q\right), \\
T^*_{e^-_R e^+_L} T_{e^-_L e^+_R} &= \left(3\cos^2\theta_q - 1\right)\cos(\eta - 2\phi_t)\left\{(|g_{tq}|^2 + |g_{qt}|^2)c_0 + (|g_{tq}|^2 - |g_{qt}|^2)d_0\right\}, \qquad (7)
\end{aligned}
$$

where a_i, b_i, c_0, d_0 coefficients are functions of the standard SM γ and Z couplings with the top and the leptons in the $t\bar{t}$ production. These equations are explicitly given in Ref. [9]. The Yukawa chiral couplings $|g_{tq}|^2$ and $|g_{qt}|^2$ are found to be proportional to the polar angle of the emitted light quark, $\cos\theta_q$ and $\cos^2\theta_q$, but those dependences differ for $|g_{tq}|^2$ and $|g_{qt}|^2$. This makes possible to control the influence of the individual chiral couplings with a suitable choice of beam polarization.

3. Asymmetries and top spin observables at the ILC

We study different distributions in the presence of chiral FCNC couplings and accordingly construct asymmetries to set limits on them. We perform our analysis considering $\sqrt{|g_{tq}|^2 + |g_{qt}|^2} = 0.16$, in accordance with the latest LHC bounds [6]. The background i.e. the SM $\bar{t}Wb$ contribution is scaled down, to be compared with the signal. For analytical results we are not applying any cuts on the final state, but a detailed analysis using all the experimental cuts is performed in the full numerical simulation.

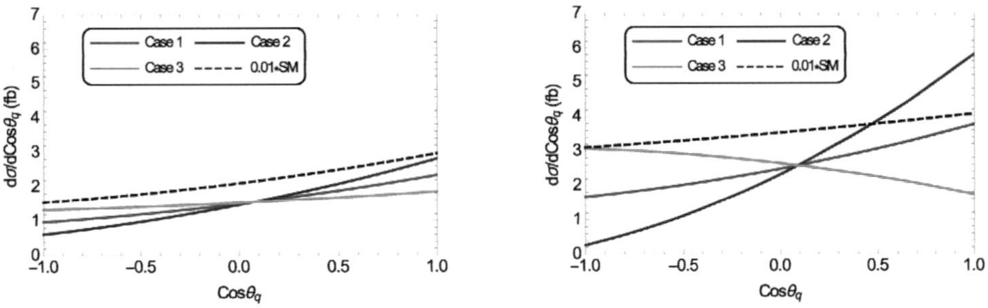

(a) (b)

Figure 1: The polar angle distribution of the quark at $\sqrt{s} = 500$ GeV, for (a) $P^L_{e^-} = P^L_{e^+} = 0$ and (b) $P^L_{e^-} = -0.8$, $P^L_{e^+} = 0.3$. The different Cases are discussed in the text.

The polar angle distribution of the emitted quark, calculated from Eqs.(5-7), is plotted in Fig. 1 for both the signal and the background, for (a) $P^L_{e^-} = P^L_{e^+} = 0$ and (b) $P^L_{e^-} = -0.8$, $P^L_{e^+} = 0.3$. As we can see the polar angle distribution is sensitive to the chirality of the Yukawa couplings and the results are given for three different cases:

- Case 1 : $\sqrt{|g_{tq}|^2 + |g_{qt}|^2} = 0.16$
- Case 2 : $\sqrt{|g_{tq}|^2 + |g_{qt}|^2} = 0.16$, with $|g_{qt}|^2 = 0$
- Case 3 : $\sqrt{|g_{tq}|^2 + |g_{qt}|^2} = 0.16$, with $|g_{tq}|^2 = 0$.

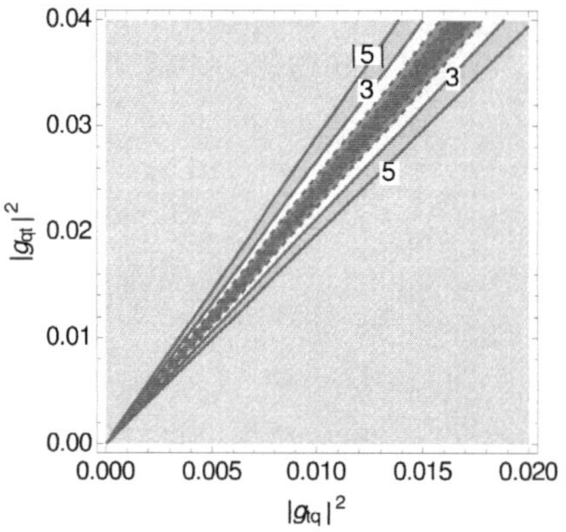

Figure 2: Contour plots of 3σ and 5σ statistical significance in the $|g_{tq}|^2 - |g_{qt}|^2|$ region from A_{fb} for $\theta_0 = 0$. The solid lines are for the unpolarized case, the dashed lines are for $P^L_{e^-}$ = -0.8, $P^L_{e^+}$ = 0.3. See text for explanation.

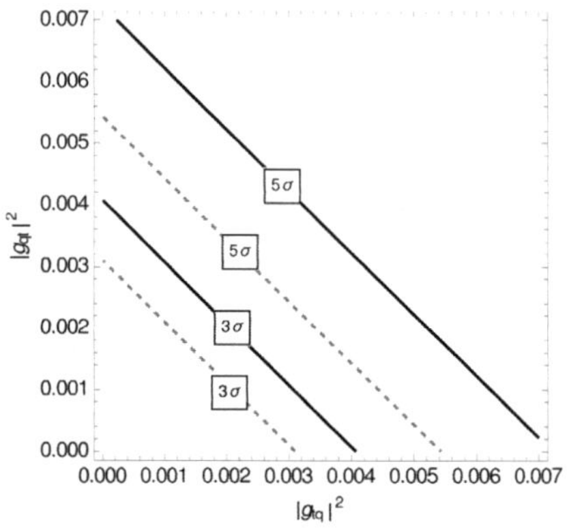

Figure 3: Contour plots in the $|g_{tq}|^2 - |g_{qt}|^2$ plane, for the statistical significance S, from the production cross section, with unpolarized beams [black] and a beam polarization of $P^L_{e^-}$ = -0.8 and $P^L_{e^+}$ = 0.3 [red-dashed].

The different Cases behave similar in the unpolarized case, Fig. 1a. For polarized beams all Cases clearly differ among each other, Fig. 1b. Therefore the manifestation of the dominance of one of the coupling, if present, will be prominent using the suitable initial beam polarization.

In the following, we also perform a full numerical simulation of the FCNC interactions in the $t \to qH$ decay at the ILC. The detailed analysis can be found in Ref. [9]. The various distributions which we consider are calculated in the $t\bar{t}-$ zero momentum frame (ZMF). The decay products, which act as spin analyzers for our case are the non-b jet (q) from the decay $t \to qH$ and the lepton (l^-) from the decay $\bar{t} \to l^-\bar{\nu}\bar{b}$. All the distribution plots are given with the number of surviving events, for $\mathcal{L} = 500$ fb^{-1} and energies $\sqrt{s} = 500$ GeV.

3.1. Constraints on the chiral FCNC couplings by angular asymmetries

Using the above fact that the couplings are sensitive to the polar angle distributions of the quark, we have also considered different asymmetries, the forward-backward asymmetry (FB) and the azimuthal asymmetry, which can be used to give simultaneous limits on both of the couplings [9]. While the FB asymmetry gives a clear distinction among the different cases, as illustrated in Sec. 3, the azimuthal asymmetry exhibits similar behavior for the signal as well as for the background and therefore cannot be an useful observable.

The limits on the FCNC couplings from the FB asymmetry,A_{fb}, are given in Fig. 2 where we show the $|g_{tq}|^2 - |g_{qt}|^2|$ region which can be probed at a statistical significance of 3σ and 5σ, with both unpolarized and polarized beams. The region in blue will be probed at 5σ and green+blue areas will be explored at 3σ with unpolarized beams. The inclusion of the beam polarization probes yellow+green+blue areas at 5σ and pink+yellow+green+blue at 3σ. The region which can not be probed by ILC with this choice of beam polarization is shown in grey. The outside area surrounding solid lines can be probed with unpolarized beams and the outside area surrounding dashed lines can be probed with a beam polarization of $P^L_{e^-} = -0.8$, $P^L_{e^+} = 0.3$. Obviously, the inclusion of beam polarization can probe a greater region in the $|g_{tq}|^2 - |g_{qt}|^2$ parameter space.

We also estimate the sensitivity that can be obtained for the FCNC tqH couplings, given by the efficient signal identification and the significant background suppression which can be achieved at the linear collider. In Fig. 3 we present the contours of 3σ and 5σ significance for our process in the $|g_{tq}|^2 - |g_{qt}|^2$ plane from the production cross section. The sensitivity of the linear collider will increase with the implementation of beam polarization with left polarized electrons and right polarized positrons. One can see that at 3σ of statistical significance and $\mathcal{L}$ = 500 fb^{-1}, $\sqrt{|g_{tq}|^2 + |g_{qt}|^2}$ can be probed to 0.063 (0.056) with unpolarized (polarized) beams. The limits obtained from the asymmetries, specially A_{fb} from Fig.2 will be more stronger and will not be symmetric in the $|g_{tq}|^2 - |g_{qt}|^2$ plane.

3.2. Constraints on the chiral FCNC couplings by top spin observables

The top(antitop)-quark polarization and spin-spin correlations can be measured using the double differential angular distribution of the top and antitop quark decay products:

$$\frac{1}{\sigma}\frac{d^2\sigma}{d\cos\theta_f d\cos\theta_{\bar f}} = \frac{1}{4}\left(1 + B_t\cos\theta_f + B_{\bar t}\cos\theta_{\bar f} - C\cos\theta_f\cos\theta_{\bar f}\right), \qquad (8)$$

where $\theta_f(\theta_{\bar f})$ is the angle between the direction of the top (antitop) spin analyser f, $(\bar f)$ in the t $(\bar t)$ rest frame.

We consider the following spin observables

$$\begin{aligned}
\mathcal{O}_1 &= \frac{4}{3}\mathbf{S}_t\cdot\mathbf{S}_{\bar t},\\
\mathcal{O}_2 &= \mathbf{S}_t\cdot\hat{\mathbf{a}}, \qquad \bar{\mathcal{O}}_2 = \mathbf{S}_{\bar t}\cdot\hat{\mathbf{b}},\\
\mathcal{O}_3 &= 4(\mathbf{S}_t\cdot\hat{\mathbf{a}})(\mathbf{S}_{\bar t}\cdot\hat{\mathbf{b}}),\\
\mathcal{O}_4 &= 4\left((\mathbf{S}_t\cdot\hat{\mathbf{p}})(\mathbf{S}_{\bar t}\cdot\hat{\mathbf{q}}) + (\mathbf{S}_t\cdot\hat{\mathbf{q}})(\mathbf{S}_{\bar t}\cdot\hat{\mathbf{p}})\right),
\end{aligned} \qquad (9)$$

giving the net spin polarization of the top-antitop system ($\mathcal{O}_1$), polarization of the top (antitop) quark ($\mathcal{O}_2(\bar{\mathcal{O}}_2)$), the top-antitop spin correlation ($\mathcal{O}_3$), with respect to spin quantization axes $\hat{\mathbf{a}}$ and $\hat{\mathbf{b}}$. The observable $\mathcal{O}_4$ is an additional top-antitop spin correlation with respect to the momentum of the incoming and the outgoing particles [12]. We have

$$\begin{aligned}
B_t &= \langle\mathcal{O}_2\rangle\,\kappa_f, \qquad B_{\bar t} = \langle\bar{\mathcal{O}}_2\rangle\,\kappa_{\bar f},\\
C &= \langle\mathcal{O}_3\rangle\,\kappa_f\kappa_{\bar f}.
\end{aligned} \qquad (10)$$

Since there is no CP violation in our case, we consider $B \equiv B_t = \mp B_{\bar t}$ for $\hat{\mathbf{a}} = \pm\hat{\mathbf{b}}$. Here $\kappa_q(\kappa_{l^-})$ are the top, anti-top spin analysers. The spin analyser for the FCNC top-Higgs decays can be either a direct t-quark daughter, i.e. H or c/u-quark, or H decay products like b or $\bar b$ in $b\bar b$ decay, or $\tau^+(\tau^-)$ in $H\to\tau^+\tau^-$ decay, or jets. On the other hand, the spin analyser for $\bar t$ are W^- or $\bar b$, or a W^- decay products $l^-, \bar\nu$ or jets. We consider the c/u quark from the top and the l^- from the anti-top as spin analysers in this work.

The top spin analyzing power of q (κ_q) from the $t\to Hq$ decay, b quark (κ_b), from the top decay to W^+b and the lepton from the top decay to $l^+\nu b$ is

$$\kappa_q = \frac{|g_{qt}|^2 - |g_{tq}|^2}{|g_{qt}|^2 + |g_{tq}|^2}, \qquad \kappa_b = \frac{m_t^2 - 2m_W^2}{m_t^2 + 2m_W^2}, \qquad \kappa_l = 1. \qquad (11)$$

The leptons emitted from the top decay, due to the $V-A$ interactions are the perfect top spin analysers, $\kappa_l = 1$, (α_s corrections are negligible), with their flight directions 100% correlated with the directions of the top spin. It is clear from Eq. (11) that with $|g_{qt}|^2 \simeq |g_{tq}|^2$, the spin information of the top will be lost ($\kappa_q \approx 0$). However, in the presence or dominance of only one of the coupling, the emitted quark acts as a perfect spin analyser ($\kappa_q \approx 1$).

The observable $\mathcal{O}_1$ can be probed using the opening angle distribution (φ), i.e. the angle between the direction of flight of the two (top and antitop) spin analysers, defined in the t and $\bar{t}$ frames, respectively, i.e $\hat{\mathbf{p}}_q \cdot \hat{\mathbf{p}}_l = \cos\varphi$.

$$\frac{1}{\sigma}\frac{d\sigma}{d\cos\varphi} = \frac{1}{2}\left(1 - D\cos\varphi\right), \qquad D = \langle \mathcal{O}_1 \rangle \, \kappa_q \kappa_{l^-} \tag{12}$$

The arbitrary unit vectors $\hat{\mathbf{a}}$ and $\hat{\mathbf{b}}$ in (9) specify different spin quantization axes which can be chosen to maximize the desired polarization and the correlation effects, and we take them as

$$\begin{aligned} \hat{\mathbf{a}} &= -\hat{\mathbf{b}} = \hat{\mathbf{q}}\,, && (\text{“helicity” basis})\,, \\ \hat{\mathbf{a}} &= \hat{\mathbf{b}} = \hat{\mathbf{p}}\,, && (\text{“beamline” basis})\,, \\ \hat{\mathbf{a}} &= \hat{\mathbf{b}} = \hat{\mathbf{d}}_{\mathbf{SM}} = \hat{\mathbf{d}}_{\rm SM}^{\rm max} = \frac{-\hat{\mathbf{p}} + (1-\gamma)z\,\hat{\mathbf{q}}_1}{\sqrt{1-(1-\gamma^2)z^2}}\,, && (\text{“off}-\text{diagonal” basis})\,, \end{aligned} \tag{13}$$

where $\hat{\mathbf{q}} = \hat{\mathbf{q}}_1$ is the direction of the outgoing top quark and $\hat{\mathbf{p}}$ is the direction of the incoming beam, both in the $t\bar{t}$ center of mass frame. The off-diagonal basis [13] is the one, where the top spins are 100% correlated ($z = \hat{\mathbf{p}} \cdot \hat{\mathbf{q}}_1 = \cos\theta$ and $\gamma = E_t/m_t = 1/\sqrt{1-\beta^2}$) and which interpolates between the beamline basis at the threshold ($\gamma \to 1$) and the helicity basis for ultrarelativistic energies ($\gamma \to \infty$). We would like to point out here that this off-diagonal basis $\hat{\mathbf{d}}_{\rm SM}$ is specific to the SM $t\bar{t}$ production, but a general procedure for finding such an off-diagonal basis for any model is given in [14, 15, 16]. The complementary basis to the "off-diagonal" one, where the correlation of the top-antitop spins is minimal can be found in Refs. [9, 16].

In Table 1 we present the values of the different spin observables in the different spin basis considered here, in the presence of beam polarizations. We note that the top (antitop) spin

Table 1: The value of the spin observables in different bases, with different choices of initial beam polarization.

Observables	Basis	$P_{e^-}^L = 0, P_{e^+}^L = 0$	$P_{e^-}^L = 0.8, P_{e^+}^L = -0.3$	$P_{e^-}^L = -0.8, P_{e^+}^L = 0.3$
$\mathcal{O}_1$		$0.333\kappa_f$	$0.333\kappa_f$	$0.333\kappa_f$
	hel	$-0.076\kappa_f$	$0.247\kappa_f$	$-0.239\kappa_f$
	beam	$-0.174\kappa_f$	$0.344\kappa_f$	$-0.436\kappa_f$
$\mathcal{O}_2$	off	$0.176\kappa_f$	$-0.351\kappa_f$	$0.443\kappa_f$
	hel	$-0.654\kappa_f$	$-0.666\kappa_f$	$-0.648\kappa_f$
	beam	$0.881\kappa_f$	$0.852\kappa_f$	$0.897\kappa_f$
$\mathcal{O}_3$	off	$0.911\kappa_f$	$0.886\kappa_f$	$0.924\kappa_f$
$\mathcal{O}_4$		$0.546\kappa_f$	$0.612\kappa_f$	$0.512\kappa_f$

polarizations are quite sensitive to the beam polarization, while this is not the case for the spin-spin correlations $\mathcal{O}_3, \mathcal{O}_4$ where the influence of the beam polarizations gets diluted. Also note that all observables are proportional to $\kappa_f = \kappa_q$ and will be equal to zero if g_{tq} and g_{qt} are equal.

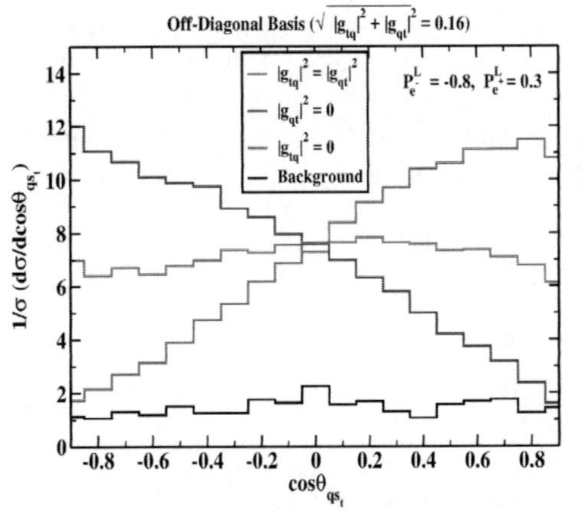

Figure 4: The normalized distribution with polarized beams, θ_{qs_t} is the angle between the direction of the top spin analyzer (q from $t \to qH$) in t rest frame and the spin quantization axis of the top (s_t) in $t\bar{t}$ZMF.

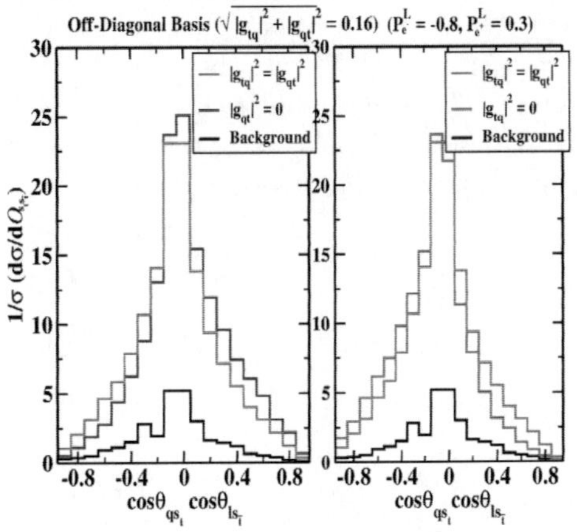

Figure 5: The normalized distribution of the product $\cos\theta_{qs_t}\cos\theta_{ls_{\bar{t}}}$, ($\theta_{qs_t} = \angle(\hat{\mathbf{p}}_q, \hat{\mathbf{a}}), \theta_{qs_{\bar{t}}} = \angle(\hat{\mathbf{p}}_l, \hat{\mathbf{b}})$), using the off-diagonal basis, at $\sqrt{s} = 500$ GeV, with polarized beams and $\mathcal{L} = 500\ \text{fb}^{-1}$.

The top spin ($\sim \langle\mathcal{O}_2\rangle$) is considered in the normalized distribution $1/\sigma(d\sigma/d\cos\theta_{qs_t})$, where θ_{qs_t} is the angle between the direction of the top spin analyzer (non-b jet) in the top rest frame and the top spin quantization axis (s_t) in the $t\bar{t}$-ZMF. The angle $\cos\theta_{qs_t}$ is the angle $\cos\theta_f$ defined in Eq. (8). The spin of the top can be chosen in the direction of any of the spin quantization axes as defined in Eq.(13). The 'beamline' basis and the 'off-diagonal' basis, Fig. 4, are most sensitive to the top polarization and therefore also to the decay dynamics of the top. The chiral nature of the FCNC coupling will be more clearly visible in these two basis, with a flat distribution in case of the equality of the two chiral coupling. The effect is further enhanced with the beam polarizations of $P^L_{e^-} = -0.8$ and $P^L_{e^+} = 0.3$, in all the spin bases considered here.

The double differential angular distribution of the top and the antitop defined in Eq. (8) provides a measurement of the spin-spin correlations ($\sim \langle\mathcal{O}_3\rangle$). It was shown in Ref. [17] that, for the experimental analysis, it is more suitable to use the one-dimensional distribution of the product of the cosines, $\mathcal{O}_{s_t,s_{\bar{t}}} = \cos\theta_f\cos\theta_{\bar{f}}$, rather than analyzing the distribution in (8). We define $\cos\theta_f\cos\theta_{\bar{f}}$ as $\cos\theta_{qs_t}\cos\theta_{ls_{\bar{t}}}$ for our analysis. The $1/\sigma(d\sigma/d\mathcal{O}_{s_t,s_{\bar{t}}})$ distribution is shown in Fig. 5, using the 'off-diagonal' basis and a longitudinal beam polarization of $P^L_{e^-} = -0.8$ and $P^L_{e^+} = 0.3$. The asymmetry of the plot around $\cos\theta_{qs_t}\cos\theta_{ls_{\bar{t}}} = 0$, although not large, signals for the spin-spin correlation. The plot for Case 2 ($|g_{qt}|^2 = 0$) shows more events for positive values for $\cos\theta_{qs_t}\cos\theta_{ls_{\bar{t}}}$, whereas for Case 3 ($|g_{tq}|^2 = 0$) one gets more events for negative values of $\cos\theta_{qs_t}\cos\theta_{ls_{\bar{t}}}$.

4. Conclusion

We have been able to show that at linear colliders one can explore the chiral structure of the flavor-violating top-Higgs interactions in the $t\bar{t}$ production, using different beam polarizations and various top spin observables. The chiral nature of these couplings is not evident from the measurements made at the LHC as both the branching ratio of the top to qH and the total production cross section are proportional to $|g_{tq}|^2 + |g_{qt}|^2$.

The possibility of the adjustment of the initial beam polarization, leads to an enhancement of the sensitivity of the measured branching ratios and the asymmetries on the FCNC parameters. The various angular distributions of the top decay products are related to the top spin observables, whose effect can be increased with a suitable choice of spin quantization axis. The quark emitted from the top FCNC decay, acts as a perfect spin analyzer ($\kappa_q = 1$)in the presence of a single chiral coupling, and the correlation is completely lost when $|g_{tq}|^2 = |g_{qt}|^2$. The off-diagonal basis along with the beamline basis is found to be most sensitive to the chirality of the coupling with its effect being enhanced with a suitable choice of beam polarization.

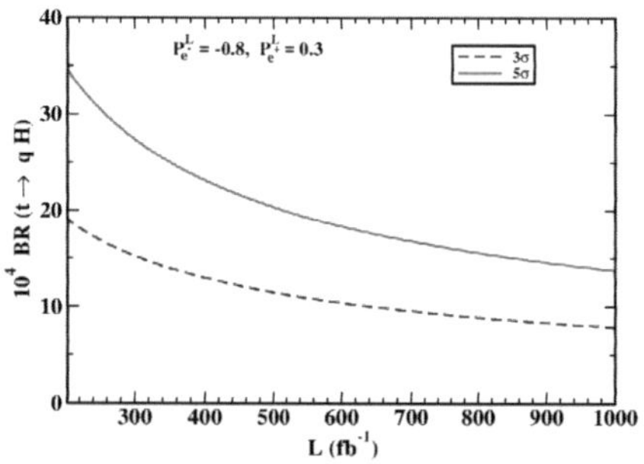

Figure 6: The sensitivity of 3σ and 5σ to BR($t \to qH$) at $\sqrt{s}$ = 500 GeV, as a function of the luminosity, $\mathcal{L}$.

We have performed a detailed analysis, applying all the relevant cuts at the linear collider and obtain a limit on the couplings from the total cross section. The 3σ and the 5σ upper bounds on the branching ratios, which can be obtained at the ILC, at $\sqrt{s} = 500$ GeV, with the choice of beam polarization $P^L_{e^-} = -0.8$, $P^L_{e^+} = 0.3$ are shown in Fig. 6.

We find that BR($t \to qH$) can be probed to $5.59 \times 10^{-3}(8.84 \times 10^{-4})$ at 3σ level at the ILC, with $\sqrt{s} = 500$ GeV, $\mathcal{L} = 500$ fb^{-1} and a beam polarization of $P^L_{e^-} = 0(-0.8), P^L_{e^+} = 0(0.3)$.

Acknowledgement
B.M. would like to thank the organizers for the invitation and organizing the pleasant conference. This work is supported by the Croatian Science Foundation (HRZZ) project PhySMaB, 'Physics of Standard Model and Beyond' as well as by the H2020 Twinning project No. 692194, RBI-T-WINNING.

References

[1] Mele B, Petrarca S and Soddu A, *Phys. Lett.* B **435**, 401 (1998) (*Preprint* hep-ph/9805498)
[2] Aguilar-Saavedra J A, *Acta Phys. Polon.* B **35**, 2695 (2004) (*Preprint* hep-ph/0409342)
[3] Blankenburg G, Ellis J and Isidori G, *Phys. Lett.* B **712**, 386 (2012) (*Preprint* arXiv:1202.5704 [hep-ph])
[4] CMS Collaboration [CMS Collaboration], (*CMS-PAS-HIG*-13-034)
[5] Aad G *et al.* [ATLAS Collaboration], JHEP **1512**, 061 (2015) (*Preprint* arXiv:1509.06047 [hep-ex])
[6] Greljo A, Kamenik J F and Kopp J, JHEP **1407**, 046 (2014) (*Preprint* arXiv:1404.1278 [hep-ph])
[7] Han T, Jiang J and Sher M, *Phys. Lett.* B **516**, 337 (2001) (*Preprint* hep-ph/0106277)
[8] Hesari H, Khanpour H and Najafabadi M M, *Phys. Rev.* D **92**, no. 11, 113012 (2015) (*Preprint* arXiv:1508.07579 [hep-ph])
[9] Melić B and Patra M, JHEP **1701**, 048 (2017) (*Preprint* arXiv:1610.02983 [hep-ph])
[10] Kleiss R, *Z. Phys.* C **33**, 433 (1987).
[11] Hikasa K , *Phys. Rev.* D **33**, 3203 (1986).
[12] Brandenburg A, Flesch M and Uwer P, *Phys. Rev.* D **59**, 014001 (1999) (*Preprint* hep-ph/9806306)
[13] Parke S J and Shadmi Y, *Phys. Lett.* B **387**, 199 (1996) (*Preprint* hep-ph/9606419)
[14] Mahlon G and Parke S J, *Phys. Lett.* B **411**, 173 (1997) (*Preprint* hep-ph/9706304)
[15] Uwer P, *Phys. Lett.* B **609**, 271 (2005) (*Preprint* hep-ph/0412097)
[16] Fajfer S, Kamenik J F and Melić B, JHEP **1208** (2012) 114 (*Preprint* arXiv:1205.0264 [hep-ph])
[17] Bernreuther W and Si Z G, *Phys. Lett.* B **725**, 115 (2013) Erratum: [*Phys. Lett.* B **744**, 413 (2015)] (*Preprint* arXiv:1305.2066 [hep-ph])

IOP Conf. Series: Journal of Physics: Conf. Series **878** (2017) 012027 doi:10.1088/1742-6596/878/1/012027

Using Integral and Differential Charge Asymmetries for BSM Searches at the LHC

Steve Muanza

CPPM, CNRS-IN2P3 & AMU, Marseille FRANCE

E-mail: muanza@in2p3.fr

Abstract. Contrary to past high energy colliders, the LHC is a charge asymmetric machine. Therefore most of the hard scattering processes producing electrically charged final states have a positive integral charge asymmetry. The latter quantity, denoted $A_C(\ell^{\pm})$, is easily measurable in event topologies bearing an odd number of hard and isolated charged leptons. We have brought to light the strong correlation between $A_C(\ell^{\pm})$ and the mass of the charged final state. This enabled us to setup a new method of indirect mass measurement [1]. For example, this method enables us to measure the mass of the $W^{\pm}$ boson with a 2% accuracy. Obviously this is not competitive with respect to the standard technique based on the $W^{\pm}$ transverse mass. However, for other processes, where more final state particles escape detection, we've demonstrated the integral charge asymmetry method to be much more effective. We illustrate this in a search for a supersymmetric production of chargino-neutralino pairs decaying in the trilepton inclusive topology, and show that we can measure $M_{\tilde{\chi}_1^{\pm}} + M_{\tilde{\chi}_2^0}$ with an accuracy better than 10%. Nevertheless, in order to apply the integral charge asymmetry method one needs to have a significant excess of signal events over the event yield of the corresponding background processes. We are currently extending this indirect mass measurement method using differential charge asymmetries. In addition to their sensitivity to the mass, the shape of these observables can also be exploited to improve the separation between a signal and its background processes. Our main physics case under study is the production of an heavy $W^{\pm\prime}$ boson which decays into a single charged lepton plus missing transverse energy. For both the use of integral and differential charge asymmetries we are also developing quantitative estimates of their sensitivity through appropriate confidence levels.

1. Introduction

Contrary to most of the previous high energy particle colliders, the LHC is a charge asymmetric machine. For charged final states, denoted $FS^{\pm}$, the integral charge asymmetry, denoted A_C, is defined by

$$A_C = \frac{N(FS^+) - N(FS^-)}{N(FS^+) + N(FS^-)} \tag{1}$$

where $N(FS^+)$ and $N(FS^-)$ represent respectively the number of events bearing a positive and a negative charge in the FS. Practically, we defined the charged final states as event topologies containing an odd number of high p_T charged and isolated leptons within the fiducial volume of the detector. For a $FS^{\pm}$ produced at the LHC in $p+p$ collisions, this quantity is positive or null, whilst it is always compatible with zero for a $FS^{\pm}$ produced at the TEVATRON in $p+\bar{p}$ collisions.

Let's consider the Drell-Yan production of $W^{\pm}$ bosons in $p+p$ collisions in order to illustrate the $A_C(\ell^{\pm})$ observable. We added the $(\ell^{\pm})$ to A_C to specify that we use the lepton charged asymmetry, not directly that of the $W^{\pm}$ boson. It is obvious for this simple $2 \to 2$ s-channel process that more W^+ than W^- are produced. Indeed, denoting y_W the rapidity of the W boson, the corresponding range of the Björken x's: $x_{1,2} = M_{W^{\pm}}/\sqrt{s} \times e^{\pm y_W}$, probe the charge asymmetric valence parton densities within the proton. This results in having more $U+\bar{D} \to W^+$ than $\bar{U}+D \to W^-$ configurations in the initial state (IS). Here U and D collectively and respectively represent the up and the down quarks. For the $W^{\pm} \to \ell^{\pm}\nu$ inclusive process, the dominant contribution to $A_C(\ell^{\pm})$ comes from the difference in rate between the $u+\bar{d}$ and the $d+\bar{u}$ quark currents in the IS. Using the usual notation $f(x, Q^2)$ for the parton density functions (PDF) and within the leading order (LO) approximation, this can be expressed as:

$$A_C(\ell^{\pm}) \approx \frac{u(x_{1,2}, M_W^2)\bar{d}(x_{2,1}, M_W^2) - \bar{u}(x_{1,2}, M_W^2)d(x_{2,1}, M_W^2)}{u(x_{1,2}, M_W^2)\bar{d}(x_{2,1}, M_W^2) + \bar{u}(x_{1,2}, M_W^2)d(x_{2,1}, M_W^2)} \quad (2)$$

where the squared four-momentum transfer Q^2 is set to M_W^2.

2. Indirect Mass Measurements

2.1. Relation Between Mass and Integral Charge Asymmetry

From equation 2, we can see that the Q^2 evolution of the parton density functions (PDFs) govern the Q^2 evolution of $A_C(\ell^{\pm})$. The former are known, up-to the NNLO in QCD, as solutions of the DGLAP equations [2]. One could therefore think of using an analytical functional form to relate $A_C(\ell^{\pm})$ to the squared mass of the s-channel propagator, here M_W^2. However, there are additional contributions to the $W^{\pm}$ inclusive production. At the Born level, some come from other flavour combinations in the IS of the s-channel, and some come from the u and the t-channels. On top of this, there are higher order corrections. These extra contributions render the analytical expression of the Q^2 dependence of $A_C(\ell^{\pm})$ much more complicated. Therefore, we choose to build process-dependent numerical mass template curves for $A_C(\ell^{\pm})$ by varying $M_{FS^{\pm}}$. These mass templates constitute inclusive and flexible tools into which all the above-mentioned contributions to $A_C(\ell^{\pm})$ can be incorporated, where they can very easily be built within a restricted domain of the signal phase space imposed by kinematic cuts.

In section 2 of this proceedings we exploit the $A_C(\ell^{\pm})$ to set a new type of constraint on the mass of the charged $FS^{\pm}$. In section 3, we present preliminary results on the use of $A_C(\ell^{\pm})$ as a new discriminating observable between a signal and its background processes.

We separate the rest of section 2 into two parts. The first one, in sub-section 2.2, is dedicated to presenting in full length the method of indirect mass measurement that we propose on a known Standard Model (SM) process. We choose the $W^{\pm} \to \ell^{\pm} + \not{E}_T$ inclusive production at the LHC to serve as a test bench. In the second part of section 2, in sub-section 2.3, we repeat the method on a "Beyond the Standard Model" (BSM) process. We choose a SUSY search process of high interest, namely $\tilde{\chi}_1^{\pm} + \tilde{\chi}_2^0 \to 3\ell^{\pm} + \not{E}_T$. For both the SM and the BSM processes, we obviously tag the sign of the FS by choosing a decay into one (or three) charged lepton(s), for which the sign is experimentally easily accessible.

2.2. A Standard Model Test Bench Process

2.2.1. Theoretical Prediction of $A_C(W^\pm \to \ell^\pm \nu)$

The theoretical $A_C(\ell^\pm)$ raw (left) and fitted (right) template curves for the MRST2007lomod PDF are displayed in Fig. 1. For the raw template curve the breakdown of the theory uncertainties are displayed as colored bands. We note the evolution of $A_C(\ell^\pm)$ as a function of the artificially varied $W^\pm$ mass is a monotonic increasing function that can be described by a polynomial of logarithms of the logarithm of $M_{W^\pm}$: $A_C[M_{W^\pm}] = \sum_{i=0}^{N} A_i \times [Log[Log[M_{W^\pm}]]]^i$. This functional form is inspired by a parametrization of the analytical solution of the DGLAP non-singlet equation.

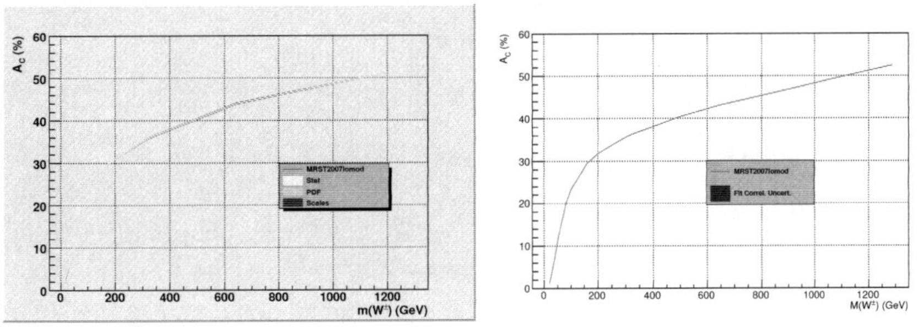

Figure 1. Theoretical $A_C(\ell^\pm)$ raw (left) and fitted (right) template curves for the MRST2007lomod PDF.

2.2.2. Experimental Measurement of $A_C(W^\pm \to \ell^\pm \nu)$

In this subsection we extract $A_C(W^\pm \to \ell^\pm \nu)$ in the presence of some background processes. To this end we generate Monte Carlo samples for the signal and the background processes in p+p colissions at $\sqrt{s} = 7$ TeV with an integrated luminosity of $L = 1\ fb^{-1}$. We use the MRST2007lomod [3] as the default PDF here and in all reported studies hereafter. We use a fast simulation of the response of the ATLAS detector. And we apply some selection cuts to optimize the signal to noise ratio. We keep the events with a hard and isolated electron or muon that has a $p_T > 25$ GeV. These events must have a missing transverse energy larger than 25 GeV and a transverse mass larger than 40 GeV. The corresponding efficiencies, events yields and expected charge asymmetries, reported in Tab. 1.

After applying the above event selection, the A_C of all surviving events has two experimenal biases. One of these biases comes from the selection cuts on the lepton kinematics, the other one comes from the background contamination. The former bias is encoded in the experimental template and not corrected for. In order to correct for the latter experimenal bias, we apply a background subtraction which relies upon the following equation:

$$A_C^{Exp}(S) = (1 + \alpha^{Exp}) \cdot A_C^{Exp}(S + B) - \alpha^{Exp} \cdot A_C^{Exp}(B) \tag{3}$$

where $\alpha^{Exp} = N_B^{Exp}/N_S^{Exp}$ is the noise (B) to signal (S) ratio. We also use this equation to propagate the experimental systematic uncertainties into $A_C^{Exp}(S)$ accounting for the correlations between α^{Exp}, $A_C^{Exp}(B)$, and $A_C^{Exp}(S+B)$. In this notation "Exp" stands for expected from the Monte Carlo (MC) simulation.

This procedure enables us to obtain $A_C^{Meas}(S)$ starting from $A_C^{Exp}(S)$ where "Meas" stands for measured and represents $A_C(S)$ after the background subtraction and the propagation of the

Process	ϵ (%)	N_{exp} (k evts)	$A_C \pm \delta A_C^{Stat}$ (%)
Signal: $W^{\pm} \to e^{\pm}\nu_e$			
$M(W^{\pm}) = 40.2$ GeV	0.81 ± 0.01	290.367	9.66 ± 1.57
$M(W^{\pm}) = 60.3$ GeV	13.69 ± 0.05	2561.508	11.22 ± 0.38
$M(W^{\pm}) = 80.4$ GeV	29.59 ± 0.04	3343.195	16.70 ± 0.18
$M(W^{\pm}) = 100.5$ GeV	39.19 ± 0.07	2926.093	20.77 ± 0.22
$M(W^{\pm}) = 120.6$ GeV	44.84 ± 0.07	2357.557	23.19 ± 0.21
$M(W^{\pm}) = 140.7$ GeV	48.66 ± 0.07	1899.820	25.29 ± 0.20
$M(W^{\pm}) = 160.8$ GeV	51.28 ± 0.07	1527.360	26.87 ± 0.19
$M(W^{\pm}) = 201.0$ GeV	54.54 ± 0.07	1.032	29.06 ± 0.18
Background	-	91.614 ± 1.706	10.07 ± 0.15
$W^{\pm} \to \mu^{\pm}\nu_{\mu}/\tau^{\pm}\nu_{\tau}/q\bar{q}'$	0.211 ± 0.003	71.350	12.92 ± 1.25
tt	5.76 ± 0.02	6.600	1.00 ± 0.37
$t + b,\ t + q(+b)$	3.59 ± 0.01	1.926	28.97 ± 0.35
$W + W,\ W + \gamma^*/Z,\ \gamma^*/Z + \gamma^*/Z$	2.94 ± 0.01	2.331	10.65 ± 0.35
$\gamma + \gamma,\ \gamma + jets,\ \gamma + W^{\pm},\ \gamma + Z$	0.201 ± 0.001	0.759	17.25 ± 0.53
γ^*/Z	0.535 ± 0.001	5.746	4.43 ± 0.23
QCD HF	$(0.44 \pm 0.17) \times 10^{-4}$	1.347	14.29 ± 37.41
QCD LF	$(0.87 \pm 0.33) \times 10^{-4}$	1.555	71.43 ± 26.45

Table 1. Selection efficiencies, event yields and expected integral charge asymmetries for the $W^{\pm} \to e^{\pm}\nu_e$ analysis.

experimental systematic uncertainties. The evolution of $A_C^{Meas}(S)$ with respect to $M_{W^{\pm}}$ enables us to construct experimental $A_C(\ell^{\pm})$ template curves. This is illustrated for the electron channel in Fig. 2.

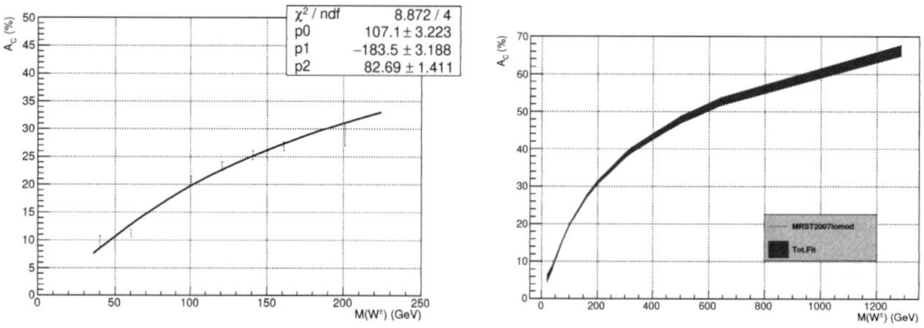

Figure 2. The A_C^{Meas} raw (left) and fitted (right) template curves for the electron channel.

2.2.3. Indirect Determination of $M_{W^{\pm}}$

The measured $A_C(\ell^{\pm})$ from the electron and the muon channels are converted into $M_{W^{\pm}}$ estimtates using the corresponding experimental templates.

- $A_C^{Meas.Fit}(S) = (16.70 \pm 0.35)\% \Rightarrow M^{Meas.Fit}(W^{\pm} \to e^{\pm}\nu_e) = 81.08^{+2.06}_{-2.01}$ GeV.
- $A_C^{Meas.Fit}(S) = (17.52 \pm 0.18)\% \Rightarrow M^{Meas.Fit}(W^{\pm} \to \mu^{\pm}\nu_{\mu}) = 79.67^{+3.56}_{-1.39}$ GeV.

These two indirect mass measurements are then combined through a weighted mean: $M^{Comb.Meas.}(W^{\pm}) = 80.30 \pm 0.96$ (Exp.Comb.) GeV. The theoretical template curve is read-up at the central value of the estimated $M_{W^{\pm}}$ and the corresponding uncertainties $^{+0.19}_{-0.21}$ GeV are summed in quadrature with the experimental ones so as to give the total uncertainties:

$$M_{W^{\pm}} = 80.30^{+0.98}_{-0.98} \text{ (MRST2007lomod) GeV.} \tag{4}$$

2.3. A SUSY Physics Case

Here we apply the previous method to a SUSY "golden channel": $p + p \to \tilde{\chi}_1^{\pm} + \tilde{\chi}_2^0 \to 3\ell^{\pm} + \not{E}_T$.

2.3.1. Theoretical Prediction of $A_C(\tilde{\chi}_1^{\pm} + \tilde{\chi}_2^0)$

The theoretical $A_C[M_{\tilde{\chi}_1^{\pm}} + M_{\tilde{\chi}_2^0}]$ raw (left) and fitted (right) template curves for the MRST2007lomod PDF are displayed in Fig. 3. The fit functional form is analogous to that of section 2: $A_C[M_{\tilde{\chi}_1^{\pm}} + M_{\tilde{\chi}_2^0}] = \sum_{i=0}^{N} A_i \times [Log[Log[M_{\tilde{\chi}_1^{\pm}} + M_{\tilde{\chi}_2^0}]]]^i$.

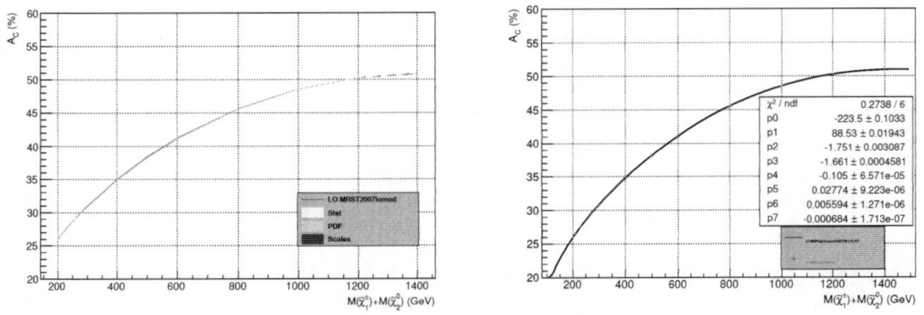

Figure 3. Theoretical A_C raw (left) and fitted (right) template curves for the MRST2007lomod PDF.

2.3.2. Experimental Measurement of $A_C(\tilde{\chi}_1^{\pm} + \tilde{\chi}_2^0 \rightarrow 3\ell^{\pm} + \not{E}_T)$

In this subsection we generate Monte Carlo samples for the signal and the background processes in p+p colissions at $\sqrt{s} = 8$ TeV with an integrated luminosity of $L = 20\ fb^{-1}$. Two different types of SUSY spectra are considered. First, there's one denoted S1 signal, where sleptons have masses intermediate between $M_{\tilde{\chi}_2^0}$ (set equal to $M_{\tilde{\chi}_1^{\pm}}$) and $M_{\tilde{\chi}_1^0}$. Second, there's another denoted S2 signal, where sleptons have masses much larger than $M_{\tilde{\chi}_2^0} = M_{\tilde{\chi}_1^{\pm}}$. Again, we use a fast simulation of the response of the ATLAS detector. And we apply some selection cuts to optimize the signal to noise ratio. We keep the events with three hard and isolated leptons (electrons and/or muons) that have a $p_T > 20, 10, 10$ GeV. These events must have a missing transverse energy larger than 35 GeV and a stransverse mass, M_{T2} [4][5], larger than 75 GeV. The A_C background subtraction is applied, the experimental uncertainties are propagated into A_C^{Meas} from which experimental template curves are derived. A_C^{Meas} are converted into estimated sums of masses: $M_{\tilde{\chi}_1^{\pm}} + M_{\tilde{\chi}_2^0}$ with the same treatment of the experimental and theoretical uncertainties as in section 2.2.1.

2.3.3. Indirect Determination of $M_{\tilde{\chi}_1^{\pm}} + M_{\tilde{\chi}_2^0}$

As in sub-section 2.2.3, the A_C^{Meas} can be translated into estimates of $M_{\tilde{\chi}_1^{\pm}} + M_{\tilde{\chi}_2^0}$. Closure tests displayed in Fig. 4 illustrate the quality (linearity, absence of offset and uncertainty) of these indirect mass measurements, especially in the mass ranges with high signal sensitivities (above 5σ for S1 and above 3σ for S2) which are highlighted.

2.3.4. Results

The accuracy of the indirect mass measurements for the W mass is better than 2%, and it is better than 6% for the S1 and S2 chargino-neutralino signal samples within their mass range of high sensitivity.

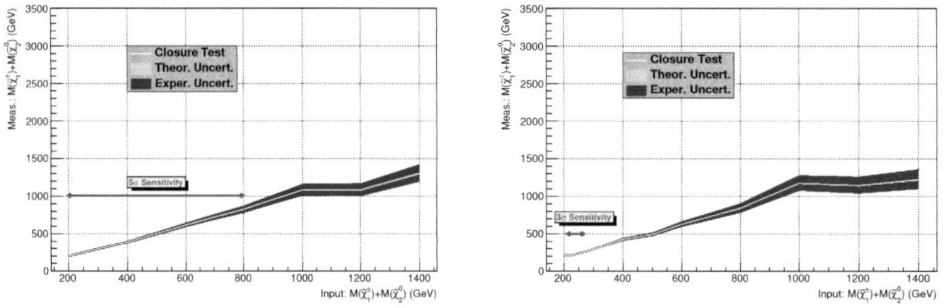

Figure 4. Closure tests for the indirect measurements of $M_{\tilde{\chi}_1^\pm} + M_{\tilde{\chi}_2^0}$ for S1 (left) and S2 (right) signal samples.

3. Improved Discrimination Between Signal and Background

In the context of BSM searches, the method of indirect mass measurement presented in section 2 can only be utilized after a significant excess of a signal produced through a charged current is discovered. This raises the following question: could the charge asymmetry observable be used in the BSM search itself? Our answer is yes, and to illustrate this we choose an exotic physics case which is the search for an additional charged electroweak boson: $p+p \to W^{\pm\prime} \to 1\ell^\pm + \not{E}_T$.

3.1. Search in the Muon Channel

More precisely, we recast an ATLAS Run 2 analysis [6] just in the muon channel so far.

3.1.1. Monte-Carlo Samples and Detector Simulation

Monte-Carlo (MC) samples are generated for the signal using Pythia 8.157 [7] and for the background processes using Herwig++ 2.5.2 [8] and Alpgen 2.14 [9] (interfaced to Pythia v8.157 for the parton shower, the hadronization and the decays). For the signal the following mass hyoptheses are considered: $M_{W^{\pm\prime}} = 1, 2, 3, 4$ TeV. All these MC samples are produced for $p+p$ collisions at $\sqrt{s} = 8$ TeV and normalized to an integrated luminosity of L=20 fb^{-1}.

3.1.2. Event Selection

The response of the ATLAS detector is crudely simulated by Delphes 3.3.2 [10]. In order to extract the signal from the background processes, we select events with one isolated muon, with $p_T(\mu^\pm) > 45$ GeV and $|\eta(\mu^\pm)| < 2.4$. Most of the background processes are suppressed by requiring $\not{E}_T > 45$ GeV and especially $M_T = \sqrt{2p_T(\mu^\pm)\not{E}_T(1 - cos\Delta\phi)} > 800$ GeV. After these cuts are applied, the resulting selection efficiencies, event yields and expected integral charge asymmetries are compiled within Tab. 2.

In the ATLAS analysis, the final discriminant is the high M_T tail (beyond 800 GeV). Instead, in our selection, the final discriminant is $A_C(\ell^\pm)$.

3.1.3. Systematic Uncertainties

The theoretical uncertanties are due to the unknown QCD scales for fixed order (here leading order) $A_C(\ell^\pm)$ calculations and to the PDF$\oplus\alpha_S$ uncertainties. The latter are calculated using the PDF4LHC15_nlo_mc_pdfas set [11][12][13][14] and following the PDF4LHC recommendations for the LHC Run 2 [15]. This is implemented using the LHAPDF 6.1.5 [16] interface.

For the experimental uncertainties, we use those quoted in Ref. [6]:

- $\not{E}_T$ scale and resolution: 0.1% (S), 0.5% (B).

IOP Conf. Series: Journal of Physics: Conf. Series **878** (2017) 012027 doi:10.1088/1742-6596/878/1/012027

$Process$	ϵ (%)	N_{exp} (evts)	$A_C \pm \delta A_C^{Stat}$ (%)
Signal: $W^{\pm\prime} \to \mu^{\pm}\nu_{\mu}$			
$M(W^{\pm\prime}) = 1$ TeV	36.36 ± 0.07	8561.59	48.56 ± 0.94
$M(W^{\pm\prime}) = 2$ TeV	64.04 ± 0.07	317.23	60.61 ± 4.47
$M(W^{\pm\prime}) = 3$ TeV	42.87 ± 0.07	12.53	60.48 ± 22.50
$M(W^{\pm\prime}) = 4$ TeV	21.15 ± 0.06	1.33	57.28 ± 71.04
Background	-	5.91	1.30 ± 41.14
$W^{\pm} \to \mu^{\pm}\nu_{\mu}/\tau^{\pm}\nu_{\tau}/q\bar{q}\prime + LF$	0.00 ± 0.00	0.00	–
$W^{\pm} \to \mu^{\pm}\nu_{\mu}/\tau^{\pm}\nu_{\tau}/q\bar{q}\prime + HF$	$5.28 \times 10^{-4} \pm 1.21 \times 10^{-5}$	1.78	82.51 ± 42.32
$t\bar{t}$	0.00 ± 0.00	0.00	–
$t + b,\ t + q(+b)$	0.00 ± 0.00	0.00	–
VV	$4.09 \times 10^{-4} \pm 1.14 \times 10^{-5}$	1.65	-100.00 ± 0.00
VVV	$5.41 \times 10^{-3} \pm 4.47 \times 10^{-5}$	2.28×10^{-2}	6.85 ± 8.26
$\gamma + \gamma,\ \gamma + jets,\ \gamma + W^{\pm},\ \gamma + Z$	0.00 ± 0.00	0.00	–
$\gamma^*/Z + LF$	$6.97 \times 10^{-2} \pm 3.71 \times 10^{-5}$	2.45	-87.15 ± 46.67
$\gamma^*/Z + HF$	0.00 ± 0.00	0.00	–
QCD HF	0.00 ± 0.00	0.00	–
QCD LF	0.00 ± 0.00	0.00	–

Table 2. Selection efficiencies, event yields and expected integral charge asymmetries for the $W^{\pm\prime} \to \mu^{\pm}\nu_{\mu}$ analysis.

- Lepton energy/momentum scale and resolution: 2.3% (S), 18.1% (B).

$Process$	$\delta A_C^{Stat} \oplus \delta A_C^{Syst}$ (B)	$\delta A_C^{Stat} \oplus \delta A_C^{Syst}$ (S+B)
$M(W^{\pm\prime}) = 1$ TeV	–	1.74 %
$M(W^{\pm\prime}) = 2$ TeV	–	9.83 %
$M(W^{\pm\prime}) = 3$ TeV	–	161.89 %
$M(W^{\pm\prime}) = 4$ TeV	–	41.31 %
Background	3.88%	–

Table 3. Total uncertainties for the S and S+B hypotheses.

Note that the very large uncertainty for $M_{W^{\pm\prime}} = 3$ TeV: $\delta A_C^{Stat} \oplus \delta A_C^{Syst} = 161.89\%$ comes from a few unphysical outlier weights in the computation of the PDF uncertainty obtained through a PDF reweighting. For the sake of being over-conservative, we left it as is in the limit calculation, though obviously it should be capped to 100% to actually make sense.

3.1.4. Statistical Interpretation

In order to distinguish the (S) and (B) hypotheses, standard log likelihood ratios were computed and used as test statistics Q following the formalism in Ref. [17]. Their distributions, including the total uncertainties, are shown in Fig. 5, in blue and in red for the (B) and the (S+B) hypotheses, respectively.

By integrating the test statistics distributions below or beyond Q_B^{Med}, we calculate each hypothesis p-value and consequently the signal confidence level, CL_S. This enables to calculate the 95% C.L. exclusion limits for the different signal mass hypotheses, as shown in Fig 6.

3.1.5. Result

Here are the respective 95% C.L. exclusion limits obtained by the ATLAS analysis and with our method:

$$M_{W^{\pm\prime}} > 2.97 \text{ TeV } (\mathrm{M_T} - \text{tail}) \tag{5}$$

$$M_{W^{\pm\prime}} > 3.0 \text{ TeV } (\mathrm{A_C}) \tag{6}$$

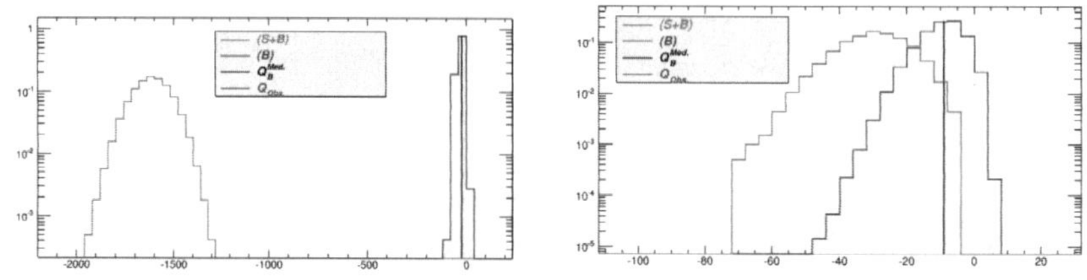

Figure 5. Distribution of the tests statistics for a $W^{\pm\prime}$ mass of 2 TeV (left) and 3 TeV (right).

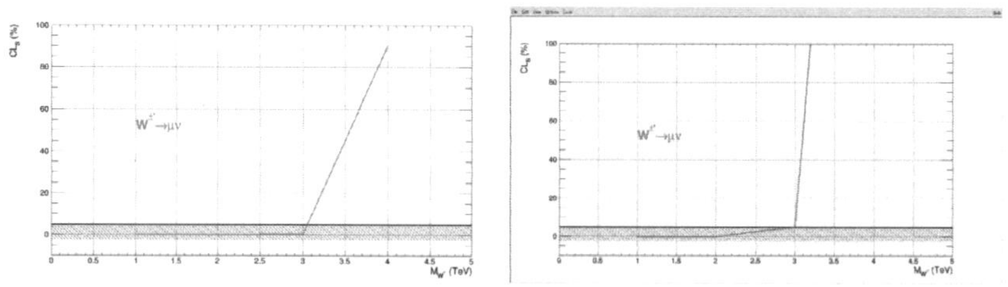

Figure 6. Confidence levels with (left) and without (right) the systematics uncertainties.

This preliminary result shows that our new method based upon A_C has a sensitivity equivalent to that of the state-of-the-art method.

3.2. *Acknowledgments*

The author would like to thank to following persons for the useful discussions regarding the calculations of the confidence levels reported in sub-section 3.1.4: F. James, A. Tilquin, K. Cranmer, G. Cowan, and Y. Coadou.

References

[1] Muanza G S and Serre T 2016 *JHEP* **04** 179 (*Preprint* `arXiv:1412.6695`)
[2] Cafarella A, Coriano C and Guzzi M 2006 *Nucl. Phys.* **B748** 253–308 (*Preprint* `hep-ph/0512358`)
[3] Sherstnev A and Thorne R S 2008 *Eur. Phys. J.* **C55** 553–575 (*Preprint* `0711.2473`)
[4] Polesello G and Tovey D R 2010 *JHEP* **03** 030 (*Preprint* `0910.0174`)
[5] Tovey D URL `http://projects.hepforge.org/mctlib`
[6] Aad G *et al.* (ATLAS) 2014 *JHEP* **09** 037 (*Preprint* `1407.7494`)
[7] Sjostrand T, Mrenna S and Skands P Z 2008 *Comput. Phys. Commun.* **178** 852–867 (*Preprint* `0710.3820`)
[8] Gieseke S, Grellscheid D, Hamilton K, Ribon A, Richardson P, Seymour M H, Stephens P and Webber B R 2006 (*Preprint* `hep-ph/0609306`)
[9] Mangano M L, Moretti M, Piccinini F, Pittau R and Polosa A D 2003 *JHEP* **07** 001 (*Preprint* `hep-ph/0206293`)
[10] de Favereau J, Delaere C, Demin P, Giammanco A, Lematre V, Mertens A and Selvaggi M (DELPHES 3) 2014 *JHEP* **02** 057 (*Preprint* `1307.6346`)
[11] Carrazza S, Latorre J I, Rojo J and Watt G 2015 *Eur. Phys. J.* **C75** 474 (*Preprint* `1504.06469`)
[12] Gao J and Nadolsky P 2014 *JHEP* **07** 035 (*Preprint* `1401.0013`)
[13] Dulat S, Hou T J, Gao J, Guzzi M, Huston J, Nadolsky P, Pumplin J, Schmidt C, Stump D and Yuan C P 2016 *Phys. Rev.* **D93** 033006 (*Preprint* `1506.07443`)
[14] Harland-Lang L A, Martin A D, Motylinski P and Thorne R S 2015 *Eur. Phys. J.* **C75** 204 (*Preprint* `1412.3989`)
[15] Butterworth J *et al.* 2016 *J. Phys.* **G43** 023001 (*Preprint* `1510.03865`)
[16] Buckley A e a 2015 *Eur. Phys. J.* **C75** 132 (*Preprint* `1412.7420`)
[17] Cowan G 2013 (*Preprint* `1307.2487`)

Light exotic Higgs bosons at the LHC

Shoaib Munir

School of Physics, Korea Institute for Advanced Study, Seoul 130-722, Republic of Korea

E-mail: smunir@kias.re.kr

Abstract. Most models of new physics contain extended Higgs sectors with multiple Higgs bosons. The observation of an additional Higgs boson, besides the $\sim$ 125 GeV '$h_{\rm obs}$', will thus serve as an irrefutable evidence of physics beyond the Standard Model (SM). However, even when fairly light, these additional Higgs bosons may still have escaped detection at the Large Electron-Positron (LEP) collider, the Tevatron and the Large Hadron Collider (LHC) hitherto, owing to their highly reduced couplings to the SM particles. Therefore, in addition to the searches based on the conventional production processes of these Higgs bosons, such as gluon or vector boson fusion, possible new search modes need to be exploited at collider experiments in order to establish their signatures. We investigate here the phenomenology of pseudoscalars, with masses ranging from $\mathcal{O}(1)$ GeV to about 150 GeV, in the Next-to-Minimal Supersymmetric SM (NMSSM) and the Type-I 2-Higgs Doublet Model (2HDM) in some such atypical search channels at the LHC Run-II.

1. Introduction

In any new physics model with an extended Higgs sector, (at least) one of the neutral Higgs bosons should have a mass and signal rates in its various decay channels consistent with those of the $h_{\rm obs}$ discovered at the LHC [1,2]. This requirement, coupled with that of the satisfaction of constraints coming from precision electroweak (EW) and b-physics experiments, almost always leads to the other two of the three neutral Higgs bosons being rather heavy in the minimal realisation of Supersymmetry (SUSY). However, in the NMSSM, where the augmentation by a singlet scalar superfield results in a total of five neutral Higgs states, scalars H_{1-3} (ordered in terms of increasing mass) and pseudoscalars $A_{1,2}$, the singlet-like states can be fairly light (for a review, see [3]). A light A_1 can prevent the over-closure of the universe when the lightest neutralino, χ_1^0, which is a crucial dark matter (DM) candidate, has a mass so low that neither the Z boson nor the $h_{\rm obs}$ can contribute sufficienty to its annihilation.

Without invoking SUSY, one can simply introduce a second Higgs doublet in the SM, with a Z_2-symmetry preventing the dangerous flavor changing neutral currents (for a review, see [4]). In a Type-I 2HDM (2HDM-I), this Z_2 symmetry is imposed in such a way that all the fermions couple only to one of the two Higgs doublets. The physical masses of the three neutral Higgs bosons can be taken to be the input parameters in this model. One can therefore assign any masses to the scalars h and H (with $m_h < m_H$) and the pseudoscalar A in order to study the phenomenological implications for different values of the other free parameters.

We analyse some of the potential discovery channels of a light pseudoscalar at the LHC Run-II in these models. In the NMSSM we consider the production of A_1A_1 and A_1Z pairs in the decays of the heavier CP-even scalars [5] and of very light A_1 and DM in the decays of the

IOP Conf. Series: Journal of Physics: Conf. Series **878** (2017) 012028 doi:10.1088/1742-6596/878/1/012028

heavier neutralinos [6]. In the 2HDM-I we discuss the electroweak (EW) production of an hA pair with a combined mass less than that of the Z boson [7].

2. Analysis methodology

For the first of our two NMSSM analyses, in order to perform a fast and efficient numerical scanning of the parameter space for obtaining A_1 solutions spanning a wide range of masses, we adopted a model version with partial universality. In this version, unified parameters m_0, $m_{1/2}$ and A_0, corresponding to the scalar masses, gaugino masses and trilinear couplings, respectively, are input at the grand unification scale. The Higgs sector soft trilinear parameters, A_λ and A_κ, though not unified with A_0, are also input at the same high scale. In contrast, the dimensionless Yukawa couplings λ and κ, and the parameters $\mu_{\rm eff} \equiv \lambda s$ and $\tan\beta \equiv v_u/v_d$, with v_u and v_d being the vacuum expectation values of the two Higgs doublets and s of the singlet, are defined at the EW scale. These input parameters were fed into the NMSSMTools [8] program to calculate the physical masses, couplings and branching ratios (BRs) of all the Higgs bosons. For the 2HDM-I, the Higgs boson couplings and BRs were obtained for the scanned ranges of the input parameters using the 2HDMC public code [9].

During the scanning process, each point was first subjected to the basic theoretical conditions like unitarity, perturbativity and vacuum stability. For every point fulfilling these conditions, we further required the H_2 (H) in the NMSSM (2HDM-I) to have its mass and signal strengths, calculated with HiggsSignals [10], consistent with the latest available measurements for the $h_{\rm obs}$ from the LHC (see, e.g., [11]). A successful model point was then tested for consistency of the remaining scalar, pseudoscalar and charged Higgs bosons with the exclusion limits from collider searches, using the HiggsBounds program [12]. It was also required to satisfy the constraints on the most important b-physics observables, the predicted values for which were calculated for a point in each of the models using the SuperIso [13] program. In addition, for the 2HDM-I points, the values of the oblique parameters S, T and U, calculated by 2HDMC, were checked against exclusion limits from the experimental measurements [14]. In the case of the NMSSM, on the other hand, a point was rejected if the relic abundance due to the χ_1^0 was computed by MicrOmegas [15] to be larger than the measurement by the PLANCK telescope [16]. The scanned ranges of the free parameters in the two models are given in table 1.

Table 1: Scanned ranges of the free parameters in the NMSSM (a) and the 2HDM-I (b).

(a)

NMSSM parameter	Scanned range
m_0 (GeV)	200 – 4000
$m_{1/2}$ (GeV)	100 – 2000
A_0 (GeV)	−5000 – 0
$\tan\beta$	1 – 40
λ	0.01 – 0.7
κ	0.01 – 0.7
$\mu_{\rm eff}$ (GeV)	100 – 2000
A_λ (GeV)	−2000 – 2000
A_κ (GeV)	−2000 – 2000

(b)

2HDM-I parameter	Scanned range
m_h (GeV)	$10 - 2M_Z/3$
m_A (GeV)	$m_h/2 - (M_Z - m_h)$
$m_{H^\pm}$ (GeV)	90 – 150
$\sin(\beta-\alpha)$	−0.25 – 0
m_{12}^2 GeV^2	$0 - m_A^2 \sin\beta\cos\beta$
$\tan\beta$	$\frac{(-0.95 - -1.1)}{\sin(\beta-\alpha)}$

The second NMSSM analysis included here is dedicated to a very specific scenario in which the A_1 and the χ_1^0 both have masses $\mathcal{O}(1)$ GeV. Thus only a couple of benchmark points (BPs) corresponding to the general NMSSM, with all the input parameters lying at the EW scale, will be discussed. For both these BPs, the role of the $h_{\rm obs}$ is once again played by the H_2, and all the relevant constraints noted above are satisfied.

3. Light pseudoscalars in the NMSSM

In the NMSSM, the tree-level mass-squared of the A_1 is written (assuming negligible singlet-doublet mixing) as

$$m_{A_1}^2 \simeq \frac{A_\lambda}{\sqrt{2}s} v^2 \lambda \sin 2\beta + \kappa(2v^2\lambda \sin 2\beta - 3sA_\kappa)\,, \quad (1)$$

where $v \equiv \sqrt{v_u^2 + v_d^2} \simeq 246$ GeV. Thus, m_{A_1} can be varied with more freedom compared to the mass of the singlet-like scalar, which is also indirectly constrained by the $h_{\rm obs}$ measurements due to the relatively stronger singlet-doublet mixing. By adjusting the trilinear couplings A_λ and A_κ, A_1 can take a broad range of values without being in conflict with the experimental data.

3.1. Heavy Higgs boson decays

For this analysis, we restricted ourselves to $m_{A_1} < 150$ GeV, so that the production cross section for the heavier Higgs bosons that decay into it did not get too suppressed kinematically. After performing the parameter space scan to find model points satisfying all the imposed conditions, we carried out a dedicated signal (S)-to-background (B) analysis for the LHC with $\sqrt{s} = 14$ TeV. We first calculated the gluon-fusion production cross section for each of the H_{1-3} using the public program SusHi [17]. The backgrounds coming from the $pp \to 4b$, $pp \to 2b2\tau$, $pp \to 4\tau$, $pp \to Z2b$ and $pp \to Z2\tau$ processes were computed using MadGraph5_aMC@NLO [18]. The hadronization and fragmentation of the signals and backgrounds was then done using Pythia 8.180 [19] interfaced with FastJet [20].

We used the two most dominant decay channels of the A_1, namely $b\bar{b}$ and $\tau^+\tau^-$. In the case of the $b\bar{b}$ decay, we employed also the jet substructure method [21], which gives enhanced sensitivity for larger masses of the decaying Higgs bosons, by assuming one fat jet from boosted b-quarks instead of two single b-jets. For three representative values of the accumulated luminosity at the LHC, $\mathcal{L} = 30$/fb, 300/fb and 3000/fb, we then estimated the signal cross sections which would give the statistical significance, $S/\sqrt{B}$, greater than 5 for a given mass of A_1 in each of the various final state combinations.

In figure 1(a) we show the cross section for the $H_1 \to A_1A_1$ process. Also shown, in this figure and the subsequent ones, are the sensitivity curve(s), which assume BR($A1 \to b\bar{b}$) = 0.9 for each $b\bar{b}$ pair and BR($A1 \to \tau^+\tau^-$) = 0.1 for each $\tau^+\tau^-$ pair, corresponding to the best final state combination for probing the given process. Here these curves are for the $2b2\tau$ final state (one corresponding to two single b-jets and the other, showing an enhanced sensitivity for smaller m_{A_1}, to one fat jet) at $\mathcal{L} = 30$/fb, and for the 4τ final state at $\mathcal{L} = 3000$/fb. In the frame (b) we show the $H_2 \to A_1A_1$ cross section. Note that in the case of the H_2, the possibility to reconstruct its mass (125 GeV) provides an important kinematical handle. We see that the $2b2\tau$ final state can be probed at the LHC with $\mathcal{L}$ as low as 30/fb, owing to the use of the jet substructure method, despite the fact that the $H_2 \to A_1A_1$ decay is tightly constrained by the $h_{\rm obs}$ signal rate measurements at the LHC. However, the maximum A_1 mass that can be accessible in this channel is $m_{H_2}/2 \sim 62.5$ GeV. Evidently, only smaller values of m_{A_1} might be accessible in the $H_2/H_1 \to ZA$ decay channels. We therefore ignore these channels and turn to the H_3 for the production of heavier A_1.

Figure 2(a) shows that the $H_3 \to A_1A_1$ channel does not carry any promise. This is due to the fact that for such high masses of H_3 (≥ 400 GeV) the production cross section gets diminished and, at the same time, other decays of H_3 dominate over this channel. The sensitivity curve in the figure corresponds to the $2b2\tau$ final state for $\mathcal{L} = 3000$/fb. Conversely, as seen in the frame (b), for the $H_3 \to A_1Z$ channel, with the Z decaying into e^+e- or $\mu^+\mu^-$ states, a number of

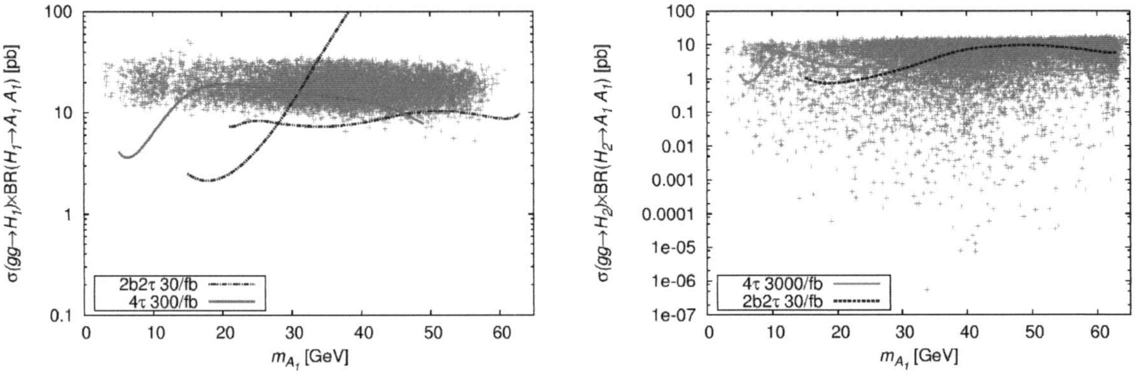

(a) (b)

Figure 1: Cross sections for (a) the $gg \to H_1 \to A_1A_1$ process, and (b) the $gg \to H_2 \to A_1A_1$ process, as functions of m_{A_1}, for points obtained from the NMSSM scan. Taken from [5].

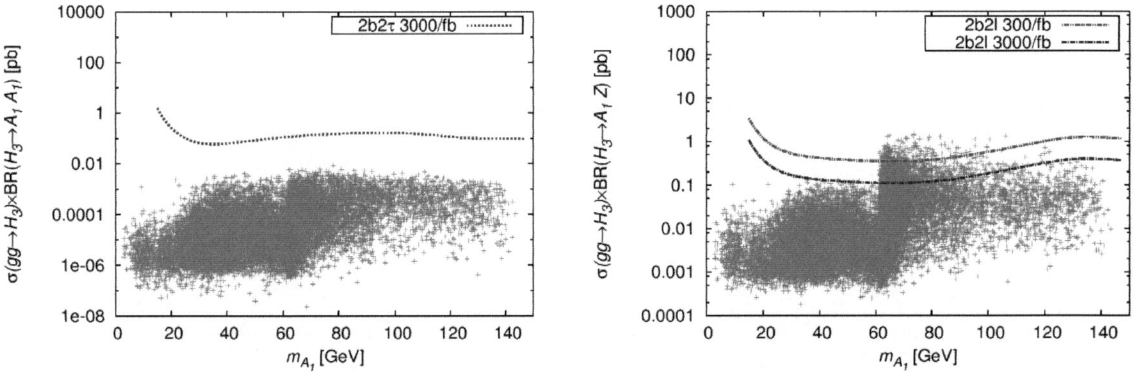

(a) (b)

Figure 2: Cross sections for (a) the $gg \to H_3 \to A_1A_1$ process, and (b) the $gg \to H_3 \to A_1Z$ process, as functions of m_{A_1}, for points obtained from the NMSSM scan. Taken from [5].

points lie above the $2b2\ell$ sensitivity curve for $\mathcal{L} = 300$/fb. Again, the use of the fat jet analysis, along with a sizable H_3A_1Z coupling resulting from a significant doublet component in A_1, make an A_1 lying in the $\sim 60 - 100$ GeV mass range discoverable in this channel.

3.2. DM associated production

Next we focus on the A_1 with mass $\mathcal{O}(1)$ GeV in the NMSSM, which also contains five neutralinos, $\widetilde{\chi}^0_{1-5}$, in total. The lightest of these states, given by the linear combination

$$\widetilde{\chi}^0_1 = N_{11}\widetilde{B}^0 + N_{12}\widetilde{W}^0_3 + N_{13}\widetilde{H}^0_d + N_{14}\widetilde{H}^0_u + N_{15}\widetilde{S}^0, \tag{2}$$

of the gaugino ($\widetilde{B}^0, \widetilde{W}^0_3$), higgsino ($\widetilde{H}^0_u, \widetilde{H}^0_d$) and singlino ($\widetilde{S}^0$) interaction eigenstates, is the DM candidate when R-parity is conserved. The presence of the singlino fraction, $Z_s = |N_{15}|^2$ in the $\widetilde{\chi}^0_1$, which is non-existent in the MSSM, leads to some interesting new possibilities in the context of DM phenomenology. A look at the NMSSM neutralino mass matrix reveals that the $[\mathcal{M}_{\widetilde{\chi}^0}]_{55}$ term, corresponding to the singlino eigenstate, is equal to $2\kappa s = 2\frac{\kappa\mu_{\rm eff}}{\lambda}$. This implies that the singlino fraction in $\tilde{\chi}_1^0$ can be increased by reducing κ and/or $\mu_{\rm eff}$ and increasing λ. Since the mass of A_1 also scales with κs, as noted above, a light A_1 can naturally accompany a light singlino-like $\tilde{\chi}_1^0$. This DM can thus undergo sufficient annihilation, via A_1 in the s-channel, to generate the correct relic abundance of the universe.

At the LHC, one of the main ways to probe the DM is in the decays of the heavier neutralinos and charginos. In particular, dedicated searches have been performed by both the CMS and ATLAS collaborations [22, 23] for the $pp \to \tilde{\chi}_{2,3}^0 + \tilde{\chi}_1^\pm$ process which is followed by the decays $\tilde{\chi}_{2,3}^0 \to Z + \tilde{\chi}_1^0 \to \ell^+\ell^- + \not{E}_T$ and $\tilde{\chi}_1^\pm \to W^\pm + \tilde{\chi}_1^0 \to \ell^\pm + \not{E}_T$, where $\not{E}_T$ implies missing transverse energy. These searches have already put strong constraints on significant regions of the NMSSM parameter space, since the $Z + \tilde{\chi}_1^0$ decay channel is by far the dominant one of $\tilde{\chi}_{2,3}^0$. However, in the scenario with a very light singlino-like DM, the $\tilde{\chi}_{2,3}^0 \to A_1 + \tilde{\chi}_1^0$ decay channel, although still subdominant, can become sizable for sufficiently large values of λ. The reason is that the $\tilde{\chi}_2^0$, $\tilde{\chi}_3^0$ and $\tilde{\chi}_1^\pm$ are predominantly higgsinos, since $\mu_{\rm eff}$ is much smaller than the gaugino soft masses M_1 and M_2, in order to maximize the singlino fraction in $\tilde{\chi}_1^0$ while minimizing its mass. The issue with the $A_1 + \tilde{\chi}_1^0$ decay mode though, is that the main leptonic decay channel, $A_1 \to \mu^+\mu^-$, is highly suppressed, with its BR never exceeding 9%. Furthermore, the muons thus produced are highly collinear and hence the isolation of this signal from the background is extremely challenging.

We show here that, for the NMSSM parameter space points yielding $\mathcal{O}(1)$ GeV $\tilde{\chi}_1^0$ and A_1, once the above complication can be overcome, the $A_1 + \tilde{\chi}_1^0$ search channel can be more promising than the $Z + \tilde{\chi}_1^0$ one at the LHC. We refer to the former as the $\mu_{\rm col}$ channel and to the latter as the trilepton (3ℓ) channel. For this purpose, from the NMSSM parameter space, we chose BP1 such that the $\tilde{\chi}_{2,3}^0 \to A_1\tilde{\chi}_1^0$ decays are typically suppressed (both having BRs of 0.004), while BP2 has relatively enhanced respective BRs of 0.089 and 0.081. The BRs corresponding to the $\tilde{\chi}_{2,3}^0 \to Z\tilde{\chi}_1^0$ decays for both the points are in excess of 60%, while the $BR(A_1 \to \mu^+\mu^-)$ is 0.039 for BP1 and 0.087 (i.e., near its maximum possible value) for BP2. For these BPs we then performed detector-level analyses of the two processes shown in figure 3.

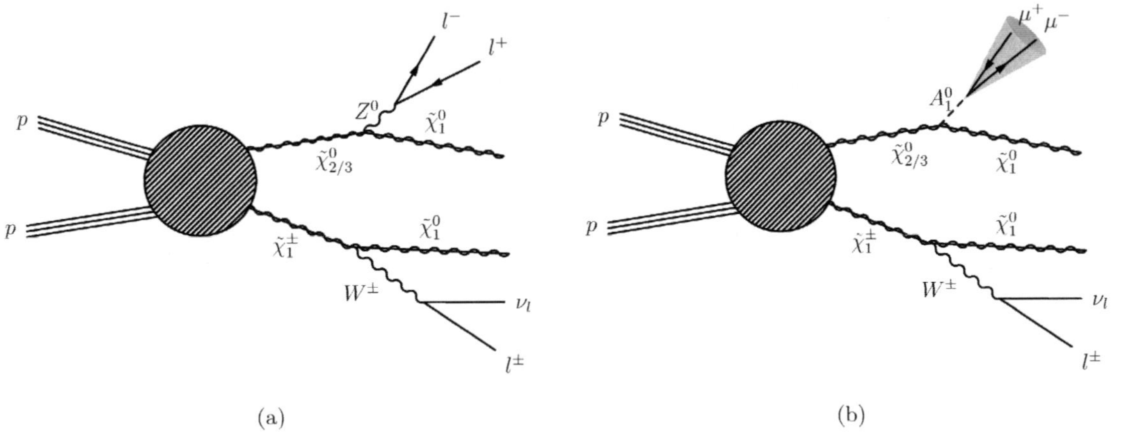

Figure 3: Diagramatic representation of the processes containing, in the final state, (a) two leptons coming form a Z boson, and (b) two collinear muons coming from an A_1.

For the 3ℓ channel, we first generated parton-level signal and background events at the 14 TeV LHC for each of the BPs using MadGraph5_aMC@NLO and passed these to Pythia 6.4 [24] for hadronization. The most dominant irreducible backgrounds for this channel come from the di-boson, tri-boson and $t\bar{t}W/Z$ productions, all of which can have three or more leptons and $\not{E}_T$ in the final states. To obtain the signal and background efficiencies, the ATLAS detector simulation was then performed with DELPHES 3 [25] via the CheckMATE program [26], wherein the six distinct signal regions defined in the ATLAS search [23] have already been implemented. By multiplying the next-to-leading order cross sections for the signal process, calculated using Prospino [27], and the backgrounds with an assumed $\mathcal{L} = 300\,\text{fb}^{-1}$, we also obtained the number of events for both in each of the signal regions.

As for the μ_{col} channel, in order to isolate the highly collimated muons, we employed the technique of clustering them together into one object, similar in concept to the construction of a lepton-jet [28]. For applying this method, the signal events generated for BP1 and BP2 were passed to Pythia 6.4 for hadronization and subsequently to DELPHES 3 for jet-clustering using Fastjet. Then, the μ_{col} object was defined by requiring the transverse momentum, p_T, for each muon in the signal to be larger than 10 GeV, and the cut $m_{\mu\bar{\mu}} < 5\,\text{GeV}$ was imposed on the invariant mass of the muon pair. It additionally satisfied the condition $I_{\text{sum}} < 3$ GeV, with I_{sum} being the scalar sum of the transverse momenta of all additional charged tracks, each with $p_T > 0.5\,\text{GeV}$, within a cone centered along the momentum vector of μ_{col} and satisfying $\Delta R = 0.4$. The main backgrounds, containing two collinear muons along with a third lepton and $\not{E}_T$, include $W(\to \ell^{\pm}v)\gamma^*$, $Z(\to \ell^+\ell^-)\gamma^*$ and $Wb\bar{b}$ and $Z\gamma^*$. These backgrounds were also generated with Pythia and Fastjet and subjected to certain cuts that maximize the isolation of the signal process from them, as explained in [6].

The yield of each of the above analysis methods applied to the two search channels is quantified in terms of S/B, which is given in table 2 for the two BPs. For the 3ℓ channel, this S/B corresponds only to the signal region that gives the highest sensitivity, and we note that it is slightly higher than the S/B in the μ_{col} channel for the BP1. For the BP2, however, the μ_{col} analysis gives a considerably larger S/B than the 3ℓ one, which is evidently a consequence of the sizable BR($\tilde{\chi}^0_{2,3} \to A_1\tilde{\chi}^0_1$) and BR($A_1 \to \mu^+\mu^-$). Thus dedicated searches in the μ_{col} channel may prove very crucial for the discovery of a very light DM in non-minimal supersymmetry at the LHC. Note that while an estimation of the statistical significance would be a more realistic indicator of the strengths of the two signal processes compared to the S/B, it is not included here since there is no consistent way of treating the systematic uncertainties.

Table 2: Measures of the strengths of the two analyses considered here, along with the masses, in GeV, relevant to them, corresponding to the two selected BPs.

BP	$m_{\tilde{\chi}^0_1}$	$m_{\tilde{\chi}^0_2}$	$m_{\tilde{\chi}^0_3}$	$m_{\tilde{\chi}^{\pm}_1}$	m_{A_1}	m_{H_2}	S/B (3ℓ)	S/B (μ_{col})
1	1.00	189.1	−201.7	195.0	2.18	124.1	0.591	0.42
2	1.41	170.1	−182.3	167.7	2.99	125.8	0.436	15

4. Scalar-pseudoscalar pair-production in the Type-I 2HDM

The Landau-Yang theorem [29, 30] prevents the contribution of an on-shell Z boson to the gluon-initiated production of a hA pair when the sum of their masses is smaller than m_Z. The $q\bar{q}$-initiated process, however, does not suffer from this limitation, and the cross section for hA pair-production can therefore get considerably enhanced due to a resonant Z boson in the s-channel. Our analysis of the Type-I 2HDM aimed at exploring this possibility, and hence the parameter space scan for this model also observed the condition $m_h + m_A < m_Z$.

In figure 4(a) we show the good points from the scan for which the $\Gamma(Z \to hA)$ additionally lies within the 2σ error on the experimental measurement of the total width of the Z boson, $\Gamma_Z = 2.4952 \pm 0.0023$ GeV [14]. The points highlighted in yellow are the three benchmark points selected for further investigation. The color map in the figure shows the production cross section for the $q\bar{q} \to hA$ process at the LHC with $\sqrt{s} = 13$ TeV, calculated using MadGraph5_aMC@NLO, which evidently grows as $m_h + m_A$ gets smaller. Near the top left corner of the figure $m_A > m_h$, and we see a high density of points. The points disappear when the $H \to AA$ decay channel opens up (for $m_A < m_H/2$) , potentially leading to a significant reduction in the signal strengths of H in the SM final states. The points start reappearing for $m_A < 35$ GeV, near the bottom right corner of the figure, when the HAA decay gets sufficiently suppressed. But they disappear again for $m_A < m_h/2$, where the $h \to AA$ decay channel, severely constrained by the LEP searches, is kinematically available.

Figure 4(b) shows that the $q\bar{q} \to hA$ production cross section at the 13 TeV LHC can exceed the $gg \to hA$ one, calculated using [31], by a few orders of magnitude, reaching up to about 90 pb. In table 3 we list the cross sections corresponding to the two production modes for the three BPs for this model. The difference between the two cross sections is much more pronounced for the BP1, wherein $m_A < m_h$, compared to that for BP2 and BP3 with $m_h < m_A$. One can also note from the table that for BP1, Z^*A is the primary decay channel of h, with the dominant mode for the subsequent decay of the A being the $b\bar{b}$ pair. Thus $Z^*b\bar{b}b\bar{b}$, $Z^*b\bar{b}\tau^+\tau^-$ and $Z^*\tau^+\tau^-\tau^+\tau^-$ could be the main signatures of interest. Similarly, for BPs 2 and 3 Z^*h is the prominent decay mode of A, so that the most common final states remain the same generally. For BP3 though, the highly fermiophobic h (owing to $\sin(\beta - \alpha) \to 0$) has a large BR into two photons, which could make the $Z^*\gamma\gamma\gamma\gamma$ final state an important unconventional probe of this scenario in the 2HDM-I.

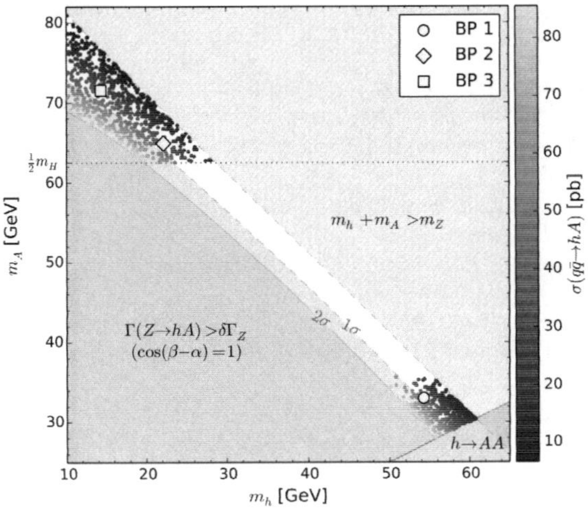

(a)

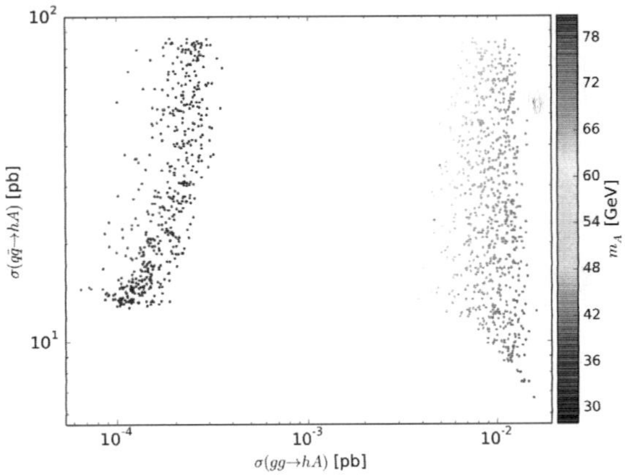

(b)

Figure 4: (a) Successful scan points with $\Gamma(Z \to hA)$ lying within the $\delta\Gamma_Z$ at the 1σ (lighter) and 2σ (darker) levels. The color map corresponds to the total cross section for the $q\bar{q} \to hA$ process and the three BPs have been highlighted in yellow. (b) QCD vs. EW production cross sections of the hA pairs, with the color map showing the mass of A. Taken from [7].

Acknowledgments
The author would like to thank his collaborators, Nils-Erik Bomark, Rikard Enberg, Chengcheng Han, Doyoun Kim, William Klemm, Stefano Moretti, Myeonghun Park and Leszek Roszkowski, each of whom were involved in one (or more) of the analyses reviewed in this contribution.

IOP Conf. Series: Journal of Physics: Conf. Series **878** (2017) 012028 doi:10.1088/1742-6596/878/1/012028

Table 3: Cross sections (in pb) for the gg- and $q\bar{q}$-initiated hA pair-production, corresponding to the three BPs. Also given are the leading BRs of h and A for each BP.

BP	m_h	m_A	$\sigma(q\bar{q})$	$\sigma(gg)$	BR($h \to Z^*A$, bb, $\gamma\gamma$, $\tau\tau$)	BR($A \to Z^*h$, bb, $\tau\tau$)
1	54.2	33.0	41.2	1.5×10^{-4}	$0.94, 0.05, < 0.01, < 0.01$	$0, 0.86, 0.07$
2	22.2	64.9	34.4	7.2×10^{-3}	$0, 0.83, 0.03, 0.07$	$0.86, 0.12, 0.01$
3	14.3	71.6	31.6	1.1×10^{-2}	$0, 0.60, 0.24, 0.07$	$0.90, 0.08, 0.01$

References

[1] Chatrchyan S *et al.* [CMS Collaboration] 2012 *Phys. Lett.* B **716** 30
[2] Aad G *et al.* [ATLAS Collaboration] 2012 *Phys. Lett.* B **716** 1
[3] Ellwanger U, Hugonie C and Teixeira A M 2010 *Phys. Rept.* **496** 1–77
[4] Branco G, Ferreira P, Lavoura L, Rebelo M, Sher M, Silva J P 2012 *Phys. Rept.* **516** 1–102
[5] Elis-Bomark N, Moretti S, Munir S and Roszkowski L 2015 *J. High Energy Phys.* JHEP07(2015)002
[6] Han C, Kim D, Munir S and Park M 2015 *J. High Energy Phys.* JHEP04(2015)132
[7] Enberg R, Klemm W, Moretti S and Munir S 2017 *Phys. Lett.* B **764** 121
[8] http://www.th.u-psud.fr/NMHDECAY/nmssmtools.html
[9] Eriksson D, Rathsman J and Stal O 2010 *Comput. Phys. Commun.* **181** 189–205
[10] Bechtle P, Heinemeyer S, Stal O, Stefaniak T and Weiglein G 2014 *Eur. Phys. J.* C **74** 2711
[11] Aad G *et al.* [CMS and ATLAS Collaborations] 2016 *J. High Energy Phys.* JHEP08(2016)045
[12] Bechtle P, Brein O, Heinemeyer S, Stal O, Stefaniak T, Weiglein G and Williams K E 2014 *Eur. Phys. J.* C **74** 2693
[13] Mahmoudi F 2009 *Comput. Phys. Commun.* **180** (2009) 1579–1613
[14] K. Olive *et al.* 2014 *Chin. Phys.* C **38** 090001
[15] Belanger G, Boudjema F, Pukhov A and Semenov A 2015 *Comput. Phys. Commun* **192** 322–329
[16] Ade P A R *et al.* [Planck Collaboration] 2016. *Astron. Astrophys.* **594** A13
[17] Harlander R V, Liebler S and Mantler H 2013 *Comp. Phys. Commun.* **184** 1605 – 1617
[18] Alwall J *et al.* 2014 *J. High Energy Phys.* JHEP07(2014)079
[19] Sjostrand T, Mrenna S, and Skands P Z 2008 *Comput. Phys. Commun.* **178** 852–867
[20] Cacciari M, Salam G P and Soyez G 2012 *Eur. Phys. J.* C **72** 1896
[21] Butterworth J M, Davison A R, Rubin M and Salam G P 2008 *Phys. Rev. Lett.* **100** 242001
[22] Khachatryan V *et al.* [CMS Collaboration] 2014 *Eur. Phys. J.* C **74** no. 9 3036
[23] Aad G *et al.* [ATLAS Collaboration] 2014 *J. High Energy Phys.* JHEP04(2014)169
[24] Sjostrand T, Mrenna S and Skands P Z 2006 *J. High Energy Phys.* JHEP05(2006)026
[25] de Favereau J *et al.* 2014 *J. High Energy Phys.* JHEP02(2014)057
[26] Drees M, Dreiner H, Schmeier D, Tattersall J and Kim J S 2014 *Comput. Phys. Commun.* **187** 227–265
[27] Beenakker W, Kramer M, Plehn T, Spira M and Zerwas P 1998 *Nucl. Phys.* B **515**
[28] Falkowski A, Ruderman J T, Volansky T and Zupan J 2010 *J. High Energy Phys.* JHEP05(2010)077
[29] Landau L D 1948 *Dokl. Akad. Nauk Ser. Fiz.* **60** no. 2 207
[30] Yang C N 1950 *Phys. Rev.* **77** 242
[31] Hespel B, Lopez-Val D and Vryonidou E 2014 *J. High Energy Phys.* JHEP09(2014)124

IOP Conf. Series: Journal of Physics: Conf. Series **878** (2017) 012029 doi:10.1088/1742-6596/878/1/012029

Medium-induced radiation beyond the Poisson approximation

Andriniaina N. Rasoanaivo and W.A. Horowitz

Department of Physics, University of Cape Town, Private Bag X3, Rondebosch 7701, South Africa.

E-mail: andriniaina@aims.ac.za

Abstract. We present a novel technique for the calculation of the probability for emission of gluons, radiated from a high-p_T quark, through a medium (QGP). Our work is an extension of the maximal helicity violating method to compute the non-Abelian correction for 2 gluon emissions.

1. Introduction

The experiments conducted at RHIC (relativistic heavy ion collision) and LHC (large hadron collider) on nucleus-nucleus collisions, also known as heavy ion collisions, provide a new arena to understand the properties of QCD (quantum chromodynamics). One of the greatest challenges in QCD is the confinement phenomenon, which explains why quarks and gluons cannot be isolated. However, data taken in heavy ion collision experiments provides strong evidence that deconfined QCD matter, also known as QGP (quark gluon plasma), has been produced [1]. Note that in this QGP state, quarks and gluons are able to move freely for a very short time.

In nucleus-nucleus collisions the three main experimental observations that confirm the observation of QGP are: *1)elliptic flow* which is the anisotropic collective flow of a fireball matter (QGP) measured in non-central heavy ion collisions[2]; *2)collective flow* in which heavy quarks, charm and bottom, are observed to be dragged along by the collective expansion that leads to a large elliptic flow [3]; *3)jet quenching* where high energetic partons passing through the medium lose large amounts of energy [4].

In this work we are going to focus on the theoretical understanding of the third observation, jet energy loss. Jets are known as the manifestation of high-p_T partons produced in nucleus-nucleus collisions. In the presence of QGP medium these high-p_T partons experience multiple interactions and lose energy through scattering, and mainly through gluon radiation prior to fragmentation and hadronisation[5, 6, 7]:

$$\text{Jet}(E) \longrightarrow \text{Jet}(E - \Delta E) + \text{Radiation}(\Delta E). \tag{1}$$

In practice we cannot detect these radiations, since they are either emitted with a very low energy compared to the resolution of the detectors (*soft*), or are emitted at a very small angle and cannot be resolved from the parent jet (*collinear*). But on the theoretical side the soft-collinear behavior

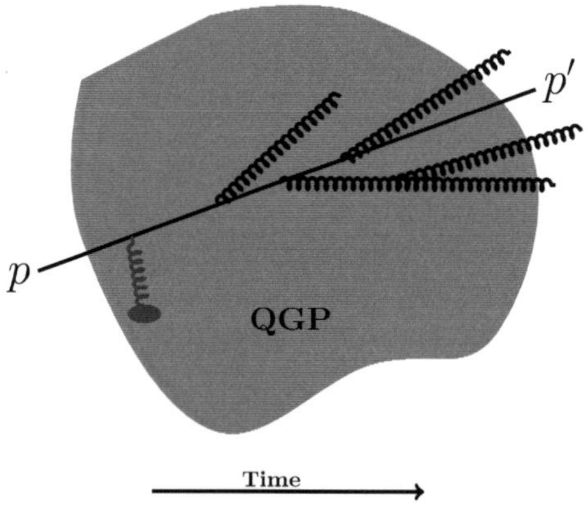

Figure 1: A patron radiating through a medium.

of these radiations provides all the necessary tools to compute the probability distribution of gluon emission. However, the non-Abelian nature of QCD complicates the computation, also the distribution is often approximated as a Poisson distribution, even if such approximation is equivalent to neglecting the gluon interferences. Therefore the main goal of this work is to present a new technique to compute the non-Abelian correction (gluon interferences) to the Poissonian approximation,

$$dP_{\text{improved}} = dP_{\text{poisson}} + dP_{\text{correction}}. \tag{2}$$

As such we want to know if the non-Abelian effect is negligible or not? For that we consider a light quark, which interacts softly with a medium by a single gluon exchange, and study the correlation between emitted gluons. In the next section we give a short introduction to the maximal helicity violating (MHV) technique that we are going to use, then we review the single emission using the new set of variables in MHV, and finally we compute the non-Abelian correction for the two gluon distribution in order to construct the correlation function.

2. Maximal Helicity Violating

This section is a short introduction for the MHV techniques that will be used through out this work in order to compute scattering amplitudes [10, 11, 12]. As physicists, it is very natural to use Feynman diagram techniques to compute amplitudes, but we know that the computation becomes complicated as the number of particles increases. In the table below we can see how the number of diagrams increases with the number of gluons emitted, for quark anti-quark production:

# Gluon	1	2	3	⋯	n
# Diagram	2	8	48	⋯	$2^n n!$

However it turns out the analytical expression of the amplitudes are extremely simple using a different set of variables, the spinor variables. It is even simpler when the conservation of helicity is maximally violated, i.e. $\Delta h = |h_i - h_f|$ is maximal. The spinor helicity formalism is one of the main ingredients of the MHV techniques, and in this formalism the momenta will be represented with a two by two matrix given by $P^{a\dot{a}} = (\sigma_\mu)^{a\dot{a}} p^\mu$. The invariant mass of a given particle will be the determinant of the 2×2 matrix that represents its momentum. For massless particles, since the invariant mass is zero, the momenta can be parametrized as follows

$$P^{a\dot{a}} = (\sigma_\mu)^{a\dot{a}} p^\mu = \lambda^a \tilde{\lambda}^{\dot{a}}, \tag{3}$$

where λ and $\widetilde{\lambda}$ are called the spinor helicity variables and transform, respectively, as left and right handed spinors. Using these new variables we can define two different Lorentz invariant products that are both antisymmetric:

$$\langle 12\rangle = \epsilon_{ab}\lambda_1^a\lambda_2^b \quad \text{and} \quad [12] = \epsilon_{\dot{a}\dot{b}}\widetilde{\lambda}_1^{\dot{a}}\widetilde{\lambda}_2^{\dot{b}}. \tag{4}$$

These products, angle and square brackets, can be related to the four vector scalar product as $2p_1.p_2 = \langle 12\rangle\,[12]$. In terms of these new variables the MHV amplitude corresponding to an exited quark that absorbs a gluon from a medium and radiates n gluons is given by the partial amplitude,

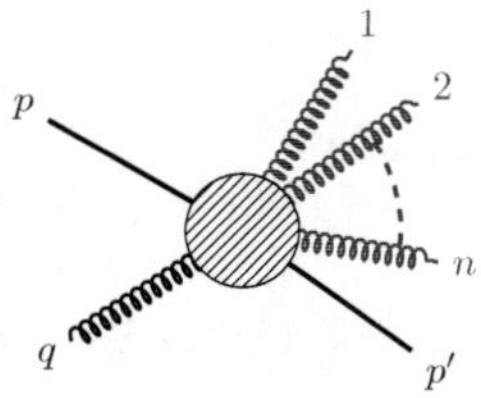

$$= g^{n+1}\frac{\langle pq\rangle^3\langle qp'\rangle}{\langle pp'\rangle\langle p'q\rangle\langle q1\rangle\langle 12\rangle\cdots\langle np\rangle}\ . \tag{5}$$

The full amplitude is obtained by summing all the permutations of the gluon indices over the color string basis $(T_{a_1}T_{a_2}\cdots T_{a_n})$, which is known as color kinematic decomposition of QCD.

3. Review of a single gluon emission

Generally speaking the momentum distribution of multiple radiations is obtained in two steps. First, we factorize the soft-collinear behavior from the Born amplitude using the following property of scattering amplitudes

$$\mathcal{M}_{n\text{-gluon}} = \mathcal{M}_{\text{born}} \times J_n(\{k_i\}), \tag{6}$$

where $\mathcal{M}_{\text{born}}$ is the Born amplitude and J_n is called the eikonal factor. The distribution will be given by the absolute square of J_n. The case of single gluon emission is a well known result, so in this section we are not going to rederive that computation, instead we are going to use the known result and translate it in the language of spinor helicity so the reader will be familiarized with the spinor variables. Recall that the eikonal factor for the single emission is given by $J_1(k_1) \sim j_\mu(k_1)\varepsilon^\mu(k_1)$ [13, 14], where $j^\mu(k_1)$ is the emission current and $\varepsilon^\mu(k_1)$ the associated gluon polarization vector. The probability distribution is defined as

$$\frac{dP^{(1)}}{d\Phi_1} = \frac{C_A}{2}\left|j_\mu(k_1)\varepsilon^\mu(k_1)\right|^2 \tag{7}$$

where $d\Phi_1$ is the differential phase space of a single emission of momentum k_1, and C_A is the Casimir operator in the adjoint representation of $SU(3)$. Let us now use the spinor helicity formalism, where any four vector V^μ will be represented by a two by two matrix $V^{a\dot{a}}$ that carries a left and right handed spinor indices. In the same way the polarization vector $\epsilon^\mu(k_1)$ are represented as

$$\varepsilon_-^{a\dot{a}}(k_1, r) := \frac{\lambda_1^a\widetilde{\lambda}_r^{\dot{a}}}{[1\,r]} \quad \text{and} \quad \varepsilon_+^{a\dot{a}}(k_1, r) := \frac{\lambda_r^a\widetilde{\lambda}_1^{\dot{a}}}{\langle 1\,r\rangle}, \tag{8}$$

where $\varepsilon_\pm^{a\dot{a}}(k_1, r)$ is the gluon polarization associated to a gluon with momentum k_1 and helicity $h = \pm 1$. The reference momentum $r^{a\dot{a}} = \lambda_r^a\widetilde{\lambda}_r^{\dot{a}}$ is arbitrary, where the two different choices of reference are related by a gauge transformation. For the case of the emission current we have

$$j^\mu(k_1) = \frac{p^\mu}{2p.k_1} - \frac{q^\mu}{2q.k_1} \longrightarrow j^{a\dot{a}}(k_1) = \frac{\lambda_p^a\widetilde{\lambda}_p^{\dot{a}}}{\langle p1\rangle\,[p1]} - \frac{\lambda_q^a\widetilde{\lambda}_q^{\dot{a}}}{\langle q1\rangle\,[q1]}, \tag{9}$$

where $p^\mu \to \lambda_p^a \tilde{\lambda}_p^{\dot{a}}$ is the momentum of the highly energetic quark, and $q^\mu \to \lambda_q^a \tilde{\lambda}_q^{\dot{a}}$ is the momentum of the gluon from the medium. For the emission of positive helicity gluons the eikonal factor is equal to

$$j_{a\dot{a}}(k_1)\,\varepsilon_+^{a\dot{a}}(k_1,r) = \frac{\langle pq\rangle}{\langle p1\rangle\,\langle 1q\rangle}, \tag{10}$$

and as we can see the left hand side of the equation depends on the reference momentum r, however the right hand side doesn't, that is, the eikonal factor is a gauge invariant quantity. In the next section we will use the MHV formula from the previous section to fully compute the two gluon emission distribution, and show that the Poisson approximation is only valid in some region of the phase space.

4. Non-Abelian correction for the two gluon emissions

To compute the full amplitude in the MHV, we need to perform a sum over the permutation of gluon indices weighted by the color string bases in which we apply the factorization. After computation the eikonal factor, J_2, can be written in two terms, which are respectively symmetric and anisymmetric under the permutation of 1 and 2,

$$J(1,2) = C_{12}^s J^s(1,2) + C_{12}^a J^a(1,2). \tag{11}$$

The symmetric part contains the independent emission, as in the Poisson approximation, and the antisymmetric contains information on the gluon interference that breaks the Poisson approximation,

$$|J_2(1,2)|^2_{\text{new}} = \frac{1}{4}[3+F(1,2)]|J_2(1,2)|^2_{\text{poisson}}. \tag{12}$$

The function $F(1,2)$ is what we call the non-Abelian correction, or gluon interference. The fact that this correction is scale invariant, i.e. conformal, allows us to express F only in terms of angles

$$F(\theta_1,\theta_2,\phi) = \frac{1-\cos\theta_1\cos\theta_2}{1-\cos\theta_1\cos\theta_2-\sin\theta_1\sin\theta_2\cos\Delta\phi}, \tag{13}$$

θ_1 and θ_2 are the emission angles of the two gluons, and $\Delta\phi$ is the separation angle between the two emission planes, as shown in Figure 2.

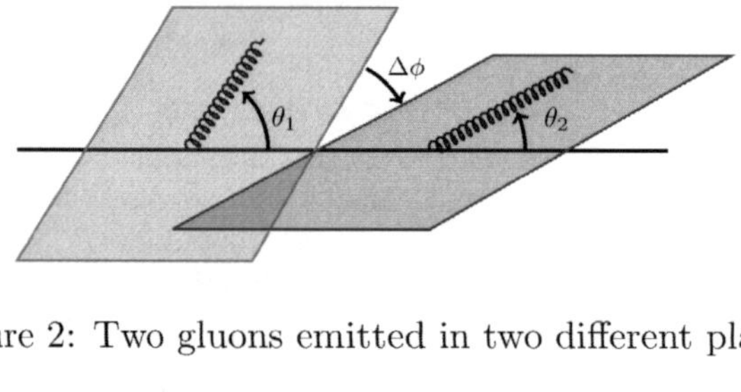

Figure 2: Two gluons emitted in two different planes.

We can also compute the two particle correlation function $C(1,2)$ between the two emitted gluons as in [15]. And in order to simplify the expression let us introduce the follow change of variables: $\eta_i = -\ln[\tan(\theta_i/2)]$, here η_i is the analogue to the pseudorapidity associated to the i-th gluon in which the quark line direction is taken as the beam direction to define η. The correlation is given by

$$C(1,2) = \frac{|J_2|^2_{\text{new}} - |J_2|^2_{\text{poisson}}}{|J_2|^2_{\text{poisson}}} = \frac{1}{4}\frac{\cos\Delta\phi}{\cosh\Delta\eta - \cos\Delta\phi}, \tag{14}$$

where $\Delta\eta = \eta_1 - \eta_2$ the relative difference between the pseudorapidity of the two gluons. In Figure 3 we can see that the two gluons are strongly correlated around $(\Delta\eta = 0, \Delta\phi = 0)$, where the correlation tends to zero as the separation angle $\Delta\eta$ increases. However, the correlation also decreases as $\Delta\phi$ increases up to $\pi/2$, then it grows negatively up to π, then increases again.

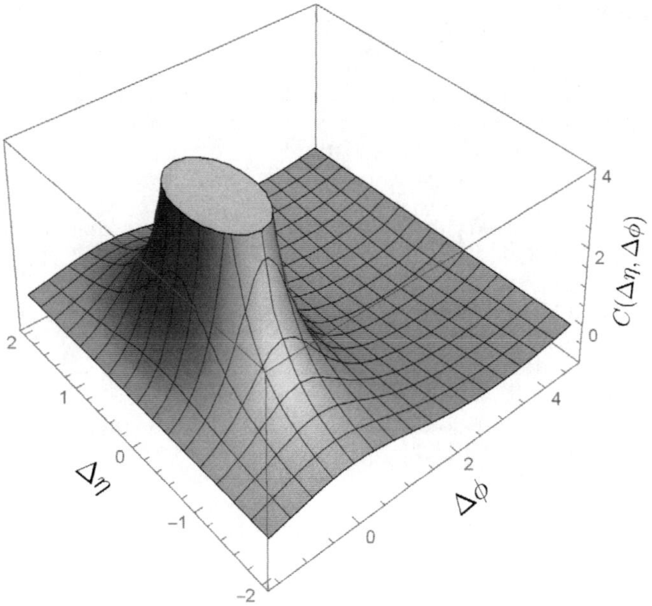

Figure 3: Two particle correlation function.

5. Conclusion

To conclude let us remind ourselves that we want to improve the momentum distribution for emitting soft-collinear gluons for a quark passing through a QGP. For that we performed a full QCD calculation using the MHV techniques in order to avoid diagrammatic complications. And for the two gluon emission case we computed the non-Abelian piece $F(1,2)$, which is a correction to the Poissonian distribution and depends only on three variables: the two emission angles and the angle between the emission planes.

We also showed that the non-Abelian correction can be translated into two particle correlations which depend only on the separation angle $\Delta\phi$, and the pseudorapidity separation $\Delta\eta$ of the two gluons. We saw that the correlation decays as the two gluons are separated in η and ϕ, and the distribution tends to be Poisson. However the two particles are strongly correlated in η, and since we are in the situation where $\theta_1 \sim \theta_2$ (collinear behavior), then we can say that the non-Abelian effect cannot be suppressed in the computation of the momentum distribution of the soft-collinear gluon radiation. This work can be extended in many different ways, one of them is to study the non-flow v_2 generated by the correlation, and see how much of the non-flow contributes in the measured v_2 in pp-collisions.

Acknowledgements

This work would have been impossible without the support of the Science Faculty PhD Fellowship of UCT. We are also grateful to the support provided by SA-CERN and the NRF-South Africa.

References

[1] Zschiesche, D. et al. *"Critical review of quark gluon plasma signals"*, Nucl.Phys. A681 (2001) 34-40 nucl-th/0007033 .

[2] K. H. Ackermann et al. (STAR Collaboration) *"Elliptic flow in Au + Au collisions at* $\sqrt{s_{NN}} = 130GeV$*"*, Phys. Rev. Lett. 86, 402 Published 15 January 2001

[3] H. A. Gustafsson et al. *"Collective Flow Observed in Relativistic Nuclear Collisions"*, Phys. Rev. Lett. 52, 1590 Published 30 April 1984

[4] Levai, P. et al. *"Discovery of Jet Quenching at RHIC and the Opacity of the Produced Gluon Plasma"*, Nucl.Phys. A698 (2002) 631-634 nucl-th/0104035

[5] Megan Connors, for the ALICE Collaboration, *"Jet Production in p-Pb Collisions"*, arXiv:1409.4655 [nucl-ex].

[6] A. Adare et al., *"Centrality-dependent modification of jet-production rates in deuteron-gold collisions at* $\sqrt{s_{NN}}$*=200 GeV"*, arXiv:1509.04657 [nucl-ex] (2015).

[7] W.A. Horowitz, *"Heavy Quark Production and Energy Loss"*, Nuclear Physics A, Volumes 904905, Pages 186c193c (2013).

[8] A. H. Mueller, Phys. Lett. 104B (1981) 161; B. I. Ermolayer and V. S. Fadin, JETP Lett. 33 (1981) 269; A. Bassetto, M. Ciafaloni, G. Marchesini and A. H. Mueller, Nucl. Phys. B207 (1982) 189.

[9] Yu. L. Dokshitzer, V. A. Khoze, A. H. Mueller and S. I. Troyan, Rev. Mod. Phys. (1987).

[10] F. Cachazo, P. Svrcek, E. Witten, *"MHV Vertices And Tree Amplitudes In Gauge Theory"*, Journal of High Energy Physics, Volume 2004, JHEP09(2004).

[11] Z. Bern, L. J. Dixon, D. A. Kosower, *"On-shell recurrence relations for one-loop QCD amplitudes"*, Phys. Rev. D 71, 105013, (2005).

[12] A. Brandhuber, B. Spence, G. Travaglini, *"Tree-level formalism"*, Journal of Physics A: Mathematical and Theoretical, Volume 44, Number 45, (2011).

[13] A. Bassetto, C. Marcello, M. Giovanni. *"Jet structure and infrared sensitive quantities in perturbative QCD."* Physics Reports 100.4 (1983): 201-272.

[14] S. Catani, L. Trentadue, *"Resummation of the QCD perturbative series for hard processes."* Nuclear Physics B 327.2 (1989): 323-352.

[15] Yuri V. Kovchegov, Kirill L. Tuchin, *"Correlation functions and cumulants in elliptic flow analysis"*. Nuclear Physics A Volume 717, Issues 34 (2003): 249-267

IOP Conf. Series: Journal of Physics: Conf. Series **878** (2017) 012030 doi:10.1088/1742-6596/878/1/012030

Exploring LHC Run 1 and 2 data using the Madala hypothesis

Stefan von Buddenbrock

School of Physics, University of the Witwatersrand, Wits 2050, South Africa

E-mail: stef.von.b@cern.ch

Abstract. The Standard Model (SM) Higgs boson, with its experimental discovery in 2012, has long been an interesting particle to study with the intention of exploring new physics ideas beyond the SM (BSM). Its properties are still not well understood, and there are several features in LHC Run 1 and Run 2 data which point at the possibility of extensions to the SM Higgs sector. This work explores the Madala hypothesis, which is the introduction of a heavy scalar (the Madala boson) to the SM, in addition to a real scalar S and dark matter (DM) candidate χ. This hypothesis has previously been used to explain several anomalous features observe in the LHC Run 1 data. This work extends the study to Run 2 data, and shows that the particle spectrum predicted in the Madala hypothesis is indeed compatible with LHC data. Further study prospects and striking signatures for searches are presented.

1. Introduction

In 2012, the Standard Model (SM) had its particle spectrum completed by the discovery of the SM Higgs boson (h) by the ATLAS [1] and CMS [2] collaborations at the Large Hadron Collider (LHC). The Madala hypothesis is an extension of the SM, and is one of the many hypotheses in the literature which predicts physics beyond the SM (BSM). At its core, the Madala hypothesis extends the SM by introducing two new scalars that are heavier than the SM Higgs boson. As discussed below, the hypothesis is merely a simplified model with the aim of explaining several particular features of the LHC Run 1 and 2 data. It can be discussed in the context of UV-complete and gauge symmetric BSM scenarios that predict extra scalars (such as 2HDMs, lefr-right symmetric models, etc.), but the focus of this short paper is to treat it as simply as possible with the purpose of exploring the data.

A first study on the hypothesis was done in 2015, where a new heavy scalar H (the *Madala*[1] *boson*) was introduced to explain several anomalous features in the LHC Run 1 data [3]. In this past work, H was considered to have Higgs-like couplings to the SM particles, as well as be a source of resonant di-Higgs production. It therefore was only considered in the mass range $2m_h < m_H < 2m_t$, since anything heavier would have been dominated by $H \to t\bar{t}$ decays and anything lighter would not allow for resonant di-Higgs decays. This also lead to the assumption that H is produced dominantly through gluon fusion (ggF), however the strength of the g-g-H interaction was considered to be a free parameter, and the SM-like value could be rescaled by a factor β_g^2. The driving force behind the existence of H was the Higgs $p_{\rm T}$ spectrum as

[1] Madala is the Zulu word for an old man, the connotation being that H is "older" and therefore heavier than the Higgs boson.

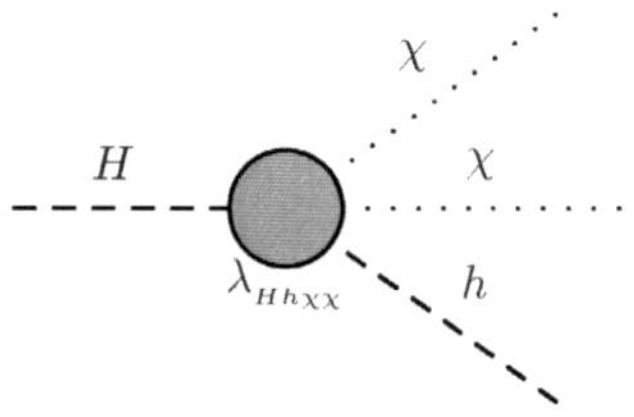

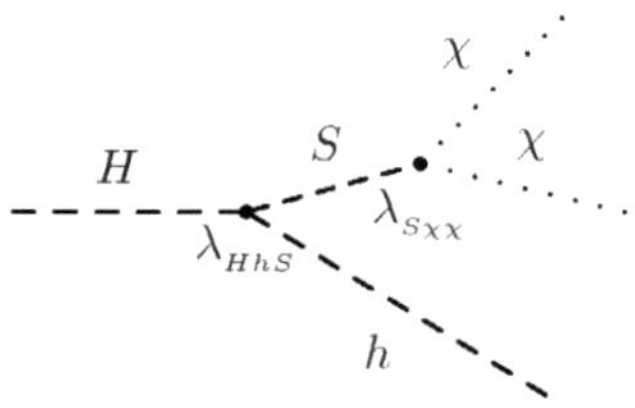

Figure 1: Higgs production in association with DM through the decay of H using (a) an effective vertex and (b) the decay of a DM mediator S.

measured by ATLAS in the $h \to \gamma\gamma$ [4] and $h \to ZZ^* \to 4\ell$ [5] decay channels. Both of these p_{T} spectra appeared to have a systematic enhancement in their intermediate bins (between 20 and 100 GeV). It was shown that this could be explained by requiring resonant associated Higgs production mode through the decay of H, i.e. $H \to h + X$. This was represented in terms of h being produced in association with a pair of dark matter (DM) candidates, denoted by χ, though an effective vertex represented in Figure 1(a). The DM candidate was a scalar, for simplicity, and fit all cosmological and detector constraints at a mass close to $\frac{1}{2}m_h$. With this model in place, a simultaneous fit was done to a collection of relevant ATLAS and CMS results, and a best fit mass of H was found at $m_H = 272^{+12}_{-9}$ GeV, with $\beta_g = 1.5 \pm 0.6$. This result was obtained as a 3σ improvement of the SM prediction.

In 2016, a study was done [6] on determining how to understand the effective vertex in Figure 1(a). The results from the 2015 study seemed to indicate that the branching ratio (BR) of the $H \to h\chi\chi$ decay mode should have to be quite large to explain the data. It is not natural for a 3-body decay to have such a large BR. For this reason, the diagram shown in Figure 1(b) was proposed to explain the nature of the effective vertex. In this case, a scalar DM mediator S was introduced, and the decays of H changed such that the dominant modes are $H \to SS, Sh, hh$. The S boson has a mass in the range $m_h < m_S < m_H - m_h$ such that it is more kinematically accessible through the decays of H as mentioned above. It was also considered that S has Higgs-like couplings to the SM, although its direct production is suppressed.[2] In this case, an S boson with a mass around 160 GeV would decay dominantly to W bosons, as can be seen in Figure 2. The BRs to SM particles would, however, have to be suppressed by the $S \to \chi\chi$ BR, which is a free parameter of the theory.

The Madala hypothesis differs from many BSM hypotheses which predict heavy scalars, in that the Madala boson should dominantly decay to pairs of h and S. The initial statistical study done in 2015 provided some insight into the potential parameter space of the model. However, since then a plethora of newer results from ATLAS and CMS have been presented (several of these at $\sqrt{s} = 13$ TeV). For this reason, it is important to identify whether or not the results which are available at the time of writing this short paper are compatible with the results from the 2015 study.

2. Statistical methodology

The experimental results which are relevant to study when considering the Madala hypothesis are shown in Table 1. It is apparent that such a diverse set of data and final states needs to be carefully combined in order for interesting information to be extracted. For this reason, a

[2] This assumption is merely a simplification to reduce the number of free parameters in the theory, and does not have a big impact on the BSM Higgs p_{T} spectrum. It would, however, have a big impact on applying the Madala hypothesis to multilepton search results.

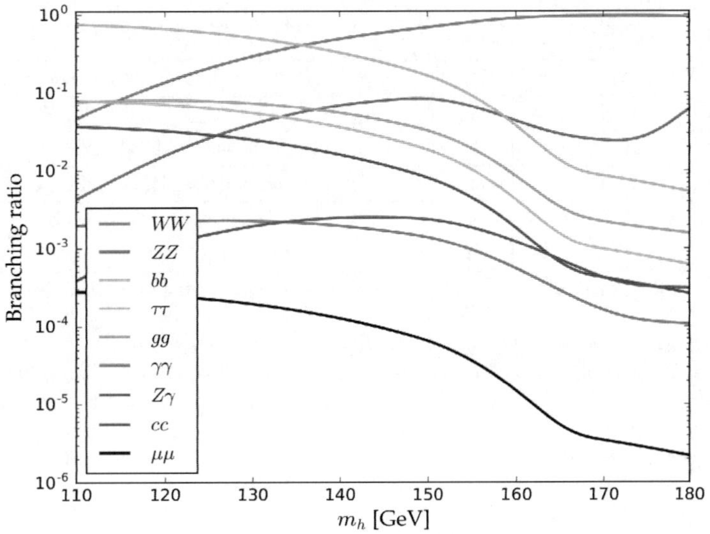

Figure 2: The BRs of a Higgs-like boson in the mass range above and below m_h, taken from the LHC Higgs cross section working group [7]. The S boson (having a mass higher than m_h) would decay dominantly to the massive vector bosons increasingly depending on its mass.

generic approach has been adopted in order to deal with statistics. That is, all experimental results are interpreted in terms of units of χ^2. This simple approach is necessary due to the fact that not enough information is presented as part of experimental results.

The experimental results considered here are usually presented in two ways. Firstly, in the case where measurements are considered, a χ^2 is calculated as Pearson's test statistic:

$$\chi^2 = \frac{(\mu^{\rm th} - \mu^{\rm exp})^2}{(\Delta\mu^{\rm th})^2 + (\Delta\mu^{\rm exp})^2}. \tag{1}$$

Here, a theoretical prediction $\mu^{\rm th}$ is compared against an experimental measurement $\mu^{\rm exp}$, along with their respective uncertainties $\Delta\mu^{\rm th}$ and $\Delta\mu^{\rm exp}$. In the denominator, the experimental and theoretical uncertainties have already been added in quadrature since they are independent of each other.

Secondly, experimental results can come in the form of limits. For searches where no significant excess is seen, a 95% CL is commonly what is presented. In this case, Pearson's test statistic is modified. The difference between the expected and observed limits are treated as a signal with a large error, and therefore limits contribute very weakly to a χ^2. The contribution is written as follows:

$$\chi^2 = \frac{(L^{\rm obs} - L^{\rm exp} - \mu^{\rm th})^2}{(L^{\rm exp}/1.96)^2}, \tag{2}$$

where $L^{\rm exp}$ and $L^{\rm obs}$ are the experimentally calculated expected and observed limits, respectively. Here again, $\mu^{\rm th}$ is a theoretical prediction and its error is considered to be negligible compared with the experimental uncertainty, which is calculated as $L^{\rm exp}/1.96$. The factor of 1.96 arises due to the fact that 95% CL corresponds to 1.96 units of standard deviation. This method of

Result type	Collaboration	Run	Final state
Higgs $p_{\rm T}$	ATLAS/CMS	1	$\gamma\gamma$, $ZZ^* \to 4\ell$, $WW^* \to e\nu\mu\nu$ [4, 8, 5, 9, 10, 11]
spectrum	ATLAS	2	$\gamma\gamma$ [12]
Di-Higgs	ATLAS	1	$bb\tau\tau$, $\gamma\gamma WW^*$, $\gamma\gamma bb$, $bbbb$ [13]
	CMS	1	$bb\tau\tau$, $\gamma\gamma bb$, multilepton [14, 15, 16]
	ATLAS	2	$\gamma\gamma bb$, $bbbb$, $\gamma\gamma WW$ [17, 18, 19]
	CMS	2	$bb\tau\tau$,$\gamma\gamma bb$, $bbbb$, $bbWW$ [20, 21, 22, 23]
Di-boson	ATLAS/CMS	1	WW, ZZ [24, 25, 26]
	ATLAS	2	$ZZ \to 4\ell, 2\ell 2\nu$, $WW \to e\nu\mu\nu$ [27, 28, 29]
	CMS	2	$ZZ \to 2\ell 2\nu$, $WW \to 2\ell 2\nu$ [30, 31]

Table 1: A list of results that are relevant for consideration in the Madala hypothesis, up until before Moriond 2017. Several other final states are relevant too, such as top associated Higgs production results, but they are the focus of a future work. The 2015 fit result used all data from Run 1, with the exception of the $h \to WW^* \to e\nu\mu\nu$ $p_{\rm T}$ spectra (for more details on the 2015 fit result, see Reference [3]).

calculating Pearson's test statistic has been tested in various cases, and found to be consistent with the standard definition given in Equation 1, assuming that the calculated limits are statistically Gaussian.

Combinations of results can be performed using this χ^2 method, and the results of the 2015 fit to data [3] that constrained the parameters of the Madala hypothesis used this procedure. With this in mind, the newer search results can undergo the same treatment in order to understand whether or not the results are compatible with the 2015 fit result.

3. Compatibility checks

The fit result of the Madala boson mass in 2015 was found to be $m_H = 272^{+12}_{-9}$ GeV, with $\beta_g = 1.5 \pm 0.6$, as mentioned in section 1. Using this as a benchmark, one can use the statistical methodology described in section 2 to combine the di-Higgs and di-boson search results in Table 1 to try and understand whether the 2015 fit result is compatible with an updated dataset. These results all contain an interpretation that a heavy resonance H is decaying to a pair of Higgs bosons (in the di-Higgs case) or a pair of massive vector bosons (WW or ZZ, in the di-boson case). Since none of these results consider resonance masses lower that 260 GeV, the full parameter space considered for the Madala boson cannot be explored.

The results of a combination of the experimental data can be seen in Figure 3. On the vertical axis of each of these plots is a best fit value of cross section times BR for the associated search channel. On the horizontal axis the mass of the Madala boson is scanned. Bands have been drawn around the combined result, which represent a 1σ uncertainty in the result. Since we would expect differences in cross section for different center of mass energies, the results are separated into whether they come from Run 1 or Run 2. As can be seen, the combined result often deviates from the null hypothesis (i.e. that no resonance exists), and this most often happens in regions below $m_H = 300$ GeV. The region around $m_H = 272$ GeV shows an enhancement in cross section times BR in every case excepting the Run 2 $H \to ZZ$ result shown in Figure 3(d). By and large, these results are compatible with the 2015 fit result, and a more detailed study could provide us with a better constraint on the best fit value for m_H.

As mentioned above in section 1, the initial driving force behind the investigation of the Madala boson is the Higgs $p_{\rm T}$ spectrum. It is therefore also important to determine whether

IOP Conf. Series: Journal of Physics: Conf. Series **878** (2017) 012030 doi:10.1088/1742-6596/878/1/012030

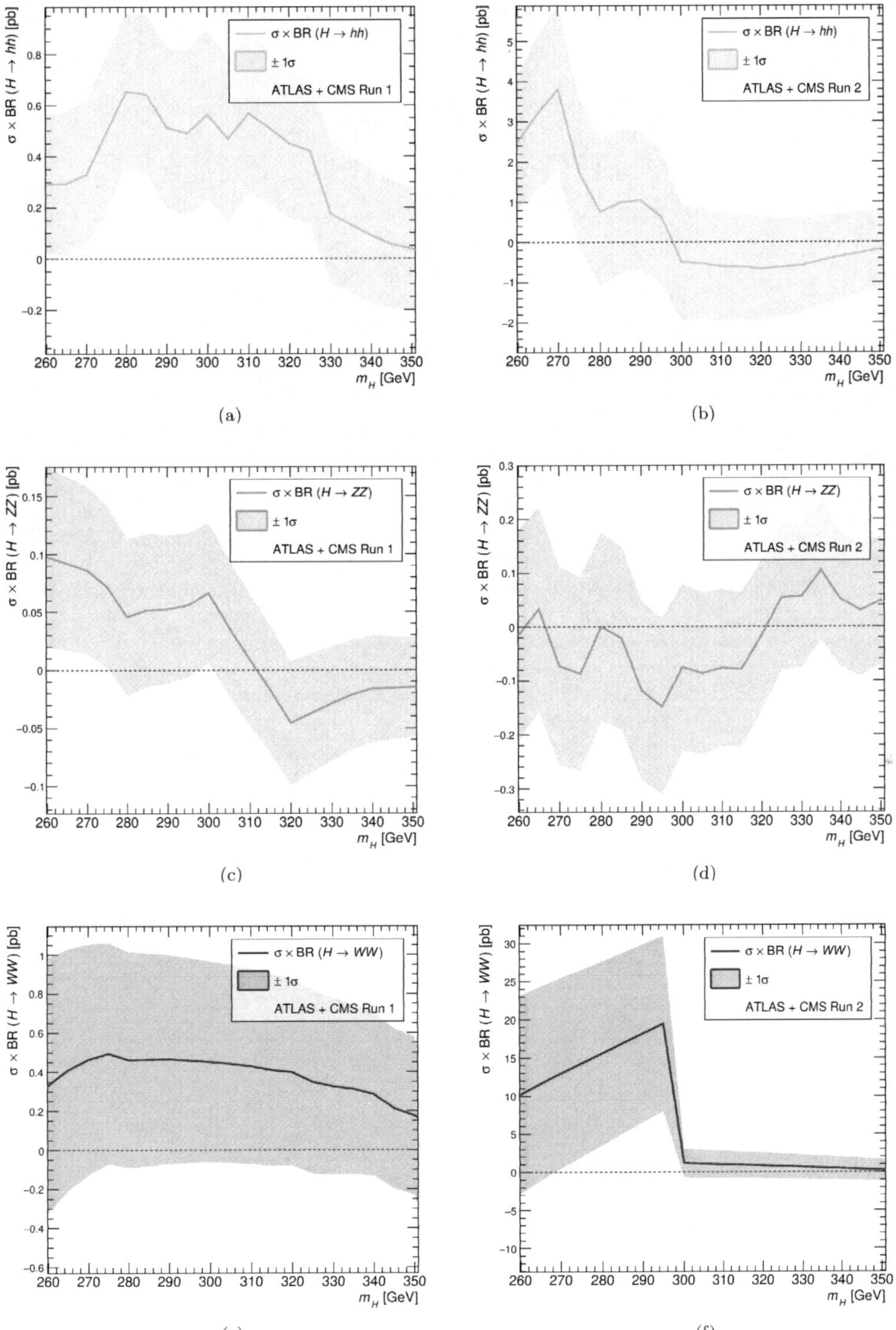

Figure 3: The best fit values of cross section times BR for H production and decay into di-Higgs (top), ZZ (middle), and WW (bottom). The values have been separated into the 8 TeV Run 1 results (left) and the 13 TeV Run 2 results (right).

Channel	β_g^2
ATLAS Run 1 $h \to WW$	1.9 ± 1.6
ATLAS Run 2 $h \to \gamma\gamma$	1.0 ± 1.7
CMS Run 1 $h \to WW$	0

Table 2: Fit results for the study done on the Higgs p_T spectrum. Here the effective vertex shown in Figure 1(a) was used, with $m_H = 270$ GeV and $m_\chi = 60$ GeV.

or not the 2015 fit result obtained with the Run 1 ATLAS and CMS $h \to \gamma\gamma$ and $ZZ^* \to 4ell$ is compatible with what is obtained for the more recent results: the Run 1 ATLAS and CMS $h \to WW^* \to e\nu\mu\nu$ results and the Run 2 ATLAS $h \to \gamma\gamma$ result.

To study the Higgs p_T, a set of Monte Carlo (MC) samples were made to reproduce the different components of it. The SM Higgs p_T spectrum was separated into its different production mechanisms. The ggF spectrum was generated using the NNLOPS procedure [32], which is accurate to next-to-next-to leading order (NNLO) in QCD. The associated production modes – vector boson fusion (VBF), Vh and tth labelled together as Xh – were generated at next to leading order (NLO) using MG5_AMC@NLO [33]. These spectra are scaled to the cross sections provided by the LHC Higgs Cross Section Working Group [7] (from which the theoretical uncertainty also comes). The events are passed through an event selection identical to the fiducial selection used by the experimental collaborations. The dominant ggF prediction was further scaled by the experimentally measured signal strength for each decay mode, $\mu_{gg\mathrm{F}}$.

The BSM prediction was considered to be the Madala hypothesis prediction of $gg \to H \to h\chi\chi$ through an effective vertex, as shown in Figure 1(a). This was generated using PYTHIA 8.2 [34], and scaled to the LHC Higgs Cross Section Working Group N^{3}LO ggF cross sections for

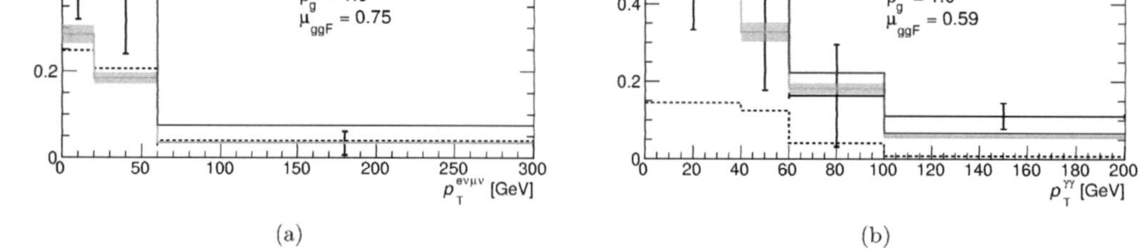

(a) (b)

Figure 4: The Higgs p_T spectra for (a) the ATLAS Run 1 $h \to WW$ channel and (b) the ATLAS Run 2 $h \to \gamma\gamma$ channel. The mass points considered here are $m_H = 270$ GeV and $m_\chi = 60$ GeV, as described in the text.

IOP Conf. Series: Journal of Physics: Conf. Series **878** (2017) 012030 doi:10.1088/1742-6596/878/1/012030

Reference	Channel	Measured μ_{tth}
CMS Run 1 [35]	Same-sign 2ℓ	$5.3^{+2.1}_{-1.8}$
	3ℓ	$3.1^{+2.4}_{-2.0}$
	4ℓ	$-4.7^{+5.0}_{-1.3}$
	Combination	$2.8^{+1.0}_{-0.9}$
ATLAS Run 1 [36]	$2\ell 0\tau_{\text{had}}$	$2.8^{+2.1}_{-1.9}$
	3ℓ	$2.8^{+2.2}_{-1.8}$
	$2\ell 1\tau_{\text{had}}$	$-0.9^{+3.1}_{-2.0}$
	4ℓ	$1.8^{+6.9}_{-2.0}$
	$1\ell 2\tau_{\text{had}}$	$-9.6^{+9.6}_{-9.7}$
	Combination	$2.1^{+1.4}_{-1.2}$
CMS Run 2 [37]	Same-sign 2ℓ	$1.7^{+0.6}_{-0.5}$
	3ℓ	$1.0^{+0.8}_{-0.7}$
	4ℓ	$0.9^{+2.3}_{-1.6}$
	Combination	$1.5^{+0.5}_{-0.5}$
ATLAS Run 2 [38]	$2\ell 0\tau_{\text{had}}$	$4.0^{+2.1}_{-1.7}$
	3ℓ	$0.5^{+1.7}_{-1.6}$
	$2\ell 1\tau_{\text{had}}$	$6.2^{+3.6}_{-2.7}$
	4ℓ	< 2.2
	Combination	$2.5^{+1.3}_{-1.1}$
Error weighted mean		1.92 ± 0.38

Table 3: The measured μ values for tth production in multileptonic analysis channels. A combination is estimated as the error weighted mean of each quoted combined result.

a high mass Higgs-like scalar. The events were passed through the fiducial selections as in the case of the SM prediction. Since the Run 1 fit result had a best fit mass of $m_H = 272$ GeV with $m_\chi = 60$ GeV, the mass points considered for this study were $m_H = 270$ GeV and $m_\chi = 60$ GeV.

The SM and BSM components were added together and then tested for compatibility with the data. A χ^2 value was calculated for each bin per channel, as in Equation 1. The normalisation of the BSM spectrum is used to minimise the χ^2, with the interpretation that it is scaled by the free parameter β_g^2. The results of this fit are shown in Table 2. Comparing against the 2015 best fit point of $\beta_g^2 = 2.25$, the ATLAS Run 1 $h \to WW$ and ATLAS Run 2 $h \to \gamma\gamma$ results are compatible with this value. The CMS Run 1 $h \to WW$ is not improved by the BSM hypothesis. For reference, the p_T spectra at their best fit values are shown in Figure 4 for the two spectra which are improved by the BSM hypothesis.

4. The future of the Madala hypothesis

The Madala hypothesis is dependent on the availability of experimental results against which it can be tested. In 2015 a limited set of results was used, and its parameters were constrained using what was available at the time. The ATLAS and CMS collaborations, however, are always hard at work producing updated results with larger datasets. This short paper has compiled the results which were available before Moriond 2017, and shown that the newer results are compatible with the fit result obtained in 2015.

The results presented in this short paper are, however, not a complete re-fit of all available data. This task will be the focus of a future work, at a time where results are presented with the full 2015 and 2016 datasets. In addition to this, several aspects of the hypothesis have not been explored in this short paper. Most notably, the Madala hypothesis predicts that an enhanced rate of top associated Higgs production should be observed, particularly in leptonic channles, due to the existence of the S boson. This could explain the enhancements we see in the data, the results of which are listed in Table 3.

As the LHC continues to run and larger datasets are analysed in the search for new physics, the Madala hypothesis will continue to be tested until such a point that we can confidently confirm it or rule it out. With several important results still to come out, more definite statements about the Madala hypothesis are left for a future work.

References

[1] Aad G *et al.* (ATLAS Collaboration) 2012 *Phys. Lett.* **B716** 1–29 (arXiv: 1207.7214)
[2] Chatrchyan S *et al.* (CMS Collaboration) 2012 *Phys. Lett.* **B716** 30–61 (arXiv: 1207.7235)
[3] von Buddenbrock S, Chakrabarty N, Cornell A S, Kar D, Kumar M, Mandal T, Mellado B, Mukhopadhyaya B and Reed R G 2015 HRI-RECAPP-2015-011, WITS-MITP-010 (arXiv: 1506.00612)
[4] Aad G *et al.* (ATLAS Collaboration) 2014 *JHEP* **09** 112 (arXiv: 1407.4222)
[5] Aad G *et al.* (ATLAS Collaboration) 2014 *Phys. Lett.* **B738** 234–253 (arXiv: 1408.3226)
[6] von Buddenbrock S, Chakrabarty N, Cornell A S, Kar D, Kumar M, Mandal T, Mellado B, Mukhopadhyaya B, Reed R G and Ruan X 2016 *Eur. Phys. J.* **C76** 580 (arXiv: 1606.01674)
[7] de Florian D *et al.* (LHC Higgs Cross Section Working Group Collaboration) 2016 FERMILAB-FN-1025-T (arXiv: 1610.07922)
[8] Khachatryan V *et al.* (CMS Collaboration) 2016 *Eur. Phys. J.* **C76** 13 (arXiv: 1508.07819)
[9] Khachatryan V *et al.* (CMS Collaboration) 2016 *JHEP* **04** 005 (arXiv: 1512.08377)
[10] Aad G *et al.* (ATLAS Collaboration) 2016 *JHEP* **08** 104 (arXiv: 1604.02997)
[11] Khachatryan V *et al.* (CMS Collaboration) 2016 *Submitted to: JHEP* CMS-HIG-15-010, CERN-EP-2016-125 (arXiv: 1606.01522)
[12] Aad G *et al.* (ATLAS Collaboration) 2016 ATLAS-CONF-2016-067
[13] Aad G *et al.* (ATLAS Collaboration) 2015 *Phys. Rev.* **D92** 092004 (arXiv: 1509.04670)
[14] Khachatryan V *et al.* (CMS Collaboration) 2015 (arXiv: 1510.01181)
[15] CMS Collaboration 2014 CMS-PAS-HIG-13-032
[16] Khachatryan V *et al.* (CMS Collaboration) 2014 *Phys. Rev.* **D90** 112013 (arXiv: 1410.2751)
[17] Aad G *et al.* (ATLAS Collaboration) 2016 ATLAS-CONF-2016-049
[18] Aad G *et al.* (ATLAS Collaboration) 2016 ATLAS-CONF-2016-071
[19] Aad G *et al.* (ATLAS Collaboration) 2016 ATLAS-CONF-2016-004
[20] Khachatryan V *et al.* (CMS Collaboration) 2016 CMS-PAS-HIG-16-029
[21] Khachatryan V *et al.* (CMS Collaboration) 2016 CMS-PAS-HIG-16-032
[22] Khachatryan V *et al.* (CMS Collaboration) 2016 CMS-PAS-HIG-16-002
[23] Khachatryan V *et al.* (CMS Collaboration) 2016 CMS-PAS-HIG-16-011
[24] Khachatryan V *et al.* (CMS Collaboration) 2015 *JHEP* **10** 144 (arXiv: 1504.00936)
[25] Aad G *et al.* (ATLAS Collaboration) 2016 *Eur. Phys. J.* **C76** 45 (arXiv: 1507.05930)
[26] Aad G *et al.* (ATLAS Collaboration) 2016 *JHEP* **01** 032 (arXiv: 1509.00389)
[27] Aad G *et al.* (ATLAS Collaboration) 2016 ATLAS-CONF-2016-074
[28] Aad G *et al.* (ATLAS Collaboration) 2016 ATLAS-CONF-2016-079
[29] Aad G *et al.* (ATLAS Collaboration) 2016 ATLAS-CONF-2016-056
[30] Khachatryan V *et al.* (CMS Collaboration) 2016 CMS-PAS-HIG-16-023
[31] Khachatryan V *et al.* (CMS Collaboration) 2016 CMS-PAS-HIG-16-001
[32] Hamilton K, Nason P and Zanderighi G 2015 *JHEP* **05** 140 (arXiv: 1501.04637)
[33] Alwall J, Frederix R, Frixione S, Hirschi V, Maltoni F, Mattelaer O, Shao H S, Stelzer T, Torrielli P and Zaro M 2014 *JHEP* **07** 079 (arXiv: 1405.0301)
[34] Sjostrand T, Ask S, Christiansen J R, Corke R, Desai N, Ilten P, Mrenna S, Prestel S, Rasmussen C O and Skands P Z 2015 *Comput. Phys. Commun.* **191** 159–177 (arXiv: 1410.3012)
[35] Khachatryan V *et al.* (CMS Collaboration) 2014 *JHEP* **09** 087 [Erratum: JHEP10,106(2014)] (arXiv: 1408.1682)
[36] Aad G *et al.* (ATLAS Collaboration) 2015 *Phys. Lett.* **B749** 519–541 (arXiv: 1506.05988)
[37] Khachatryan V *et al.* (CMS Collaboration) 2017 CMS-PAS-HIG-17-004
[38] Aad G *et al.* (ATLAS Collaboration) 2016 ATLAS-CONF-2016-058

Concluding Remarks: Connecting Relativistic Heavy Ion Collisions and Neutron Star Mergers by the Equation of State of Dense Hadron- and Quark Matter as signalled by Gravitational Waves

Matthias Hanauske[1,2], Jan Steinheimer[2], Luke Bovard[1], Ayon Mukherjee[2], Stefan Schramm[2], Kentaro Takami[1,4], Jens Papenfort[1], Natascha Wechselberger[1], Luciano Rezzolla[1,2] and Horst Stöcker[*,1,2,3]

[1] Institut für Theoretische Physik, Max-von-Laue-Straße 1, 60438 Frankfurt, Germany
[2] Frankfurt Institute for Advanced Studies, Ruth-Moufang-Straße 1, 60438 Frankfurt, Germany
[3] GSI Helmholtzzentrum für Schwerionenforschung GmbH, 64291 Darmstadt, Germany
[4] Kobe City College of Technology, 651-2194 Kobe, Japan

Abstract. The underlying open questions in the fields of general relativistic astrophysics and elementary particle and nuclear physics are strongly connected and their results are interdependent. Although the physical systems are quite different, the 4D-simulation of a merger of a binary system of two neutron stars and the properties of the hot and dense matter created in high energy heavy ion collisions, strongly depend on the equation of state of fundamental elementary matter. Neutron star mergers represent optimal astrophysical laboratories to investigate the QCD phase structure using a spectrogram of the post-merger phase of the emitted gravitational waves. These studies can be supplemented by observations from heavy ion collisions to possibly reach a conclusive picture on the QCD phase structure at high density and temperature. As gravitational waves (GWs) emitted from merging neutron star binaries are on the verge of their first detection, it is important to understand the main characteristics of the underlying merging system in order to predict the expected GW signal. Based on numerical-relativity simulations of merging neutron star binaries, the emitted GW and the interior structure of the generated hypermassive neutron stars (HMNS) have been analyzed in detail. This article will focus on the internal and rotational HMNS properties and their connection with the emitted GW signal. Especially, the appearance of the hadon-quark phase transition in the interior region of the HMNS and its conjunction with the spectral properties of the emitted GW will be addressed and confronted with the simulation results of high energy heavy ion collisions.

1. Introduction

One hundred years after Albert Einstein developed the field equations of general relativity and predicted the existence of gravitational waves, his theory triumphantly corroborates all experimental and observational tests it has been put through to date. Among the ≈ 2500 known neutron stars (NSs), there are some which are in binary systems where the companion of the NS is either a normal star, a planet, a white dwarf or again a NS. The most impressive binary neutron star system, so far[1], is the so called *Double Pulsar: PSR J0737-3039A/B*, which has been discovered in 2003 [2]. The two neutron

[1] An even more relativistic binary pulsar system has been discovered recently and the corresponding experimental results will be reported in the near future [1].

stars in this binary system, which are only separated by 800000 km, orbit around each other with an orbital period of 147 minutes and a mean velocity of one million km/h. This system is excellent for testing Einstein's theory of general relativity and alternative theories of gravity in the strong-field regime [3]. Due to the emission of gravitational waves, the distance between the two NSs decreases with time and finally the two objects will merge accompanied by neutrino radiation, a strong gravitational radiation chirp, and short gamma-ray burst.

Gravitational waves have been recently observed from a pair of merging black holes (BHs) by the LIGO detectors [4, 5] and GWs emitted from merging neutron star binaries are on the verge of their first detection. The main difference between GWs originating from a merger of two BHs or NSs is the possibility of an existence of a post-merger phase after the collisions of the two objects, which can be only present in the case of a NS-merger. Depending on a variety of parameters, e.g. the initial mass of the two stars, the product right after the merger could be a prompt collapse to a BH, a metastable HMNS or a stable supramassive NS. The GWs produced by a merger of NSs are by far more interesting as the GWs resulting from a BH-merger, as, in the case of existence of a post-merger phase, the equation of state (EOS) of elementary matter might be deduced by a frequency analysis of the GW [6, 7, 8]. This is insofar interesting, as the EOS until now is mainly understood by high energy heavy ion collisions and only coarse constraints are coming from astrophysical observations, like the current observational constraint on the observed maximum mass in neutron stars, i.e., $2.01 \pm 0.04 M_\odot$ [9]. In this article we will compare the results of numerical simulations of merging NS binaries with simulations of high energy heavy ion collisions. We will discuss how one can create a similar state of hot and dense nuclear matter in two seemingly different 'experimental' setups, namely the mergers of two neutron stars and relativistic heavy ion collisions. By studying the properties of this QCD natter in a single consistent approach we can finally address one of the most relevant challenges of high-energy nuclear theory. This is to determine the properties and phase structure of QCD at large densities and temperature.

2. Numerical general relativity of neutron star mergers

Einstein's theory of general relativity and the resulting general relativistic conservation laws for energy-momentum in connection with the rest mass conservation are the theoretical groundings of neutron star binary mergers:

$$R_{\mu\nu} - \frac{1}{2} g_{\mu\nu} R \; = \; 8\pi \, T_{\mu\nu} \, , \;\; \nabla_\mu T^{\mu\nu} \; = \; 0 \, , \;\; \nabla_\mu \left(\rho \, u^\mu \right) \; = \; 0 \qquad (1)$$

$T_{\mu\nu}$ describes the energy-momentum tensor, $R_{\mu\nu}$ is the Ricci tensor, which contains first and second derivatives of the space-time metric $g_{\mu\nu}$, ∇_μ is the covariant derivative and u^μ is the four velocity of the star's fluid. The Einstein equation (first equation in Eq. (1)) describes in which way the space-time structure need to bend (left hand side of the equation) if energy-momentum is present (right hand side of the equation). These highly non-linear differential equations describe on the one hand how matter moves in a curved space-time and on the other hand formulates in which way the amounts of energy-momentum curves the space-time structure. In the ideal-fluid energy-momentum tensor $T_{\mu\nu} = (e+p)\, u_\mu u_\nu + p\, g_{\mu\nu}$ enters the energy and pressure densities of the nuclear and elementary particle physics contributions of the underlying neutron star matter and $u^\mu = dx^\mu / d\tau$ describes the four velocity of the star's fluid which is defined as the derivative of the coordinates $x^\mu = (t, x, y, z)$ by the proper time τ.

The interior metric of a single, spherically symmetric and static neutron star can be calculated by assuming the following Ansatz of the metric $g_{\mu\nu}$ (infinitesimal line element ds)

$$ds^2 \; = \; g_{\mu\nu} \, dx^\mu dx^\nu \; = \; - \, e^{2\nu(r)} dt^2 + \left(1 - \frac{2m(r)}{r} \right)^{-1} dr^2 + r^2 d\theta^2 + r^2 \sin^2\theta \, d\phi^2 \quad . \qquad (2)$$

By inserting this Ansatz of the metric and the expression for the energy momentum tensor into Eq. (1) one derives the Tollman-Openheimer-Volkov (TOV) equations (see e.g. [10])

$$\frac{dm}{dr} \; = \; 4\pi r^2 \, e \, , \quad \frac{d\nu}{dr} \; = \; \frac{m + 4\pi r^3 p}{r\,(2m - r)} \, , \quad \frac{dp}{dr} \; = \; - \, (e+p) \frac{d\nu}{dr} \quad . \qquad (3)$$

For a given equation of state (EOS: a function $p(e)$) the TOV equation can be solved numerically by fixing the central pressure p_c and integrating outwards to the star's surface where the pressure is zero. It can be shown (Birkov Theorem, see e.g. [11], page 843f) that the metrical structure outside the neutron star is identical to the Schwarzschild metric of a black hole. It is impressive that Karl Schwarzschild found this analytical solution of the Einstein equation only three month after Einstein's article and it is very unfortunate that he died soon after his remarkable result. The Schwarzschild solution is only valid in a spherically symmetric approximation and even for a single rotating neutron star the metric outside the body cannot be described analytically, as a similar theorem like the Birkov Theorem does not hold for the Kerr solution of a rotating black hole. However, all of the known NSs, including the individual stars of the famous double PSR J0737-3039A/B, can be described in good approximation with the TOV equations and only in the case of a very cramped binary system tidal deformations become important.

In order to solve the evolution of a merging neutron star binary system numerically, Eq. (1) needs to be rewritten, because its structure is not well posed. To reformulate Eq. (1), the so called $(3+1)$-split is used, which starts by slicing the 4-dimensional manifold $\mathcal{M}$ into 3-dimensional space-like hypersurfaces Σ_t. The space-time metric $g_{\mu\nu}$ is then sub-classified into a purely spatial metric γ_{ij}, a lapse function α and a shift vector β_i ($\mu, \nu = 0,1,2,3$ and $i,j = 1,2,3$):

$$g_{\mu\nu} = \begin{pmatrix} -\alpha^2 + \beta_i \beta^i & \beta_i \\ \beta_i & \gamma_{ij} \end{pmatrix} . \tag{4}$$

The lapse function α describes the difference between the coordinate time t and the proper time of a fluid particle τ ($d\tau = \alpha\, dt$). The shift vector β_i measures how the coordinates are shifted on the spatial slice if the fluid particle moves an infinitesimal time step further. In the case of a rotating neutron star, this shift vector formulates one important difference between Newtonian gravity and general relativity - the dragging of local inertial frames in the case of a rotating body. The Lensing-Thirring effect, which has its nature in this property, has been experimentally proved [12] and the gravitomagnetic effect, which corresponds to a general relativistic counterpart of the Lorentz force has been indirectly verified in the binary system PSR 1913+16.

By inserting the metric (4) into the Einstein equation (1) one can restate the equations into a system of first order differential equations, the so called Arnowitt-Deser-Misner (ADM) equations. As the ADM equations are still not 'well posed' (for details see [10]), they need to be further transformed using a conformal traceless formulation. In this article we follow the 'well posed' Baumgarte-Shapiro-Shibata-Nakamura-Oohara-Kojima (BSSNOK) formulation [13, 14] of general relativity. This BSSNOK formulation of general relativity together with the relativistic hydrodynamical equations are finally used as the grounding equations in our computer program [15, 16].

In the following, the results of two simulations will be presented. The EOS used within the first framework is composed of a cold nuclear-physics part and a thermal ideal fluid component (for details see [17, 7]). The cold part have been modelled by a hybrid star matter model (`ALF2-EOS` [18]), where a phase transition to color-flavor-locked quark matter has been implemented. Within this model, the hadronic particles begin to deconfine to quark matter above a certain transition rest-mass density $\rho_{\rm trans} = 3\,\rho_0$, where $\rho_0 := 2.705 \times 10^{14}$ g/cm^3 is the nuclear-matter rest-mass density. Assuming a moderate surface tension of the quark matter droplets, a phase transition is implemented by using a Gibbs-construction (for details see e.g. [19]). As charge neutrality is only globally conserved within this construction, a mixed-matter phase exists in the rest-mass density range $3\,\rho_0 \leq \rho \leq 7.8\,\rho_0$. The EOS used within the second simulation is the purely hadronic but temperature dependent Lattimer-Swesty EOS (`LS220-EOS` [20]). The initial coordinate separation of the stellar centers has been set to $45\,\rm km$ for both models. For each EOS, two equal-mass binaries have been considered, where the (gravitational) mass of each star at infinite separation has the value of $M_{\rm In} = 1.35 M_\odot$ for the `ALF2-M135` run and $1.32 M_\odot$ for the `LS220-M132` simulation.

The results of 4-dimensional numerical simulations of merging neutron star binaries in full general relativity (see Fig. 1) show that the emitted GWs of the merger and post-merger phase are strongly

IOP Conf. Series: Journal of Physics: Conf. Series **878** (2017) 012031 doi:10.1088/1742-6596/878/1/012031

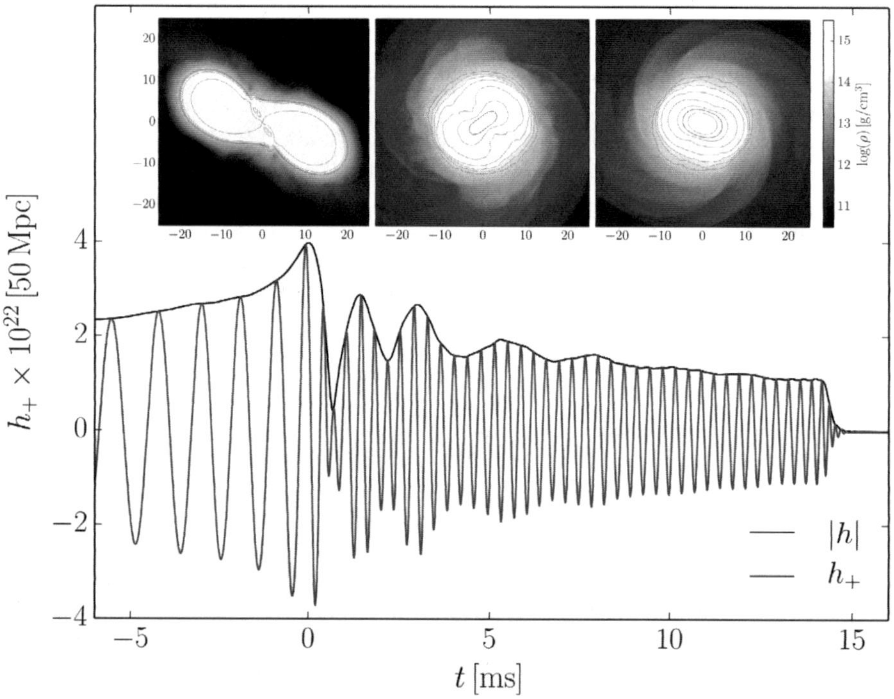

Figure 1. Rest mass density profiles on the equatorial plane and gravitational wave amplitude $|h|$ and h_+ at a distance of 50 Mpc for the ALF2-EOS with $M_{\rm In} = 1.35 M_\odot$.

determined by the high density region of the EOS. The underlying general-relativistic hydrodynamical evolution of the produced HMNS depends on the hadronic and quark matter properties, which also enters in heavy ion collisions (see Sec. 3). Fig. 1 depicts the GW-amplitude of of the ALF2-M135 run. The absolute maximum of $|h|$ marks the time of the merger (i.e., $t = 0$ ms) and the last peak of h_+ corresponds to the time where the HMNS collapses to a black hole (i.e., $t = 14.16$ ms). The upper part of Fig.1 shows the logarithm of the rest mass density profiles for three different time snap-shots ($t = -0.17, 4.05, 13.16$ ms), whereat the boundary of the hadron-quark phase transition (HQPT) is marked with a red curve. For the used ALF2-EOS with $M_{\rm In} = 1.35 M_\odot$, the maximum density reached within the inspiral phase is below the onset of the HQPT, but soon after the merger the density reaches values above $\rho_{\rm trans}$, forming a mixed phase inner region of deconfined quark matter. The detection of GWs from merging neutron star binaries can be used to determine the high density regime of the EOS. The power spectral density profile of the post-merger emission is characterized by distinct frequency peaks and with the knowledge of the total mass the system, the GW signal can set tight constraints on the EOS (see [6, 7, 8] for a definition and discussion of the various frequencies of the post-merger signal).

While the produced HMNS of the ALF2-M135 run collapses to a BH, no gravitational collapse happens for the LS220-M132 run within our simulation time domain. Fig. 2 shows a comparison between the two simulations in respect to the evolution of the minimum value of the lapse function $\alpha_{\rm min}$ and the maximum magnitude of the rest mass density $\rho_{\rm max}$. The collapse to the final BH in the ALF2-M135 case, can be easily seen as a sudden increase in $\rho_{\rm max}$ (sudden decrease in $\alpha_{\rm min}$). In contrast, the LS220-M132 run does not collapse to a BH but evolves to a quasi-stable HMNS. Fig. 3 shows the density and temperature profiles of the HMNS at a post-merger time of $t = 6.34$ ms. Additionally, in Fig. 3, the flowlines tracked by massless tracer particles that are advected in the flow are visualized in

IOP Conf. Series: Journal of Physics: Conf. Series **878** (2017) 012031 doi:10.1088/1742-6596/878/1/012031

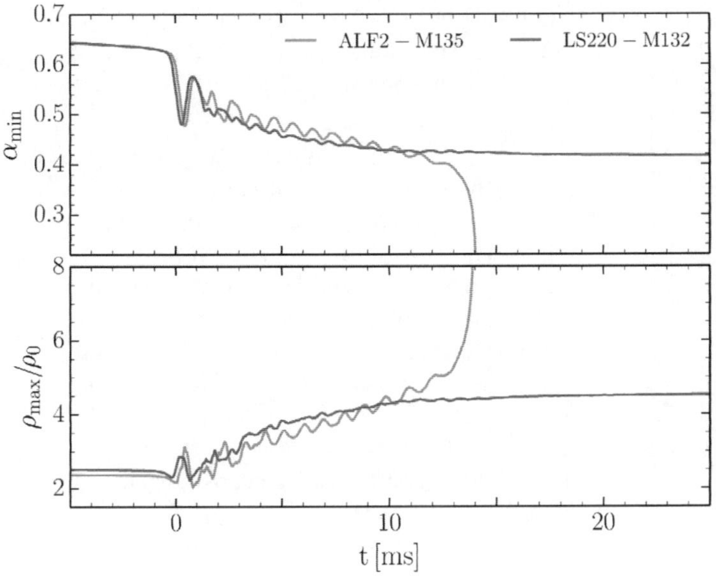

Figure 2. Minimum value of the lapse function $\alpha_{\rm min}$ (upper panel) and maximum of the rest-mass density $\rho_{\rm max}$ in units of the nuclear-matter rest-mass density ρ_0 (lower panel) versus time in milliseconds after the merger for the `ALF2-M135` and `LS220-M132` simulations. Before the merger, the maximum (central) value of the rest-mass density is essentially constant. In the transient post-merger phase strong fluctuations of $\rho_{\rm max}$ and $\alpha_{\rm min}$ occur, while during the "post-transient" phase, $\rho_{\rm max}$ ($\alpha_{\rm min}$) show a quite regular oscillating behaviour with an average increasing (decreasing) value.

the corotating frame of the rest-mass density (for details on the implementation of tracers see [21, 22]). The fluid trajectories indicate that the temperature hot spots also represent vortices around which fluid elements rotate. In order to understand the temperature profile and underlying reason of the curious tracer paths, the angular velocity distribution Ω is visualized in Fig. 4. The temperature and Ω distributions bear a remarkable similarity and the position of the hot spots overlaps closely with the position of the maxima

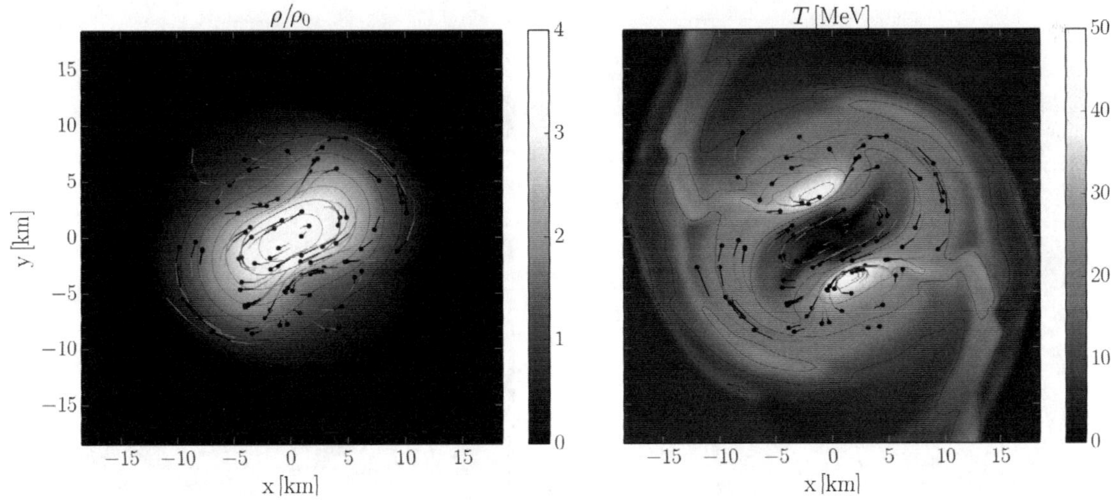

Figure 3. Distributions of the rest-mass density ρ in units of ρ_0 (left panel) and the temperature (right panel) on the equatorial plane at a post-merger time of $t = 6.34$ ms for the `LS220-M132` binary. Also shown are portions of the flowlines of several tracer particles that remain close to the (x, y)-plane and for which we show only the final part of the flowlines. In the left picture, the black isocontours have been drawn at $\rho/\rho_0 = 0.5n$ ($n \in \mathbb{N}$), while the red isocontour indicates $\rho = 3\,\rho_0$. The temperature isocontours (right picture) have been drawn at $T = 10n$ MeV.

IOP Conf. Series: Journal of Physics: Conf. Series **878** (2017) 012031 doi:10.1088/1742-6596/878/1/012031

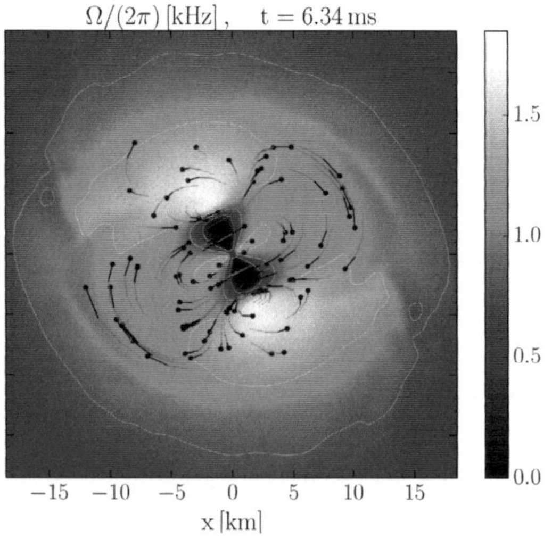

Figure 4. Angular velocity distribution Ω on the equatorial plane at a post-merger time of $t = 6.34$ ms for the `LS220-M132` binary. Ω, which is defined as $\Omega := \frac{d\phi}{dt} = \frac{u^\phi}{u^t} = \alpha\ v^\phi - \beta^\phi$ consists of a lapse-corrected part of the ϕ-component of the three-velocity, minus a frame-dragging term provided by the ϕ-component of the shift vector. The figure shows that the HMNS is a strongly differentially rotating object whereat the two maxima appear at a radial distance of $\simeq 6$ km, while the inner core is rather slowly rotating. The largest gradients in the Ω-profile take place at the positions of the temperature hot spots. The white isocontours have been drawn at $\Omega = [0.4, 0.8, 1.2, 1.6]$ kHz.

in the angular-velocity distribution. Furthermore, the profiles for ρ and Ω show a $m = 2$ deformation, however, they have a phase offset of $\simeq 90$ degrees between them. The physical cause of this phase offset and the underlying reason of the hot spots can be explained rather simply with the Bernoulli's theorem for which areas of low rest-mass density (pressure) are accompanied by regions of large velocity (for details see [23]).

For later times the HMNS reaches a stationary state, the two hot spots disappear and the temperature distribution reaches an axisymmetric pattern. Interestingly, the high-temperature region is not the central one, which is slowly rotating and comparatively colder, but appears at an annular region at about 7-8 km from the center. The figures A1 and A2 in the Appendix A show the rest-mass density ρ and the

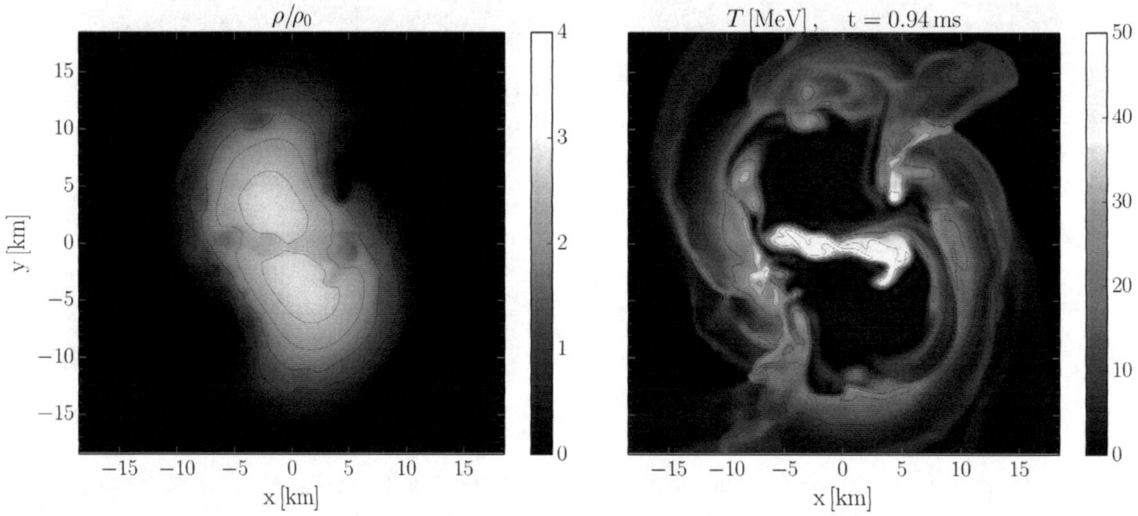

Figure 5. Distributions of the rest-mass density ρ in units of ρ_0 (left panel) and the temperature (right panel) on the equatorial plane at a post-merger time of $t = 0.94$ ms for the `LS220-M135` binary. In the left picture, the black isocontours have been drawn at $\rho/\rho_0 = 0.5n$ ($n \in \mathbb{N}$). The temperature isocontours (right picture) have been drawn at $T = 10n$ MeV.

IOP Conf. Series: Journal of Physics: Conf. Series **878** (2017) 012031 doi:10.1088/1742-6596/878/1/012031

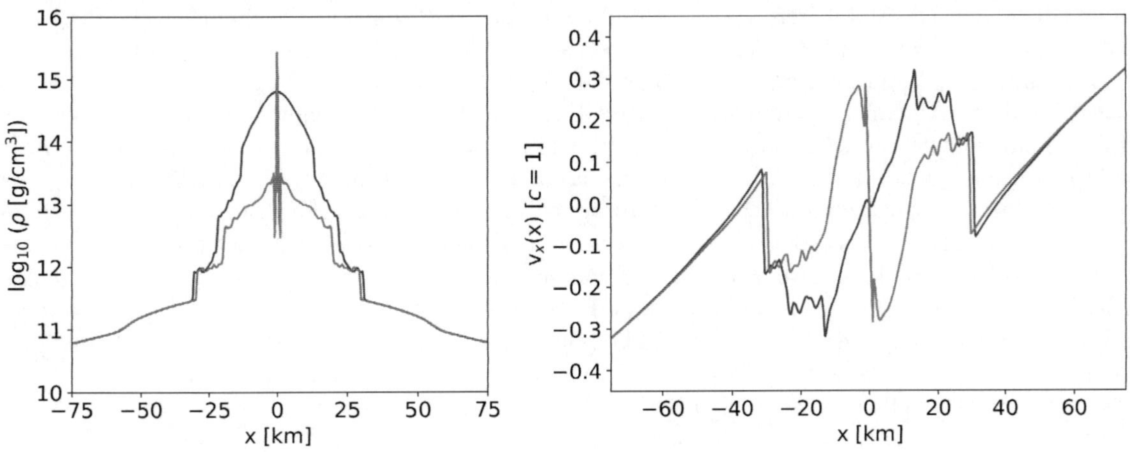

Figure 6. Density (left panel) and velocity (right panel) profiles for head on mergers of NSs. The snapshots are taken at a time $t = 1.4$ ms. We compare collisions where a black hole is formed (red line) and collisions where a NS remnant remains (blue line).

temperature profile at a post-merger time of $t = 11.54$ ms and $t = 23.83$ ms.

However, within the transient post-merger phase ($0 < t < 3$ ms) the temperature hot spots appear in the central area of the inner core of the HMNS and could reach values up to $T \approx 100$ MeV. Fig. 5 shows the density and temperature distributions 0.94 ms after merger. The equatorial density profile (left picture) indicate clearly that the HMNS, at this early phase, has still a double core structure and the high temperature region is located between these density maxima. The underlying simulation of Fig. 5 has been set up with slightly different initial conditions ($M_{\rm In} = 1.35 M_\odot$, for details see [23]) but the main characteristics of the described density and temperature structures of the different post merger phases are independent on the EOS or initial mass of the binary, if the merger product does not immediately collapse to a BH due to a to heavy initial mass.

Figure 6 shows the density and velocity profiles of neutron star matter in a head on merger of two NSs. Note the considerable radial acceleration of the matter at the shock fronts of the colliding neutron star matter slabs, which yields in a millisecond time step relative flow velocities above 0.5 c - and densities of several times the central density of the separated NSs. These velocities are close to the velocities and densities achieved in relativistic heavy ion collisions at the GSI and FAiR accelerators, but below those at RHIC and the LHC.

In the following section we will present, for those accelerator energies, the compressions achieved in relativistic nuclear collisions. By using different models for the dynamical description of the compressional phase, with a realistic EOS for hot and dense nuclear matter, we can make a connection from heavy ion collisions to the matter created in NS mergers.

3. The hot and dense QCD equation of state for heavy ion collisions and neutron star mergers

As we have seen, the densities created in the mergers of compact stars can exceed several times the nuclear ground state density. Furthermore we have shown that in the early time of the merger high temperatures $T \leq 100$ MeV are obtained. The properties of matter at such high temperature and density are very different from what we expect from cold nuclear matter. In fact we know that similar densities and temperatures can be created in the relativistic collisions of heavy nuclei at different particle accelerators. In such heavy ion experiments the heavy nuclei are accelerated to relativistic velocities. As they collide, they create a small system (of several fm in size and a lifetime of approximately 20 fm/c) which is expected to have a temperature of $T \geq 80$ MeV and densities several times the nuclear ground state density. It is therefore very intriguing to study QCD matter at similar temperatures and densities in two rather different 'experimental' setups, in neutron star mergers and heavy ion collisions. By combining the findings from both observations one may be able to deduce information on the properties of the QCD matter at high densities and finally on the phase structure of QCD. The properties of the equation of state of QCD are the link connecting the neutron star mergers and relativistic nuclear collisions. Consequently the goal of such studies has to be to find a description for the EOS that is able to describe neutron star merger and nuclear collision observables and therefore establish the connection.

In the following we will introduce such a model for hot and dense QCD matter and discuss how it is constructed. But first we have to establish common features and differences of the systems created in heavy ion collisions and neutron star mergers. For that we estimated the expected maximal compression reached in nuclear collisions at different colliding beam energies. Since the very early stage of a nuclear collision is a very rapid and violent process, expected to take place out of thermal equilibrium, estimating the maximal compression is no unambiguous task. We compare the energy and net baryon densities reached, as function of the colliding beam energy per nucleon pair $\sqrt{s_{NN}}$ in figures 7. Here we use different methods, which do not depend on the EOS, to estimate these densities.

The dashed lines follow from a simple geometric overlap model where one assumes that the total energy and baryon number of the nuclei, colliding head on, is completely stopped in a volume which is equal to the Lorentz contracted volume of a singe nucleus. The expected densities then can be written as:

$$\rho_{\rm ini} = 2\,\gamma_{\rm c.m.}\,\rho_0 \quad \text{and} \quad \epsilon_{\rm ini} = 2\,m_N\,\rho_0\,\gamma^2_{\rm c.m.} \tag{5}$$

where ρ is the net-baryon density, ϵ is the energy density, $\gamma_{\rm c.m.}$ is the Lorentz gamma of the nuclei in the center of mass (c.m.) frame of the collision and m_N is the nucleon mass. The densities from the geometric overlap model serve as a lower bound of the expected densities since this simple approach does not take into account the additional compression which occurs as the two nuclei penetrate each other.
To get a more realistic estimate for the initial compression we also show results where a microscopic transport model is used to simulate the initial non-equilibrium compression stage (grey band in figures

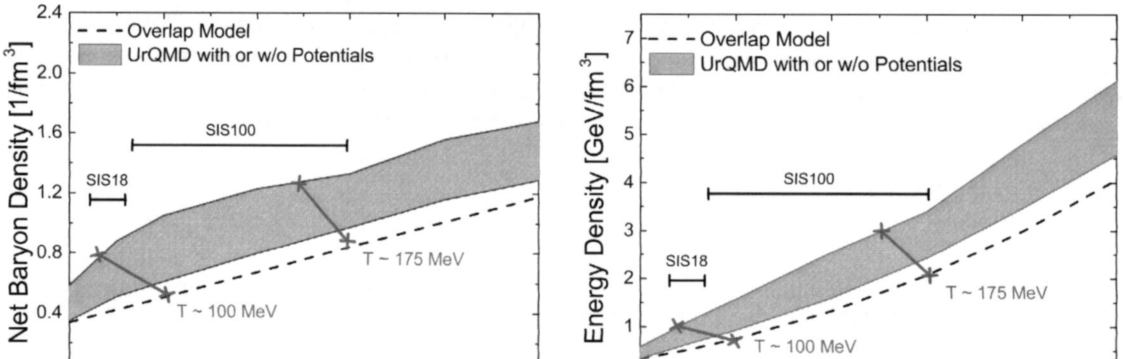

Figure 7. Left: Largest net-baryon densities achieved in central collisions of Au+Au nuclei at different colliding beam energies. Right: Largest energy density achieved in central collisions of Au+Au nuclei at different colliding beam energies. For both figures we compare results from an overlap model (dashed black line) with results where we used the UrQMD model to estimate the initial compression. The green lines with crosses indicate the beam energies where we expect the maximal temperature to exceed 100 or 175 MeV. The temperatures are calculated using the QχP model described in the text.

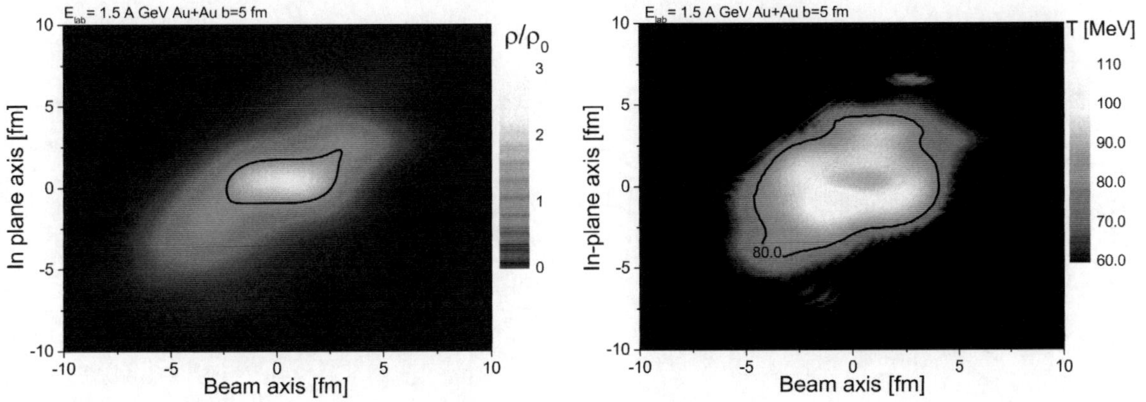

Figure 8. Left: Net baryon density contour in the reaction plane of a non-central (b= 5 fm) nuclear collision of Au+Au nuclei at a beam energy of $E_{\rm lab} = 1.5$ A GeV. The contour where the density exceeds two times nuclear ground state density is highlighted (black dashed line). The densities where calculated using the UrQMD transport model. Right: Same as left but for the temperature. The temperature has been calculated from the density and energy density using the QχP model for the equation of state.

7). The Ultrarelativistic Quantum Molecular Dynamics Model (UrQMD) is used in its cascade mode as well as a setup that includes nuclear interactions via potentials [24, 25]. To obtain a smooth density distribution from the microscopic model we run a large number of events and average densities over this event-ensemble. The so obtained values of the densities are generally larger, by up to a factor of 2, than the values of the overlap model. We find that is we want to study systems that have densities of approximately 4 times the nuclear ground state density $\rho \approx 0.6 fm^3$ we have to study systems created at beam energies of $\sqrt{s_{\rm NN}} \approx 2.5 - 3.0$ GeV. This is the energy region of the current SIS18 accelerator at GSI, as can be seen in figures 7.

As in the case of neutron star mergers the spatial density distribution in nuclear collisions is far from uniform. While in the center of the collision zone very high densities and temperatures can be obtained, we also observe a steep gradient of the densities. To illustrate this we show in figures 8 contour plots of the net-baryon density and the corresponding temperatures for collisions of Au+Au nuclei at a fixed target beam energy of 1.5 A GeV, as expected for the SIS18 accelerator. This snap-shot of the densities as taken at a time $t = 15 fm/c$, a time where one expects the system to be at least partially in local equilibrium. Again we observe that the bulk of the system reaches densities ranging from 1 to 4 time nuclear ground state density and temperatures from 50 to 100 MeV.

It is important to note at this point that in order to extract the temperature of the system at given densities, one requires knowledge on the effective degrees of freedom of the system, encoded in the equation of state. So depending on the EOS the temperatures reached in these relativistic collisions may vary significantly. It is therefore most important to employ an EOS that entails a realistic set of degrees of freedom as well as interactions. In the following we will present a model for such an EOS which can be employed to describe the matter produced in neutron star mergers as well as heavy ion collisions, thus an EOS which is able to link the properties of collision events in drastically different environments.

3.1. The QχP model

The model we employed is the so called Quark-Hadron Chiral Parity Doublet Model (QχP) [26, 27]. In this approach, an explicit mass term for baryons in the Lagrangian is possible, which preserves

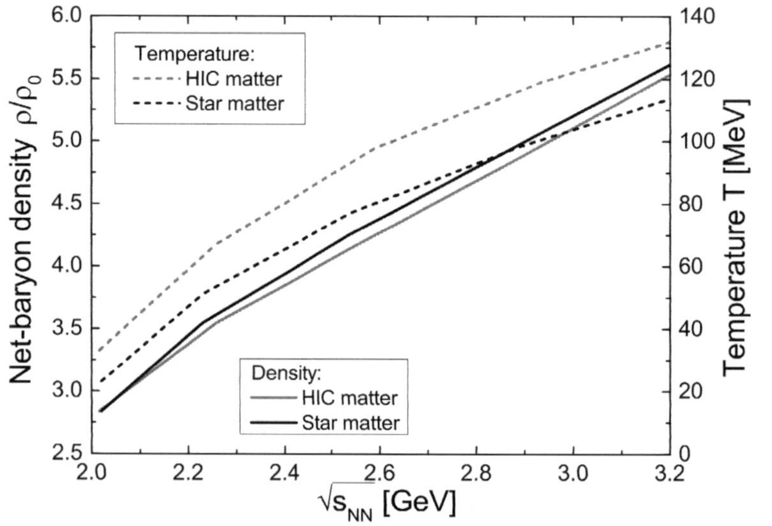

Figure 9. Largest net-baryon density (solid lines) and temperatures (dashed lines) achieved in collisions of heavy ions and compact stars at a given center of mass beam energy $\sqrt{s_{\rm NN}} = 2 \cdot \gamma_{\rm c.m.} \cdot m_N$. To calculate the densities and temperatures we used the Taub adiabat (see text) with the QχP EOS. Due to the different properties of the EOS as function of iso-spin the temperatures in heavy ion collisions are larger and densities slightly smaller, at the same relative velocities.

chiral symmetry. Chiral symmetry is considered one of the most important features of QCD and the restoration of chiral symmetry may be deeply connected with the deconfinement phase transition. The chiral condensate serves as order parameter of the chiral transition and is the only well defined order parameter in QCD thermodynamics.
In the (QχP) model, the signature for chiral symmetry restoration is the degeneracy of the usual baryons and their respective negative-parity partner states. The positive and negative parity states of the SU$(3)_f$ baryons are grouped in doublets $N = (N^+, N^-)$ as discussed in [28, 29].

Taking into account the scalar and vector condensates in mean-field approximation, the resulting Lagrangian $\mathcal{L}_\mathcal{B}$ includes [27]

$$\begin{aligned} \mathcal{L}_\mathcal{B} &= \sum_i (\bar{B}_{\rm i} i \partial\!\!\!/ B_{\rm i}) + \sum_i \left(\bar{B}_{\rm i} m_{\rm i}^* B_{\rm i}\right) \\ &+ \sum_i \left(\bar{B}_{\rm i} \gamma_\mu (g_{\omega {\rm i}} \omega^\mu + g_{\rho {\rm i}} \rho^\mu + g_{\phi {\rm i}} \phi^\mu) B_{\rm i}\right) , \end{aligned} \tag{6}$$

summing over the states of the baryon octet. Furthermore the scalar meson interaction, driving the spontaneous breaking of the chiral symmetry, is expressed in terms of SU(3) invariants $I_2 = (\sigma^2 + \zeta^2)$, $I_4 = -(\sigma^4/2 + \zeta^4)$ and $I_6 = (\sigma^6 + 4\zeta^6)$ as:

$$V = V_0 + \frac{1}{2} k_0 I_2 - k_1 I_2^2 - k_2 I_4 + k_6 I_6 , \tag{7}$$

where V_0 is fixed by demanding a vanishing potential in the vacuum. The quark and gluonic degrees of freedom are introduced as done in the PNJL approach [30, 31]. This model uses the Polyakov loop Φ field to describe an effective thermal de-confinement. To suppress hadrons in the deconfined phase we also introduced a simple excluded volume for the hadrons. The various parameters of this model are fixed by demanding a reasonable description of nuclear ground state properties like the saturation density, binding energy and symmetry energy. Furthermore if this model is extended to finite temperature and vanishing chemical potentials it gives a reasonable qualitative description of lattice QCD thermodynamics. A more detailed description of this model, can be found in [26, 27].

This model is highly qualified to study the properties of matter at high densities and intermediate temperatures, as expected in nuclear collisions as well as neutron star mergers. Within the same

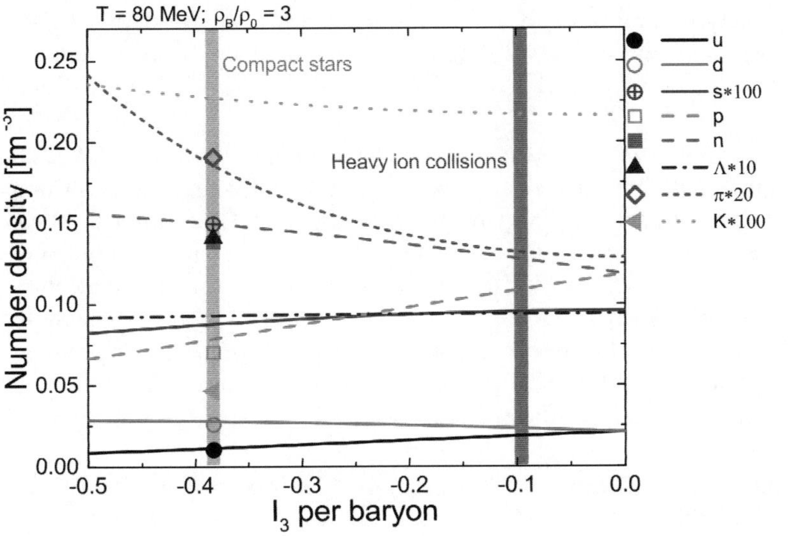

Figure 10. Number densities of different hadronic and free quark species (rescaled for visibility) as function of the iso-spin per baryon at a fixed temperature T=80 MeV and net-baryon density $\rho = 3\rho_0$. The lines correspond to matter with conserved net strangeness, as expected for heavy ion collisions while the symbols represent results where the matter is on β-equilibrium (as expected for neutron star matter).

parameter set we can use this model to calculate the EOS and chemical properties of QCD matter. A straight forward way of connecting the features of the EOS with the maximally achievable compression of a relativistic collision is by employing the so called Rankine-Huguniot-Taub-Adiabat [32]. The Taub adiabiat is essentially a shock wave solution of two colliding infinite slaps of matter. If the EOS, i.e. the connection between pressure, energy density and baryon density is know (as p(ϵ,ρ)), then one can calculate the maximum compression in a collision by solving the following Taub-equation:

$$(\rho_0\, X_0)^2 - (\rho\, X)^2 - (p_0 - p)(X_0 + X) = 0 \tag{8}$$

with $X = (\epsilon + p/\rho^2)$, the generalized volume. For simplicity we assume $p_0 = 0$. One can furthermore connect the center of mass gamma factor $\gamma_{\rm c.m.}$ of the colliding slabs to the densities created:

$$\gamma^2_{\rm c.m.} = \left(\frac{\epsilon\, \rho_0}{\rho\, \epsilon_0}\right)^2 \tag{9}$$

The resulting beam energy dependence of the net-baryon density and temperatures reached is shown in figure 9 for two different scenarios:

(i) The EOS for heavy ion collisions, i.e. with conserved strangeness and no beta-equilibrium
(ii) The EOS for compact stars, i.e. in beta-equilibrium

Again we observe that the densities and temperatures achieved in this consistent approach are similar to those discussed earlier. An important observation is that the density compression in iso-spin symmetric matter and NS matter is very similar the actual temperature is quite different. The reason for this difference are the additional degrees of freedom, i.e. leptons in beta equilibrium and non-conserved strangeness, which decrease the temperature at a given compression. Again this highlights the importance to employ a consistent and realistic temperature dependent EOS for the description of NS matter. Eventually this EOS can also be used in full 3+1D fluid dynamical simulations of heavy ion collisions. As in the NS merger simulations the goal here is to possibly find observables for a (non-equilibrium) first order phase transition in dense QCD matter [33].

To highlight similarities and differences in the chemical composition of the systems created in these collisions we show in figure 10 the number densities of different hadronic species at a fixed temperature and net-baryon density, as function of the iso-spin per baryon of the system. This stresses the different composition of systems created in collisions of Au+Au nuclei (where the iso-spin per baryon is -0.1) and of neutron star matter. The matter in neutron stars not only has a an iso-spin per baryon of -0.38 but also, according to beta equilibrated strangeness, a significantly different composition of strange particles.

4. Conclusions

In this article we show that the properties of elementary matter at high temperatures ($T \approx 100$ MeV) and densities ($\rho \approx 3\,\rho_0$) can be studied in two different physical scenarios. High energy heavy ion collision experiments try do determine the phase structure of the iso-spin symmetric QCD equation of state, and the knowledge of the iso-spin asymmetric QCD-EOS is needed in a general relativistic computer simulation of binary neutron star mergers. These two different fields of physics, namely elementary particle physics and astrophysics, combine when two neutron stars collide. It is therefore possible to study the properties of dense QCD for systems of different size, time-scales and chemical composition, which will eventually lead to an understanding of the properties of this elementary form of matter.

Acknowledgements
Support comes from "NewCompStar", COST Action MP1304, from the LOEWE-Program in HIC for FAIR, and from the European Union's Horizon 2020 Research and Innovation Program (Grant 671698) (call FETHPC-1-2014, project ExaHyPE). MH gratefully acknowledges support from the Frankfurt Institute for Advanced Studies (FIAS) and the Goethe University Frankfurt, while HS acknowledges the Judah M. Eisenberg laureatus Professur endowment. The simulations were performed on SuperMUC at LRZ-Munich, on LOEWE at CSC-Frankfurt and on Hazelhen at HLRS in Stuttgart.

Appendix A. Evolution of the density and temperature distributions in the late post merger phase

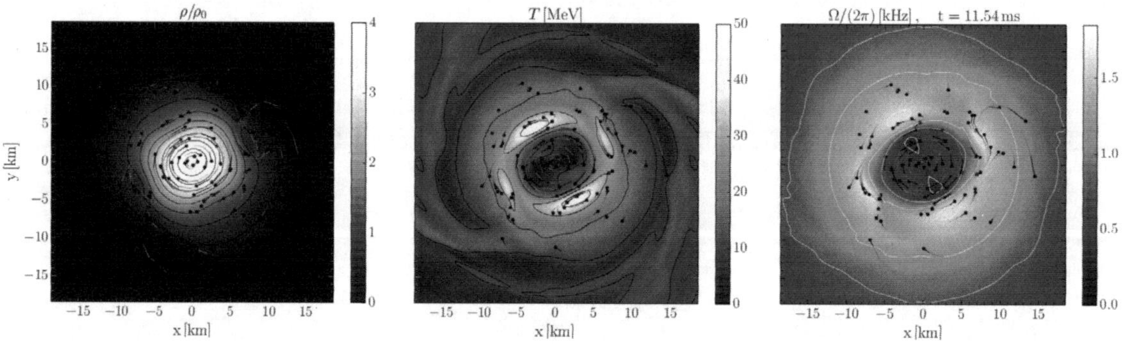

Figure A1. Same as in Figs. 3 and 4 but at a post-merger time of $t = 11.54$ ms.

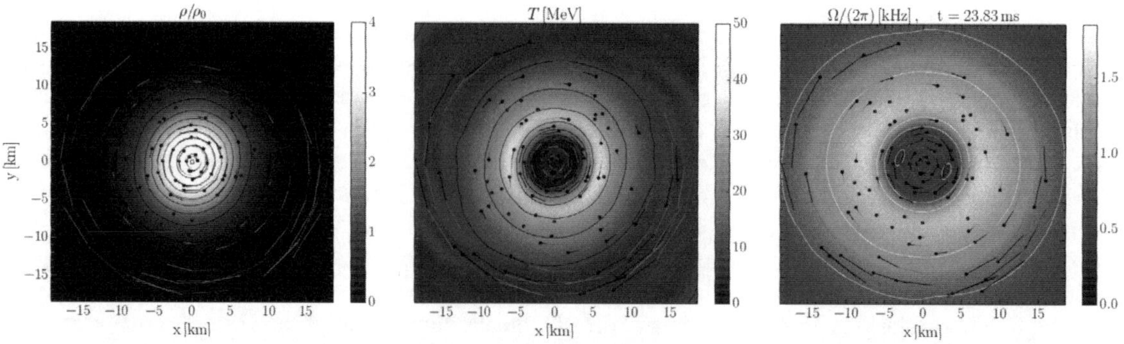

Figure A2. Same as in Figs. 3 and 4 but at a post-merger time of $t = 23.83$ ms.

References

[1] Wex N 2017 *Presentation at the DPG-Spring Meeting in Bremen, Gemany*

[2] Lyne A G, Burgay M, Kramer M, Possenti A, Manchester R N, Camilo F, McLaughlin M A, Lorimer D R, D'Amico N, Joshi B C, Reynolds J and Freire P C C 2004 *Science* **303** 1153–1157 (*Preprint* arXiv:astro-ph/0401086)

[3] Kramer M and Wex N 2009 *Class. Quantum Grav.* **26** 073001

[4] The LIGO Scientific Collaboration and the Virgo Collaboration 2016 *Phys. Rev. Lett.* **116** 061102 (*Preprint* 1602.03837)

[5] Abbott B P, Abbott R, Abbott T D, Abernathy M R, Acernese F, Ackley K, Adams C, Adams T, Addesso P, Adhikari R X and et al 2016 *Astrophys. J. Lett.* **818** L22 (*Preprint* 1602.03846)

[6] Takami K, Rezzolla L and Baiotti L 2014 *Phys. Rev. Lett.* **113** 091104 (*Preprint* 1403.5672)

[7] Takami K, Rezzolla L and Baiotti L 2015 *Phys. Rev. D* **91** 064001 (*Preprint* 1412.3240)

[8] Rezzolla L and Takami K 2016 *Phys. Rev. D* **93** 124051 (*Preprint* 1604.00246)

[9] Antoniadis J, Freire P C C, Wex N, Tauris T M, Lynch R S, van Kerkwijk M H, Kramer M, Bassa C, Dhillon V S, Driebe T, Hessels J W T, Kaspi V M, Kondratiev V I, Langer N, Marsh T R, McLaughlin M A, Pennucci T T, Ransom S M, Stairs I H, van Leeuwen J, Verbiest J P W and Whelan D G 2013 *Science* **340** 448 (*Preprint* 1304.6875)

[10] Rezzolla L and Zanotti O 2013 *Relativistic Hydrodynamics* (Oxford, UK: Oxford University Press) ISBN 9780198528906

[11] Misner C W, Thorne K S and Wheeler J A 1973 *Gravitation* (San Francisco: W. H. Freeman)

[12] Everitt C W F, Debra D B, Parkinson B W, Turneaure J P, Conklin J W, Heifetz M I, Keiser G M, Silbergleit A S, Holmes T, Kolodziejczak J, Al-Meshari M, Mester J C, Muhlfelder B, Solomonik V G, Stahl K, Worden Jr P W, Bencze W, Buchman S, Clarke B, Al-Jadaan A, Al-Jibreen H, Li J, Lipa J A, Lockhart J M, Al-Suwaidan B, Taber M and Wang S 2011 *Physical Review Letters* **106** 221101 (*Preprint* 1105.3456)

[13] Shibata M and Nakamura T 1995 *Phys. Rev. D* **52** 5428–5444

[14] Baumgarte T W and Shapiro S L 1999 *Phys. Rev. D* **59** 024007 (*Preprint* gr-qc/9810065)

[15] The Einstein Toolkit Consortium: einsteintoolkit.org

[16] Radice D, Rezzolla L and Galeazzi F 2014 *Class. Quantum Grav.* **31** 075012 (*Preprint* 1312.5004)

[17] Read J S, Lackey B D, Owen B J and Friedman J L 2009 *Phys. Rev. D* **79** 124032 (*Preprint* 0812.2163)

[18] Alford M, Braby M, Paris M and Reddy S 2005 *Astrophys. J.* **629** 969–978 (*Preprint* nucl-th/0411016)

[19] Hanauske M 2004 *Properties of Compact Stars within QCD-motivated Models* Ph.D. thesis J.W. Goethe-University Frankfurt, Germany

[20] Lattimer J M and Swesty F D 1991 *Nucl. Phys. A* **535** 331–376

[21] Mewes V, Galeazzi F, Font J A, Montero P J and Stergioulas N 2016 *Mon. Not. R. Astron. Soc.* **461** 2480–2489 (*Preprint* 1605.02629)

[22] Bovard L, Radice D and Rezzolla L 2016 *in preparation*

[23] Hanauske M, Takami K, Bovard L, Rezzolla L, Font J A, Galeazzi F and Stöcker H 2016 *arXiv preprint arXiv:1611.07152*

[24] Bleicher M *et al.* 1999 *J. Phys.* **G25** 1859–1896 (*Preprint* hep-ph/9909407)

[25] Bass S A *et al.* 1998 *Prog. Part. Nucl. Phys.* **41** 255–369 [Prog. Part. Nucl. Phys.41,225(1998)] (*Preprint* nucl-th/9803035)

[26] Mukherjee A, Steinheimer J and Schramm S 2016 (*Preprint* 1611.10144)

[27] Steinheimer J, Schramm S and Stocker H 2011 *Phys. Rev.* **C84** 045208 (*Preprint* 1108.2596)

[28] Detar C E and Kunihiro T 1989 *Phys. Rev.* **D39** 2805

[29] Hatsuda T and Prakash M 1989 *Phys. Lett.* **B224** 11–15

[30] Fukushima K 2004 *Phys. Lett.* **B591** 277–284 (*Preprint* hep-ph/0310121)

[31] Ratti C, Thaler M A and Weise W 2006 *Phys. Rev.* **D73** 014019 (*Preprint* hep-ph/0506234)

[32] Taub A H 1948 *Phys. Rev.* **74** 328–334

[33] Steinheimer J and Randrup J 2012 *Phys. Rev. Lett.* **109** 212301 (*Preprint* 1209.2462)